T-1 Networking

How to Buy, Install, and Use T-1 From Desktop to DS-3

T-1 Networking

How to Buy, Install, and Use T-1 From Desktop to DS-3

Fifth Edition

William A. Flanagan

try's central source of
u may receive a FREE
copy of the latest catalogs by calling 212-691-8215
or by writing to the address below.

QUANTITY PURCHASES

If you wish to purchase this, or any other Telecom Books in quantity, please contact:

Book Manager
Telecom Books
12 West 21 Street
New York NY 10010
1-800-LIBRARY or 212-691-8215

This book was designed and produced by Telecom Books
Published in the United States by Telecom Books and Flatiron Publishing, Divisions of Miller Freeman, Inc.

In Memoriam

STAN ELLIN

Who showed me how to write a book.

Contents

1 What T-1 Does For a Network 1
What is T-1? 1
Who Provides It 2
Advantages of Private T-1 Networks 3
Is T-1 for Me? 14
If I think We May Need a Private T-1 Network 15

2 T-1 Defined Historically 21
A Short History of T-1 21
FDM 24
Line Noise 25
TDM 25
The Digital Hierarchy 28
T-1 Everywhere 29

3 Digital Voice 33
Analog versus Digital 34
Pulse Code Modulation 35
The Voice Channel Bank 39
The Nature of Voice Signals 42
A Question of Voice Quality 52
Digital Speech Interpolation 52
Packetized Voice 53
Sub-band Coding 54
VQL: Variable Quantizing Level 54
LPC: Linear Predictive Coding 55
Telephone Signaling 55

Voice Applications 59
Direct Digital Interfaces 59
Emulating Analog Ties 59
OPX: Private Centrex 61
D4 Framing: A Fundamental Concept 63
The Single Frame 63
The M24 Superframe 64
Robbed Bit Signaling 65
ESF: Extended Superframe 67
Facilities Data Link 69
False Framing 73
Proprietary Framing 73

4 T-1 Circuits 77
Terrestrial 77
CSU: Channel Service Unit 83
Loopback 83
Keep Alive 84
ESF Statistics 86
1's Density Enforcement 87
DSX-1 Interface 88
User Clocking 89
Carrier Clocking 91
DSU: Digital Service Unit 92
Clear Channels 95
Satellite 100
Who Does What 102
Other T-1 Transmission Media 105
Digital Subscriber Line 109
Base Technologies 110
2B1Q From ISDN 111
Carrierless Amplitude/Phase (CAP) 112
Discrete MultiTone (DMT) 113
Time Shift Keying (TSK) 114
Product Categories (Applications) 115
ADSL: Asymmetrical 115
HDSL: High-speed 117
IDSL: ISDN (Basic Rate) 117
MDSL: Medium-speed 117
RADSL: Rate-Adaptive 117
SDSL: Symmetrical 118
VDSL: Very-high-speed 118

Specialized Circuits 118
Access and Cross-Connect Systems 118
Fractional T-1/T-3 121
Customer Controlled Reconfiguration 123
DACS Control Limitations 125
T-1/E-1 Conversion 128
Switched T-1 128
Inverse Multiplexing 129
Software Defined and Virtual Private Networks 129
Outlook For T-1 and ISDN 131
Why ISDN Is Inevitable 132

5 Equipment You Will Need At Your Office 133
Interfaces 133
How Many User Ports? 134
Link Efficiency 135
Many Kinds of Inputs 138
Channel Routing Flexibility 148
Getting Data Through a Node 148
Data Bypass or Drop and Insert 148
Multiple Data Links 151
Bypass vs. D/I 152
Migrating to a BIG Node 154
How Is Switching Done? 155
The T-1 Node as Voice/Data Switch 157
Distributed vs. Centralized Switching 159
Reliability and Redundancy 160
The Effect of Nodal Architecture 168
Where Is Control? 169
Internal Bus Architecture 169
Access Multiplexers 170
Internet Access 170
Fractional T-1 Access 171
Local Dial Tone 172

6 How To Build A T-1 Network 173
Making Tradeoffs 173
T-1 or Not T-1? 175
Dollars & Necessity 176
Should Your Network Be Public or Private? 176
Whose Job? 178
Should data be mixed with voice? 179
Which Network Topology? 179

Design Effects on Network Topology 190
Reliability and Availability 190
CPE Multiplexing, Switching, Recovery 194
Similar Central office Functions 195
Single-Ended T-1 Access 200
Where are they being used? 201
Internet Access 201
Frame Relay Access 201
Dial Tone 202
Integrated Access 203
Project Management For T-1 Installations 205

7 Beyond T-1: Desktop to DS-3 211
The new networking 212
Saving Money 213
Unifying Control 214
Integrating Equipment 215
Up to DS-3 and Beyond 216
DS-3 Options 216
Networking At DS-3 and Faster 219
Down to the Desktop 221
Modem 221
ISDN Terminal Adapter 225
Cable-Based LAN 227
Statistical Multiplexer 230
Data PBX 232

8 Network Management And Control 235
Network Control 237
Distributed Control 238
Centralized Control 241
Center-Weighted Control (CWC) 243
Control Functions 246
Making Connections 246
Shared Features Among Control Types 247
Automatic Alternate Routing (AAR) 249
Following Priorities 250
Remote Reconfiguration 251
Testing 254
Advanced Diagnostic Features 256
Network Management 258
Configuration Setup 258
Network Design Tools 261

Appendices

A. T-1 Case Examples 265
Manufacturer I 266
Manufacturer II 267
Three nodes Intra-State 272

B. Basic Multiplexing 277
Time division multiplexing 278
Packet Switching 279
Frequency division multiplexing 281
Bit vs. Byte Interleaving 281
Fixed Vs. Flexible framing 283

C. Global Framing Formats 287
ITU Format For 2.048 Mbit/s 287
HDB-3 Coding 288
Frame 289
Time Slot 290
Multiframe 290
Possible Variations 291
T-1 in Japan 292

D. Networking Acronyms 295

Index 331

Preface to the Fifth Edition

The number of T-1 lines in the US will reach 1 million about the time this book is printed. Wider use has been driven by costs for both lines below the analog and 56K equivalents, and much cheaper hardware. There is no end in sight.

My earlier predictions about growth in the T-1 market have come true, and much more. (Prefaces from previous editions are reprinted below.)

When the 4th edition went to press, the Internet was still an obscure way for researchers and government people to communicate among themselves. The public hadn't heard of it. Who would have guessed that Internet access would be a very popular application for T-1? Not even my most optimistic predictions included Internet and frame relay access at 45 Mbit/s. But they are here now. Some RBOCs are beginning to offer cell relay (ATM) over T-1 access.

In other words, it's time for an update on T-1 Networking. This fifth edition includes more new applications and new types of equipment.

While T-1 remains a backbone transmission technology, it is also taking on the duties of a familiar standard interface. It's almost like V.35 in that regard.

The V.35 recommendation was deleted by ITU years ago. T-1 refers to a combination of twisted pairs and signal regenerators more than 30 years old and not used in new installations.

Yet both terms persist in common usage because they mean something useful. T-1 will be around a while. This book is designed to help you cope with T-1, wherever you find it.

An expanded Glossary of Acronyms is almost double the size 7 years ago. We look at new services related to T-1, like fractional DS-3, Digital Subscriber Line (DSL), and ISDN Primary Rate Interface. Some topics have been expanded or revised in response to reader comments, for example clocking and ESF quality reporting.

The huge developments in voice compression are mentioned here, but not explored in detail as they appear in my 1977 book, Voice Over Frame Relay. That volume also contains more information on analog signaling.

We hope you like the new page size and find it more convenient to find a spot on your bookshelf. Again, your comments are always welcome.

—William A. Flanagan
August 1997

Preface to the Fourth Edition

Changes in the last year have been very significant for T1 network operators. In particular, they have seen new services and lower costs.

Fractional T-1 grabbed the headlines for many months, and promises to live up to its publicity in having a huge impact on network design. The low cost of fractional T-1, showing a breakeven in the U.S. with as few as three voice grade lines, should speed the migration of private networks from analog to digital transmission technologies. Several sections of this book now include the impact of fractional T-1. But rapid change continues and will require "real time" reference to tariffs, including those for full T-1's.

This year alone AT&T twice cut the per mile charge for T-1 lines; it is now half the $15/mile quoted in the examples. Payback times dropped by about a third, again.

The distinction between public and private networks began to blur. Carriers started to offer essentially private networks, but based on T-1 multiplexers installed in central offices as well as on customer premises. Here, the customer gets all the benefits of private networking, including low cost and direct control, but the carrier assumes responsibility for operation and maintenance. This could be a very large market worldwide in the next decade.

Another new service much talked about in the U.S. is T-3. It is definitely real and looks like an important factor for the near future. Several vendors have announced DS-3 interfaces to T-1 multiplexers for delivery before the end of 1990. At least three T-1 vendors have bought T-3 vendors. Market analysts forecast amazing revenue growth in DS-3 products. The text in this book was adjusted at several points to reflect the latest situation.

Oceanic Fiber cables came on line this year. More will be cut over in the next few years. These efforts reinforce my forecast for larger networks and global coverage.

Past references in the text to imminent events like the introduction of new services had to be updated now that many of them have happened. In most cases there was little or no impact on the meaning.

Other than that, there has been little change in the fundamental outlook for digital networking. The preface to the third edition, reprinted below, still reflects my outlook.

One new standard has been adopted in this volume: ANSI has settled the question of how to abbreviate "thousands of bits per second." What we had printed as kbps is now kbit/s. Likewise, megabits per second becomes Mbit/s. There will never again be any confusion between bits and bytes per second.

How to use this book

There are two ways:

1. As a textbook in a college course or for self-guided study.
2. As a tool to help a telecommunications professional choose, install, use, and maintain a high-speed voice, data, and video telecommunications network.

In a college course, read the book in the order we've published it.

If you're using this book for practical network-making, then read the chapters in this order: 1, 5, 6, 7, 8.

Then if you want to get in-depth understanding of the technology, read Chapters Two, Three, Four, and the Appendices. Don't overlook the acronym list in Appendix B and the index as ways to reach particular topics directly.

I hope this Fourth Edition better meets your current needs and helps you plan for the future. As before, I solicit reader comments.

William A. Flanagan
September 1989

Preface to the Third Edition

Although T-1 technology was created in the 1960s, its applications really started in private networks with the 1983 T-1 tariff. Much has changed since then. In just the two years between the 1986 and 1988 editions of this book:

- New vendors have introduced new equipment that creates new ways of building networks.
- New carriers have brought lower tariffs and shorter delivery times for T-1 lines in many areas.
- A new and higher tariffed speed, DS-3 service at 45 Mbit/s, is becoming common.

Together, these factors put a different slant on how networks are planned, installed, and used. Thus, this third edition. Revised and expanded, it offers a look at how digital networking can extend throughout an enterprise spread over many sites. Certain sections in the original book have been clarified and expanded. New material has been added. Some chapters are reorganized to clarify the topics they cover.

Not that the first edition didn't foresee change. In fact, the earlier editions contained an explicit Outlook for CPE from the viewpoint of 1985:

For the next few years the trends in CPE (Customer Premise Equipment) seem to be:

- DACS compatibility on the T-1 link. DS-0 channelization will dominate the carrier offerings. To participate in sophisticated services the end user will have to provide a compatible format from his/her CPE.

- Customer Controlled Reconfiguration will migrate, probably in more than one step, to ISDN. The controlling factor will be the speed with which the carriers can install digital central office equipment to replace what was almost entirely analog in 1983.
- Bandwidth "needs" will grow rapidly. More than one corporation has installed a fiber optic link running at 45 million bit/s. This will become a common speed within a few years.
- More and more functions will be brought inside the T-1 node. These advanced nodes, in the past called "T-1 multiplexers," will properly be renamed to distinguish them from common TDMs. The name could be something like Network Exchange.
- The node of a few years from now will:
  — Multiplex any kind of information,
  — Cross connect DS-0s and slower data channels,
  — Allocate bandwidth on demand,
  — Offer test access, and
  — Function as a matrix switch at standard data rates.

The "nodes of a few years from now" have begun to ship. All those 1985 trends are on track in 1988, with two exceptions:

1. "Bandwidth Manager" seems to be winning out over "Network Exchange" as the generic name for a T-1 node.
2. CCR itself has not advanced much in functionality (although AT&T has announced Bandwidth Management Service, which is closer to ISDN).

With some experience to back up the process, let's refine the list and again look ahead a few years with a new Outlook for CPE (1988):

Larger nodes; look for installed capacity beyond 200 T-1 ports, or the equivalent in DS-3 ports.

Larger networks; expect to see 1000 nodes on a digital backbone. Many nodes could be small, for example if an insurance company ties all it's offices together.

Digital replacing analog; not only in switches and PBXs, but in the lines to smaller locations. Tail circuits—voice and data on 56 kbit/s—will be part of the digital backbone.

Rapid advances in network management; essential to deal with larger nodes and networks. Most will involve powerful computers dedicated to running the network, on-line, rather than enhancements within the nodes. Graphical interfaces, between operators and the machines, will become essential for "mere humans" to comprehend the huge volume of information needed to control a large network.

Much lower T-1 line costs; many new optical fiber installations will bring down international rates as well as domestic. Multinational networks will be restrained for some years by monopolistic practices of the PTTs in most countries.

Much faster market growth in the U.S.; lower costs for lines and CPE will make many more applications economic and practical, fueling more sales than predicted as late as 1987.

We shall see.

William A. Flanagan
May 1988

Acknowledgements

(Third Edition)

During the rapid change of the past two years many people have helped keep me up to date. Coworkers, consultants, other vendors, and even competitors have shared their knowledge quite freely. Again a special thanks to Jim Michaels, particularly for his insights into network clocking.

A special note of thanks to Harry Newton, my publisher and editor. His continuing support was a key element in the production of this revised and expanded third edition.

In addition, readers of the first edition have offered many valuable suggestions. My thanks to all of you.

(First Edition)

Many people have helped, in particular Vicki Brown and Harry Thomas. They gave me support and encouragement on the original project, and permission to reproduce many of the diagrams in this book.

The interest in T-1 networking has grown steadily since the 1.544 Mbit/s digital service was first tariffed by AT&T in January 1983. As T-1 lines became more readily available, and new multiplexer and switching equipment was shipped, large users of communications realized this technology was impossible to ignore. And difficult to learn about.

One of the earliest sources of T-1 information was a T-1 seminar series, introduced in late 1983. Al Spiegleman was the original presenter of these seminars. A special thanks to him, Jim Michaels, and George Hendrix for their technical insights.

Before it was discontinued, that seminar was modified continually in response to comments from hundreds of attenders. In the same spirit, I solicit reader comments on this book.

William A. Flanagan

What T-1 Does For a Network

▶ Why should a communications manager take the time to read a fairly technical book on T-1 communications services? The answer is simple. Putting T-1 circuits into a network can:

- save large amounts of money;
- give you the user the flexibility to reconfigure connections quickly and at will — without waiting for the service from the local or long distance telephone company;
- improve voice quality, compared to long distances over analog lines;
- improve reliability by simplifying the network and its maintenance; and
- provide the bandwidth (data throughput) needed by growing businesses.

To explain these advantages requires a basic understanding of what "T-1" is.

What is T-1?

In simplest terms, T-1 service offers a two-way connection at 1.5 million bits per second. That is, one T-1 line will carry the equivalent of 160 computer ports running at 9600 bit/s each.

The most common use for T-1 circuits through the 1980s was inside phone companies, usually between central offices. In a phone company, one T-1 circuit on two normal twisted pairs of copper wire carries 24 phone conversations in digital form.

In the strictest sense, the rate of 1.544 Mbit/s (megabits per second) is "digital signal level one," DS-1. With the same strictness, "T-1" refers to a system of copper wire cables and amplifiers or regenerators that reinforce the digital signal at intervals of about one mile. The nature of the cable, regenerator, and signal are defined very tightly for the United States. In popular use, T-1 has come to mean any transmission line or connection running at 1.5 Mbit/s.

In Europe, the U.K., Mexico, and other countries that adhere to ITU standards, the equivalent is E-1, a 2.048 Mbit/s service (see Appendix C).

Who Provides It

Like long distance telephone service, T-1 circuits in the U.S. are available from many sources. Most often the local telephone company (telco) provides the copper wire from each of your offices to the nearest central office. These are called local loops.

If your chosen T-1 circuit is entirely within a local area — called in telephonese the Local Access and Transport Area, LATA — only one telco will be involved. A LATA is usually one metropolitan area or one state.

For longer circuits, there will be different telcos at either end, in general. They provide access to the central offices of a long haul phone company which carries the signal between LATAs. Three carriers per T-1 circuit is very common. As many as five or six carriers may be needed to build a T-1 line.

Sometimes the user may provide his own T-1 using microwave, optical fiber, copper wire, or other medium.

Outside the U.S. the local PTT (the national Postal, Telephone, and Telegraph authority) provides the entire circuit within a country. Most PTTs are monopolies, though competition is being admitted (slowly in most cases) to more advanced countries. PTTs interconnect their circuits at national borders for multinational networks.

Advantages of Private T-1 Networks

Dollars

The economic incentive to use T-1 can be overwhelming despite the apparent high start-up cost of equipment and continuing costs for leased lines. Here are some typical costs:

- T-1 multiplexers and bandwidth managers can cost from less than $10,000 to more than $100,000 per location;
- Leased T-1 circuits from the major carriers range up to $16,000 per month (New York to Los Angeles), down from $25,000 per month a decade ago. Smaller carriers charge half that figure or less.

Still, an investment in a new network based on T-1 can break even within the same fiscal year. Annual reductions in phone bills at one company reach into the millions of dollars.

Network Control

Greater than dollar savings to many users is the promise of network control. The end user organization operating a private T-1 network enjoys:

- The diagnostic power to identify and isolate faults, often before they cause a catastrophic failure.
- The ability to set up and change connections between points on the network at will, and at little or no incremental cost.
- Simplification that makes control more practical, and improves reliability.

Rerouting of connections is part of network control. Ideally, the network operator or communications manager should be able to perform any function on the network from a single point, usually the central or largest processing site. Single point of control implies that any tests or changes at remote sites, short of physically changing out printed circuit cards or other components, can be handled without a remote operator.

Network users consistently report that reliability and uptime are their most pressing concerns. Therefore diagnostics are often considered the most important control function. The central site can invoke all types of tests and gather usage and quality reports from any node. With perhaps three to five carriers responsible for sections of a single circuit, quick positive fault isolation reduces finger pointing among vendors and speeds restoration of service.

In practice, the ability to reconfigure the network often proves of more immediate value than diagnostics. A network can be restored to full capacity almost immediately by the multiplexer or network management system automatically rerouting essential connections over the remaining links.

Diagnostic Control

Divestiture of AT&T in 1984 split the responsibility for a single circuit among as many as six carriers. Their coordination of troubleshooting is not what it used to be. T-1 nodes, modems, computers, and terminal equipment can add another 4 to 6 vendors. When the connection is not working properly, who's at fault? How do you get it repaired?

An integrated T-1 network with a central "war room" monitoring the progress of the digital T-1 network gives the user control of his own communications resources. Control means the ability to locate faults quickly and reliably. Once you locate the fault, you can dispatch the proper maintenance organization to restore service. This reduces finger pointing among vendors as to whose fault it is, and sharply reduces the time needed to get circuits working again.

Configuration Control

Configuration control also means the ability to manipulate network resources. The network operator assigns them where needed most, to compensate for failures or changing demands.

Has a snow storm prevented people from reaching the western regional office? Set up additional connections to accommodate temporary help in other offices. Tomorrow, everything can go back to normal.

Does the president tell you at 10:30 a.m. he wants to hold a video teleconference right after lunch? During lunch, rearrange bandwidth allocation to provide the connection. When he is finished, give the bandwidth back to other applications, like normal long distance voice and/or data traffic.

Has teller productivity in the branch offices dropped due to slow terminal response? Try increasing the speed on the data link between the mainframe (actually the front end or communications processor) and the remote site. Test different speeds to different offices, at different times of day, on specific days of the week. There's no wait for the service changes, and no charge from the telco. The cost of the experiments is the cost of your own time.

In short, the configuration flexibility of a T-1 network maximizes the reliability and efficiency of the communications resource within a corporation.

Simplification

T-1 technology simplifies, and thereby improves, corporate communications. Combining multiple voice and data channels over a single high-speed digital circuit eliminates many separate networks and many forms of network management. Once the digital backbone is in place it is possible also to carry compressed video, for teleconferencing; facsimile; and wideband or hifi audio. Many other applications can be integrated on one private network.

Not only does a T-1 backbone reduce the number of circuits that must be monitored, it affects the network at a conceptional level. A backbone, perhaps with branches, is easier to understand, diagnose, and (inevitably) to reconfigure than a mesh of low speed lines.

Reliability Improvement

As any engineer will attest, simplicity is harder to design than complexity. It's worth the effort because "simple" usually is more reliable.

The same is true of telecommunication networks. Fortunately, the difficulty of designing a T-1 network is most often handled by the carrier or multiplexer vendor. The benefits accrue to the user, including reliability and ease of troubleshooting.

With fewer circuits and a clear topology, it is easier to isolate faults when they occur. Of course each circuit carries many individual channels, and when a circuit is lost so are all the channels. However, it is more practical to arrange for one back-up circuit in case of total failure, compared to arranging alternatives for multiple leased lines.

One particularly attractive feature of T-1 circuits of most carriers is inherent redundancy on the long-haul portion of the route. That is, even in the event of a total loss of your T-1 circuit, your connection will be re-established automatically. Restoration usually occurs within seconds or even milliseconds.

To complete this redundancy between the end sites requires a hot spare T-1 circuit from each customer location to the local phone company's central office. The T-1 multiplexer or switch detects the loss of circuit and transfers to the backup. If an alternate to each circuit is not practical, then the latest T-1 multiplexers and bandwidth managers provide another form of backup: automatic alternate routing. If a connection is broken by loss of a circuit, this feature lets the network re-establish the connection via another path within the private network, if one exists. The nodes or the central network management system use information about network topology to seek out available bandwidth, or even create it (if users can program ahead of time which applications or connections may be "bumped").

But what about failures within the multiplexer? Here is it important to have redundant components for all functions shared by multiple users, the "common" components. For example, redundant master control cards, T-1 line interfaces, data switching components, and power supplies must continue without interruption despite the failure of any single module or circuit board. This redundancy is available for all common control and support modules. Individual In/Out ports on the user side are not usually protected with redundancy, though this is desirable in some applications. The loss of one I/O board will affect only a few users and is easily replaced without disturbing other users.

For optimum economics, the redundancy in the communications network should match that in the host computer. A customer running a garden variety minicomputer might not invest in more than load sharing power supplies (the most common failure point). If the computer is a fault tolerant design with multiple hot spares, the network nodes probably should be fully redundant.

Quality

By eliminating voice-grade lines (formerly called 3002 by AT&T), a T-1 network eliminates the variability in quality that can affect analog facilities. Noise and crosstalk on PBX tie lines may make conversations more difficult to understand. When carrying a data channel, however, an impaired line can reduce data throughput, increase response time, or even render a circuit useless. Conditioning VG lines to carry data used to be common practice. There is nothing comparable for the T-1 user. Usually a digital circuit either works or it doesn't. When it works, the quality is high. When it doesn't, you fix it.

An all-digital path between PBXs ensures that voice quality of the transmission is consistently high. For the carriage of real human voices, a line that contributes less noise and distortion improves the overall quality. Even after compressing the voice signal digitally (Chapter 3), there can be a net overall gain in perceived voice quality. The improvement was especially noticeable over distances of more than 1500 miles in the days when repeated amplifications of analog signals increased the noise level.

For data channels, the all-digital path usually means virtually error-free operation. Intermittent problems can occur — noise spikes for example, or short periods of disruption (error bursts lasting up to about 2 seconds). But typically, failures on T-1 lines are total. Close monitoring can spot the narrow transition zone of increasing error rate as a line goes from good to bad. Upon a failure, the carrier can invoke its backup, or the user can reroute essential channels over those circuits that remain functioning.

It is possible for speed alone to improve the quality of service. A major bank in New York state installed a T-1 network. They then doubled the speed of the connections between the mainframe's front end processor and the remote terminal cluster controllers — from 9,600 to 19,200 bit/s. Response times at teller stations dropped from as much as 5 seconds to below 2 seconds. They now often run at 1 second. These relatively small savings of a few seconds per transaction dramatically improved overall productivity of the tellers, and made the customers happier too.

New Capabilities

With the availability of large amounts of bandwidth, under immediate control, the network operator is able to offer the user community or corporation new services not practical in the past. Additional circuits are readily available through reconfiguration.

For example, due to changing offices, a CAD/CAM group needs three new 56,000 bit/s lines between two cities. Ordered as separate circuits, they will cost thousands of dollars for the installation alone and take weeks or months to get. A private T-1 network with spare bandwidth (or low priority traffic that can be interrupted) could set up those CAD/CAM connections in minutes. There would be no additional cost per month above the normal fixed cost of the network. The installation charge would be at most the cost of two expansion modules, typically about $2000 each.

Wide bandwidth channels can be set up in minutes for video teleconference;ing, facsimile transmissions, or computer-to-computer transfers of bulk data. With the flexibility of configuring by time of day, the T-1 network can replace the physical shipment of computer tape reels with a high-speed connection. Entire data bases can be backed up at remote locations in hours, or even minutes. As news of the new network's capabilities spreads through an organization, entirely new functions come to light. "Traffic expands to fill the bandwidth available." Experience shows the new traffic often saves money, improves productivity, and gives better customer service.

Examples

Simplification

To see what T-1 can bring to a network, consider a fairly large volume of voice and data traffic between Dallas and Denver. While hypothetical, the situation is very representative of real applications.

Fig. 1-1 shows a mix of voice and several kinds of data. Through simple statistical multiplexing, lines serving some of the remote terminals (async contending and interactive BSC applications) have already been consolidated. This example is conservative in not starting from the crudest possible arrangement. In the US today, many BSC terminals still operate over dedicated leased lines with a pair of modems devoted to each.

In the example, a multiplexer has put as many as 30 or 40 terminals on one 56,000 bit/s DDS line, saving a large number of individual lines. The same consolidation will be applied to the 56K lines by putting them on a single T-1 circuit, see Fig. 1-2.

Notice that the voice tie lines and the facsimile line are multiplexed onto the same T-1. Even with that much traffic the capacity of the T-1 hasn't been filled half way. There is still bandwidth of 896,000 bit/s available for future expansion.

The significant benefit of simplicity comes from the single point of control offered by the T-1 CPE (customer premise equipment). Instead of monitoring 20 separate circuits, there is only one. The lack of complexity makes it easier to track quality of service.

This example concerns only two nodes (locations). Many successful conversions to T-1 networking in the 1980s involved a dozen or more nodes. Some networks have approached locations and the forecast is for continued expansion. Rationalizing a network that large represents a simplification of far greater magnitude—and a correspondingly greater cost reduction.

Cost Reduction

Easier maintenance and higher quality have a price. Fortunately, most often the price is lower.

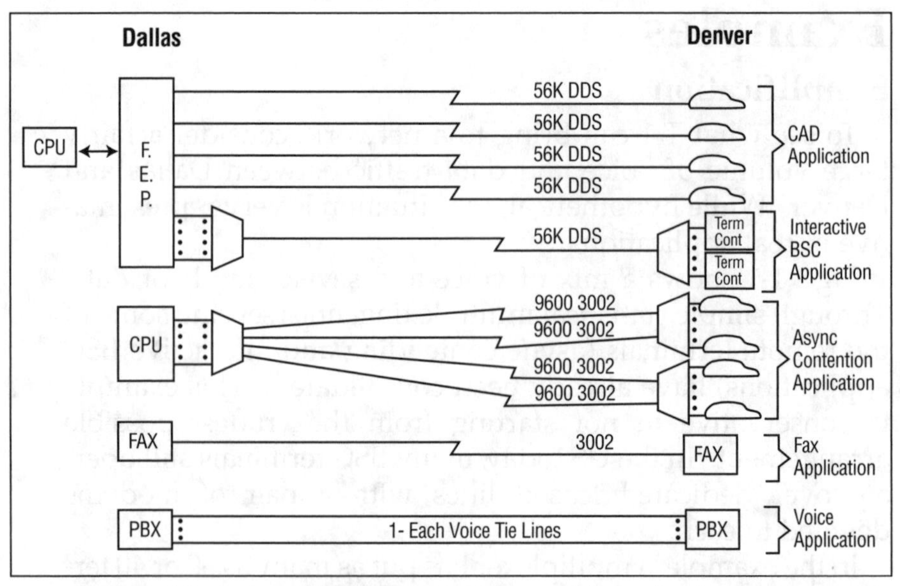

Figure 1-1 *Typical multi-line network reflects need for on-line interaction between corporate locations. While hypothetical, it resembles many real installations.*

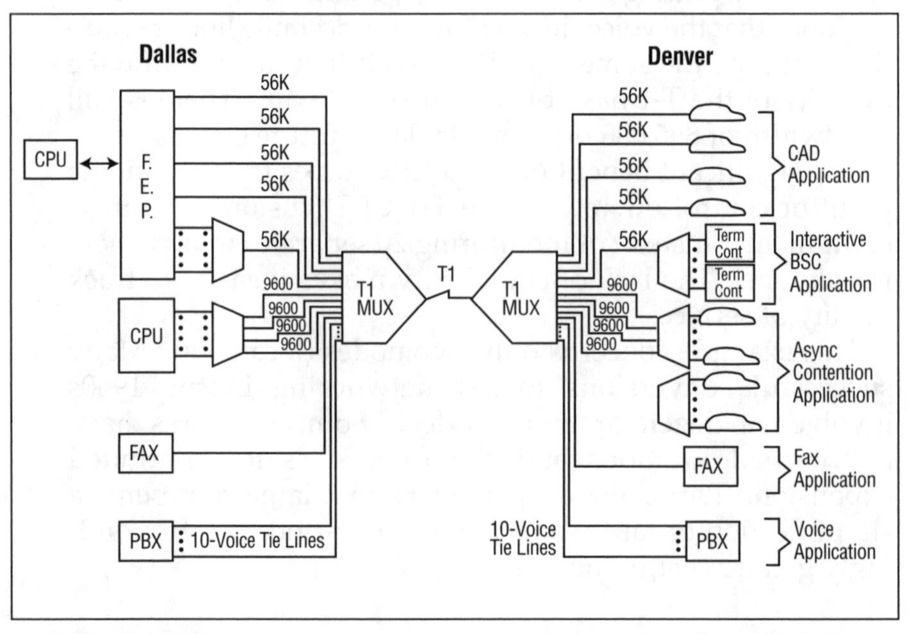

Figure 1-2 *Consolidated network combines all existing functions on a single T-1 circuit — and occupies less than half of the bandwidth. More that 800,000 bit/s remains for future expansion, which will cost nothing more for circuit capacity.*

The cost for the example in Fig. 1-1 was $27,710 per month in 1985 (Table 1-1). The new T-1 network costed out that year at $20,584 per month, or about $85,000 less per year. Since then the tariffs have moved to favor T-1 even more. The annual savings have increased to over $150,000 by 1988 (Table 1-2). In 1989 the per mile charge again dropped to $7.50, on top of a lower fixed charge of only $2025. Reductions in 56 kbit/s line charges were also seen in the early 1990s. By 1993 the per-mile charge for T-1 dropped to $3.50 per month. However, for the next few years the price trend was upward, almost doubling by 1997. Tariffs are all local, and change constantly.

The overall effect is still to push the balance more and more toward completely digital solutions and away from analog lines.

The yearly saving (Table 1-3) more than pays for a pair of T-1 multiplexers—and still saves money year after year. And for the relatively modest additional cost of expansion modules in those multiplexers, the single T-1 link will accommodate a doubling of the current traffic volume at no additional cost for lines.

This price analysis, while based on actual tariffs, is very conservative in its estimates of savings. Analog and DDS costs were calculated from the 1986 rates, disregarding a 3 to 5% increase in the following two years. T-1 rates reflect the reduction in January 1988, the second price cut in two years, but not the sharp reductions in 1989 and 1992. The trend of tariffs continues to be toward increased prices of leased voice-grade lines, with reduced costs for T-1 and other digital circuits.

Clearly the new T-1 services offer advantages to both Telco and user. With some new central office (CO) equipment, the phone company can lease existing local loops in the form of T-1 for 8-10 times what the wire would fetch as voice grade circuits. The user gets at least 24 (digital) voice channels for less than the cost of that number of analog lines.

Voice compression of the simplest sort (see Chapter 3) increases the number of voice channels on a T-1 to at least 44. Newer compression algorithms deliver many more with little or no reduction in sound quality. For that many voice channels, T-1 service is economical for any distance when compared to an equal number of analog leased lines.

A much bigger advantage can be seen with terminal data. A typical 9600 bit/s data channel occupies a full analog line, or 1/24th of a T-1. Now it need occupy only 9600 bit/s of bandwidth on the T-1 line. In other words, on the bandwidth used by 24 voice channels, a modern T-1 network can transport at least 150 data channels, each of which used to take a full voice grade line.

The really big payoff comes when voice and data share the same T-1. The convenience of a single circuit to monitor, flexibility, user control, and much wider bandwidth for high speed transmissions, adds to the value.

Line Costs For Figure 1-1

(1988 AT&T Tariffs)

Voice Grade:			
Inside Wire	2 @	$8.	$16.
Access coordination	2 @	20.	40.
Local loops:			
Intra-LATA	1 @	73.	73.
Inter-LATA	1 @	103.	103.
Central office connections	2 @	15.50	31.
Inter-office channel:			
Fixed charge		305.	305
Mileage	660 @	.30	198.
Total for lines			$766.
Modems (for data links)	2 @	109.	218.
			$984./month
56K DDS:			
Inside Wire	2 @	$8.	$16.
Access coordination	2 @	27.	54.
Local loops:			
Intra-LATA	1 @	109.	109.
Inter-LATA (10 miles)	1 @	328.	328.
Central office connections	2 @	50.	100.
Inter-office channel:			
Fixed charge			1325.
Mileage	660 @	1.60	1056.
Total for lines			$2988.
CSU/DSU	2 @	35.	70.
			$3058./month
Cost of Configuration in Fig. 1-1:			
5 56K DDS @ 3050.			$15,250.
5 voice grade data lines	@ 976.		7,880.
10 voice tie lines @ 758.			7,580.
Monthly line charges			$27,710.

Table 1-1

Line Costs For Fig. 1-2
(1989 AT&T Tariffs)

T-1 Circuit:		
Inside Wire (2)		$16.
Access coordination (2)		42.
Local loops:		
1 @ 1 mile (Denver)		807.
1 @ 10 miles (Dallas)		1163.
Central office connections (2)		124.
Inter-office channel:		
Fixed charge		2,600.
Mileage	660 @ 7.50	4,950.
Total for lines		9127.
ESF CSU/DSU (2)		170.
Total monthly line charge		$9,297.

Table 1-2

Line Costs Comparison

Multi-line network		$27,710. per month
T-1 network		– $9,297. per month
	Savings:	$18,413. per month
	x12	$220,956. per year

Table 1-3

Fast Internet Access

Just as the facsimile machine became an essential business tool, the company site on the World Wide Web is becoming a measure of the company itself. Net "surfers" will judge a company's commitment to customer service and even the usefulness of its products from the quality of the WWWeb display and its contents.

As congestion on the Internet increases the time to retrieve a page, each WWWeb site is being pressed to perform faster. A key element is the speed of the connection between the Web server and the Internet. Where 56 kbit/s has been the standard, a T-1 access line is part of many new installations and is a common upgrade for older servers.

There are drop-in interface cards for T-1 access that support routing protocols, frame relay, and X.25 services. Small routers are available with integral T-1 CSUs.

Since the Internet Service Provider (ISP) most often has a

local office, the T-1 is also local and not very expensive. Increasing graphics content of WWWeb sites, meet the declining cost of T-1 lines.

Merry Christmas.

Is T-1 for Me?

Here are some questions and answers that cover a simple (conceptual) approach to network design. They should help you determine whether a T-1 network makes any sense for your company or government organization:

1. Do I Seriously Need a Private Digital Network?

- Are there "technical" people at the top of my company? Does my company seriously want control of its own destiny?
- Are the "users" in my company (especially my president) requesting changes and additional connections I can't meet because the "phone company" (which ever phone company) delivers circuits too slowly?
- Has the management of multiple voice-grade and digital lines become a proverbial nightmare?
- Do I have difficulty finding people to do the management?
- Does my network map look like spaghetti, incomprehensible to figure, and impossible to diagnose?
- Has it become difficult to train new operators?
- Are "old-timers" in our company finding our telephone network impossible to learn? Do they feel they have a "right" to "dial around" our network? (i.e. use normal — but more expensive — dial-9, dial-up phone lines).
- Have my sundry and various phone bills for lines become incomprehensible and probably wrong?
- Have the expenses for dial-up phone calls between fixed points on my network become so high, it's obvious we should go to leased lines? (This is a simple break-even calculation.)

2. Do I Need a T-1 Communications Network?

Issues (perhaps not all will apply):

- In my company are communications becoming a key competitive weapon? For example, are we a bank running automated teller machines? Are we pulling orders in from our branches on data lines? Can our customers dial into our computers and place orders? Do we use teleconferencing to alert them to new products and/or train our sales people?
- Do I need many connections between a smaller number of sites?
- Do I need voice, data and perhaps video connections among my company's various offices and factories?
- Do I need high reliability? Are there "big" dollars at stake if our network goes down?
- Do I need the ability to monitor the quality and reliability of my own circuits? Do I feel uncomfortable about relying on outside vendors to do this monitoring?
- Have I asked several long distance vendors to check on the "logic" of my swapping my present "slow-speed" voice lines to higher-speed T-1 lines? Are they all hammering on my door eager to sell me T-1 lines?

3. Do I need Fast Access to a Service?

- Is Internet response too slow to tolerate?
- Has my data throughput needs outgrown a 56 kbit/s line?
- Are more applications moving to frame relay, more than will fit one or two 56 kbit/s lines?
- Does Basic Rate Interface service to the ISDN not meet capacity needs? (Primary Rate Interface ISDN, 12 times the capacity of a BRI, is a T-1 local loop.)
- Do I want to use Asynchronous Transfer Mode (ATM) at small offices that cannot justify an optical fiber access? (Carriers have started to tariff cell relay service, ATM, over T-1 local loops.)

If I think We May Need a Private T-1 Network...

...Here's what I should do now.

Do The Homework

- Consider the skill level of the present staff: Will they be able to move up to the more sophisticated management of a digital backbone network?
- Consider the budget: While monthly savings from reduced phone expenses should add up quickly, there is an initial capital outlay for new equipment and high installation charges for new lines.

The break-even calculation is simplest with existing analog leased lines vs. new digital circuits. The phone bills for the last three months will give a good average for the dial-up time between points. You may need a computer program to do the calculations.

Hint: Look for patterns in connections that could be concentrated between hubs. For example, there may not be more than two hours of connection per month between any one Midwest location and New York, but ten Midwest locations could put 20 hours on a backbone from NYC to Chicago or St. Louis.

- Consider the space: While not huge, new T-1 telecommunications equipment may fill more than one cabinet. (The smallest T-1 multiplexer is slightly larger than an attache case and hangs on the wall. The largest will fill about eight 6-ft. cabinets.) Where will I locate the Network Management Center?

4. Make the Decision

- Involve the telecommunications manager if there are any voice connections and the head of MIS (Management Information Systems) if there are any data connections. Involve the corporate audio/visual or presentations department, if there is one, to consider video teleconferencing and other image services.
- Involve the CEO: He or she will want to know that the new network will improve the reputation of the company and its finances.
- Involve the telecommunications staff. Get them trained on the new technology. Go to the manufacturer's school.

- Involve the PBX vendor; you may find your PBX offers a direct T-1 interface that could simplify network design significantly. (See the index.)

And when it's a go...

Design the Network

- Ask equipment vendors and carriers for help. They will offer proposals in considerable detail, and are usually glad for the chance to bid. Note that a carrier's design will probably stress lines and central office services. The hardware vendor will stress customer premises equipment. Take the best of both.
- Do you need a consultant? This book will provide enough information to make you dangerous. It won't make you an instant, seasoned expert. Consultants should be bought initially by the day, and then by the project, as you see how they work out. Remember consultants are vendors, eager to sell their ongoing services.
- Get it all down on paper. In pencil. Be prepared to change it. Nothing today is more volatile than a communications network.

Negotiate the Contracts

- You can get long-term lower, stabilized prices from some long distance and local carriers if you guarantee that you'll keep the circuits for x number of months. The longer you guarantee, the lower the price. Make sure you get some guarantee of quality built in, so you can escape the contract if the quality of the circuits becomes intolerable.
- The "list" price for hardware is for one item of equipment. As soon as you buy two (and you'll always need two) price is negotiable.
- Most T-1 equipment is very reliable. Most manufacturers guarantee their products for 90 days; some for a year. The guarantee should not start, however, until after you, the customer, have "accepted" the network. Most vendors in this business are actually very fair about their guarantees. They will fix, at their expense, equipment

on which there is honest doubt about whose fault it is. Fact is the stuff is electronic and most fails (if it is to fail) within the first 60 days. After that, if it's treated fine (housed in a happy not-too-hot, not-too-cold, not-too-humid home) it will often work forever.

Build the Network.

- Order the circuits. T-1 circuits have been known to take as long as 18 months. Sometimes they may not even be available. Even though T-1 uses two "normal" twisted pairs of normal telephone line, in some areas, the "normal" twisted pairs are not normal, having been repaired, modified, spliced, eroded and kluged beyond belief. Ordering and then checking circuits may be (correction — is likely to be) the biggest agony of implementing your T-1 network.
- Order the equipment. You can get common T-1 equipment on 90 days' notice. Some vendors keep hardware in stock for delivery in a few days.
- Build a "War-Room." This is a central room from where you can monitor your network's performance. This room should be a minimum of 20ft x 20ft. You should be able to house plenty of equipment in that room. This means air-conditioning, power, and lots of space for cables and manuals. More than one operator keeps a small futon mattress in the corner for those occasional emergencies.

Plan to Maintain It

- Documentation and training are the keys. There are great network management systems available now, that essentially put all your documentation "on line" in a small desk top computer. This allows the entire network to be manipulated from one point. Typically this software comes from a hardware vendor and runs on an IBM PC or compatible. The trend, however, is toward more powerful central processors. Engineering workstations, and even entire IBM mainframes, are being dedicated to the network management system.

- Service contracts with the hardware vendors are useful. Decide how critical your network is and buy service contracts based on hours of service per day needed. Vendors give preferential treatment to equipment under contract.

While most offer service on a per call basis, they may not get to you as quickly as you like without a service contract. Most repair can, however, be diagnosed from a central site and these days most repair consists of changing a board. If you're located remotely, or your network is vital, you may want to consider having several critical spare boards around. Put them under lock and key and keep them in their static-free containers. Better yet, install them in spare slots right in the equipment. The nodes should be able to monitor them for failures even when they are not on-line.

Start Planning Again

The next change and expansion in your network won't be long in the future. Successful networks are never static, but constantly changing and growing.

T-1 Defined Historically

A Short History of T-1

T-1 circuits are still new to many end users. The first widely tariffed T-1 service was offered in 1983. But in the voice world, inside the telephone companies, T-1 has been around since the mid 1960s.

The very first telephone service sold to the public connected only a few hundred phones wired to one central office. Each subscriber had a pair of wires that linked his instrument to the switchboard in the central office. The two wires were twisted to reduce crosstalk and extend the range.

To establish a call, the operator manually inserted plugs in jacks to set up one continuous loop of copper wire from the battery in the CO, to one subscriber, back through the CO to the second subscriber, and returning to the battery (Fig. 2-1). That portion of the circuit between each customer and the CO is still called the "local loop." Note that the path was solid copper all the way.

Isolated central offices yielded almost immediately to the demand for interconnection. Interoffice trunks were run (Fig. 2-2). Again they were solid copper pairs. The call connections were still set up by establishing a continuous copper loop between subscribers. Only now the loop could be

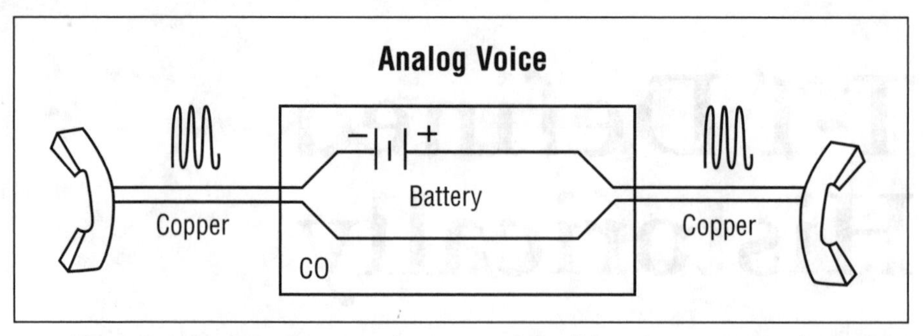

Figure 2-1. *Analog voice started as a two-wire solid copper connection in the form of one big loop between two subscribers through the battery in the central office. Hence the name "local loop" for the wires between the customer site and the CO.*

extended between central offices over the trunks. Each interoffice connection required a separate copper pair.

Within a few decades after Alexander Graham Bell commercialized the telephone, the number of local loops and interoffice trunks increased rapidly. The technology then was aerial wires on knob insulators. Soon the skies above cities were filled with poles, cross arms, and countless wires (Fig. 2-3). There was no more room for additional wire.

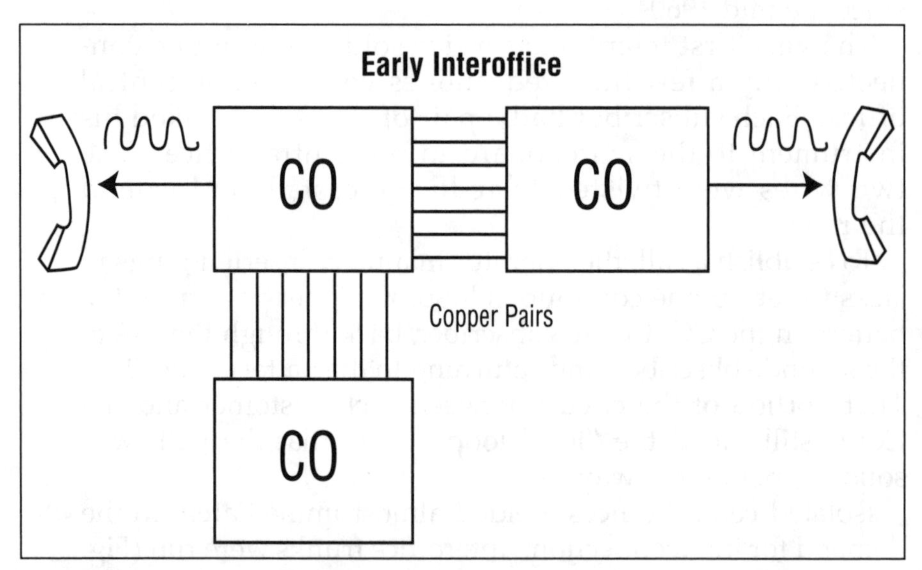

Figure 2-2. *Demand for extended connections required inter-office trunks. Originally these were solid copper pairs like the local loops.*

Figure 2-3. *The number of telephones in service expanded rapidly before the end of the 19th century. Every city and town was overrun by poles and wires.*

The vacuum tube arrived just in time to save the situation. With amplifiers and oscillators, telco engineers invented frequency division multiplexing (Fig. 2-4).

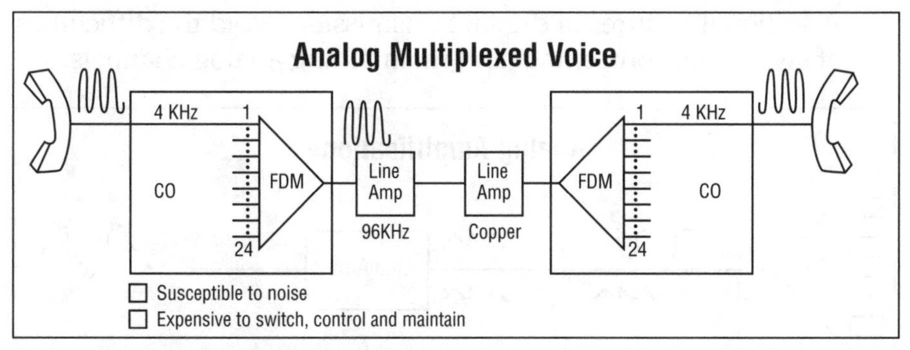

Figure 2-4. *Analog multiplexing reduced the need for wires. The technique assigned each voice channel to a different 4-kHz frequency band. When added together, 24 voice channels totaled 96 kHz, well within the capacity of a twisted pair.*

FDM

The solid copper loop can carry a much wider bandwidth signal than the 4,000 Hertz required for a single voice conversation. A pure metallic path can carry frequencies above 1 MHz. LANs, short haul modems, and Digital Subscriber Lines (see Chapter 4) prove this.

In the early stage of vacuum tube designs, however, conservative practice led to a top figure of 96,000 Hz. This allowed 24 voice channels of 4,000 Hz each to be multiplexed on two pairs of wire (one for each direction). These were the same wires that previously carried two conversations.

The Frequency Division Multiplexer (FDM) accepted the voice inputs at audible frequencies, 0-4,000 Hz. Channel 1 was transmitted unchanged. Channel 2 was translated electronically to the frequency range 4,000 to 8,000 Hz; channel 3, became 8,000-12,000 Hz; and so on. Adding all channels together produced a broadband signal that could be treated as one channel for transmission over interoffice trunks. (A fuller explanation appears later.)

FDMs were still in wide use up to the 1980s for long-haul circuits over microwave transmission links (remember when MCI blew up their towers?). They have been replaced with digital facilities. The main reasons FDM is gave way to digital signal processing is the susceptibility of analog paths to noise. Additional features of digital transmission avoid the difficulties of switching, controlling, and maintaining analog channels.

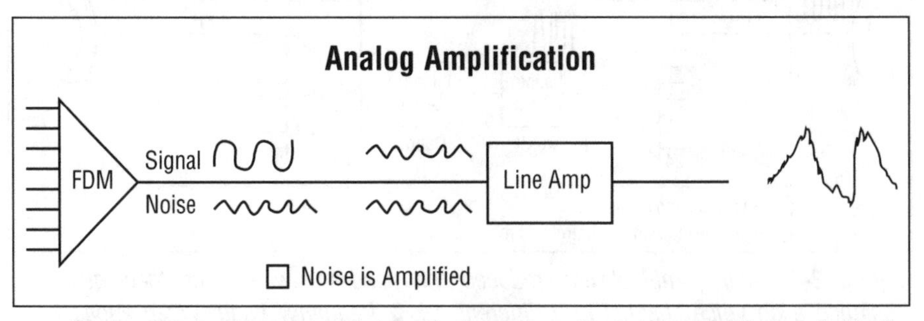

Figure 2-5*. Analog signals fade with distance and must be amplified periodically. Noise, present on all lines, is amplified also and accumulates.*

Line Noise

Every wire acts like an antenna. It receives signals from electrical radiation sources that exist almost everywhere. An electrical motor may spark, sending out radio waves. Even if it doesn't spark, the motor's magnetic fields disturb every circuit in the vicinity. Fluorescent lights, motor starters, switches, and even other phone lines contribute. They ensure every circuit has on it electrical signals that are not wanted: noise.

Noise exists uniformly along most lines. The desirable signal starts out strong at the sending end, but fades as it travels. The user's signal is weaker at the receiver (Fig. 2-5). The voice signal can be amplified easily, but the amplifier cannot separate voice and noise. Consequently, the noise is amplified too.

If the distance is great, the voice signal may have to be amplified many times. It will fade between repeater stations. A new component of noise is added by each subsequent line segment. The noise portion of the signal is added to and amplified each time. Eventually the volume of the noise exceeds the signal and the conversation becomes difficult or impossible to understand.

To be switched or accessed for diagnostics, operator intervention, etc. analog signals must be de-multiplexed. That is, they must be converted back to the 0-4,000 Hz range and placed on separate wire pairs. This is costly in equipment and power, and can be done a limited number of times before distortion exceeds acceptable limits.

TDM

To bring the problem of noise under control, the telephone industry developed digital transmission techniques. Digital methods address directly the problems caused by allowing the signal to take on any value. This is the case for continuously varying analog voice signals. In contrast, digital signals are allowed only a very limited number of digital values. The most often cited values are 0 and 1, but digital

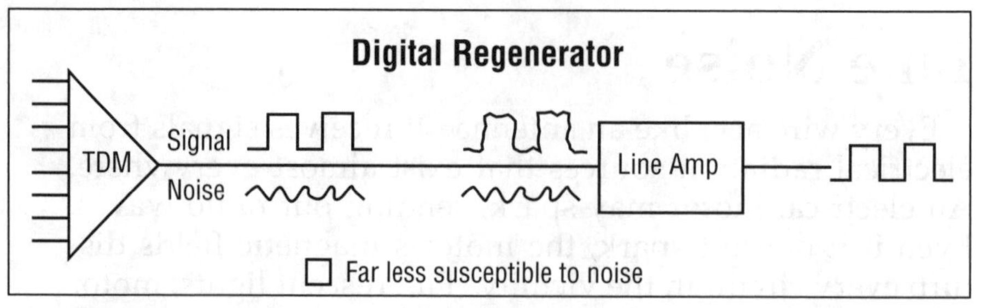

Figure 2-6. *Digital regeneration restores the original signal and thus eliminates most noise.*

transmission actually uses -1, 0, and +1; or +/-1.

The great difference between analog and digital is that the digital receiver knows to a great extent what to expect. An analog signal legitimately can be any value in a range, so a slight variation due to noise is difficult to detect and almost impossible to remove.

A digital signal can have only a few true values. Deviations from the allowed states represent fading or noise. A digital amplifier, therefore, can distinguish between the signal and noise. The line amplifier (Fig. 2-6) is really a regenerator. It examines the input at periodic time intervals, and decides what value the input probably represents.

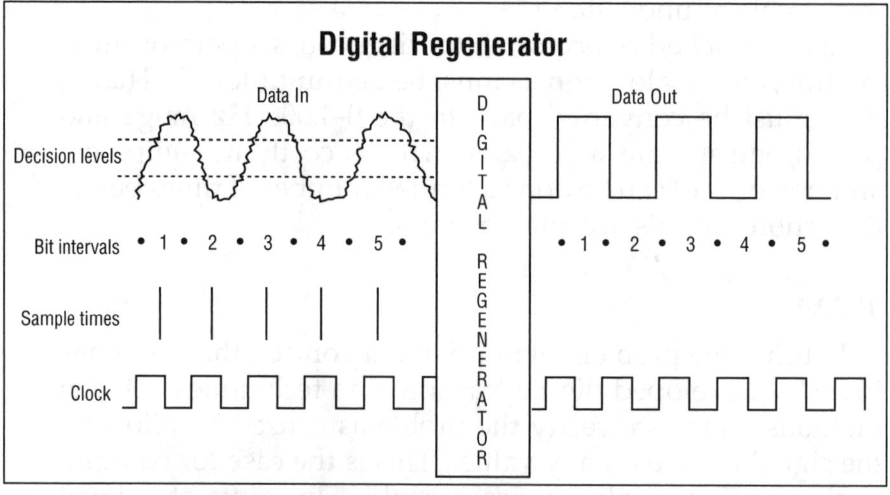

Figure 2-7*. Sampling at clock times, the regenerator decides if the line voltage is above the upper, or below the lower, decision levels. If so, the signal is regenerated exactly. If not, noise is too great to tolerate and causes an error.*

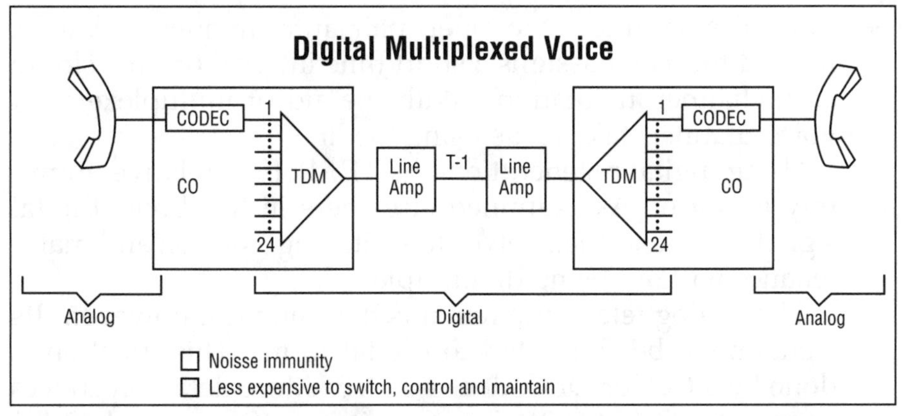

Figure 2-8. *A channel bank digitizes analog voice signals, then multiplexes the information into a composite signal on a digital line.*

Exactly that value is then sent on the output side.

The regenerator samples the voltage on the line at time intervals controlled by the system clock. The sampled value is compared to two preset decision values (Fig. 2-7). Clearly if the sample is above the upper level, or below the lower level, the choice is easy and the signal can be regenerated without error. Greater than expected noise will produce a sample between the levels. Here the regenerator cannot decide and may simply indicate an error.

The transistor and digital logic made possible a new device: the voice Time Division Multiplexer (Fig. 2-8). This is also known as a channel bank. The TDM doesn't divide a

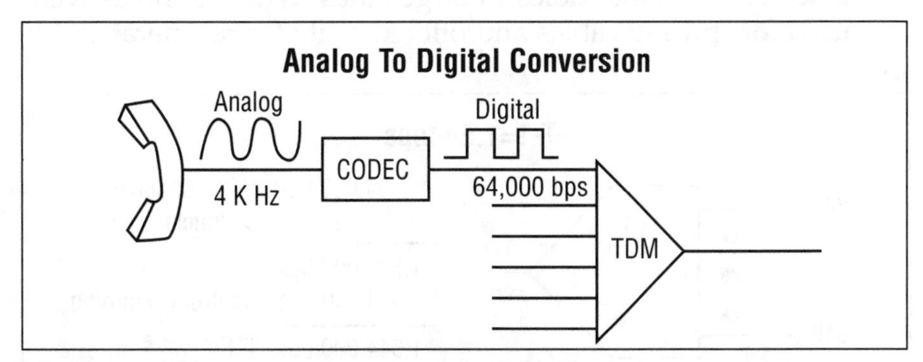

Figure 2-9. *Inside a channel bank, a COder/DECoder converts analog voice into a digital bit stream. Operation is full duplex, but there are usually separate sections for signals traveling in each direction.*

transmission line or copper pair into frequency bands. Instead the TDM assigns a short time interval or time slot to each channel in rotation. With the aid of multiplexers, 24 voice channels take turns using the line.

Using digital regeneration, the TDM offers relative immunity to noise. As examined elsewhere in this book, digital signals also lend themselves to switching, control, and maintenance without being demultiplexed.

The analog telephone remains the standard, however. Its signal must be converted to digital form. This function is done by a COder-DECoder or codec. More detail on codecs appears in Chapter 3. For now, it is sufficient to describe them by their input and output (Fig. 2-9):

IN = an analog voice signal of 300-3,300 Hz bandwidth.
OUT= a 64,000 bit per second digital stream.

The codec is bidirectional, so "in" and "out" refer to the end user's view of the network.

The Digital Hierarchy

A single voice signal in digital form takes 64,000 bit/s. This is known as digital signal level zero, or DS-0. When T-1 originated, it turned out that 1.5 Mbit/s (megabits) was about the highest rate that could be supported reliably for the 1-mile distance between manholes in large cities. The manholes were used for splicing cables and offered a site for regenerators. It

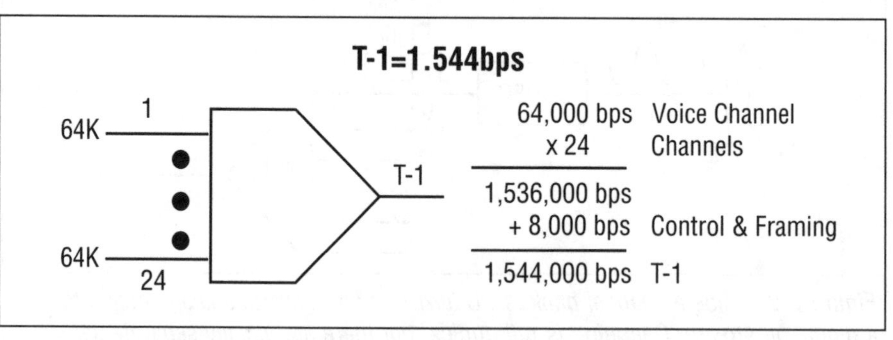

Figure 2-10. Derivation of the T-1 rate involves 24 voice channels plus 8,000 bit/s for control and synchronization.

was natural to retain the convention that originated with FDM designs and multiplex 24 channels on the new TDMs.

At 64,000 bit/s each, the aggregate data rate of 24 channels became 1,536,000 bit/s (bits per second). To keep the two multiplexers synchronized, an additional 8,000 bit/s was added. The total then is 1.544 million bit/s. This is the standard T-1 rate in North America, Japan, and Australia (Fig. 2-10). In European and other CCITT countries, 30 voice channels and two signaling channels are multiplexed at a rate of 2.048 Mbit/s (2,048,000 bits per second). Both are called data signal level one, or DS-1.

The term "T-1" originated with the phone company as a very specific type of physical equipment: selected cable pairs and digital regenerators at 6,000 ft intervals. The term has become somewhat corrupted in common usage, and now T-1 refers to the DS-1 rate more than the cable system.

Note that the 8,000 bit/s of control or framing bits are not available for user data. The fastest serial interface, like a V.35, that can feed into a T-1 is what will fill the 24 time slots, or 1.536 Mbit/s.

Later developments improved on the capacity of the original T-1 repeaters. It was found that by placing the regenerators at 3,000 ft intervals the bit rate could be doubled to 3.152 Mbit/s. This is called DS-1C. As newer electronics appeared, the rates climbed, creating an entire range of digital signal levels, and corresponding regenerators and telephone cable (Table 2-1).

T-1 Everywhere

Initially the T-1 facilities carried trunks between central offices. As channel banks became smaller and more reliable, telcos (telephone companies) realized the customer who needed 20 outside lines could be served as well by two twisted pairs. A channel bank at the customer site (Fig. 2-11) would save up to 22 local loops. The number of wires saved would be higher if the subscriber wanted E&M trunks (see Chapter 3).

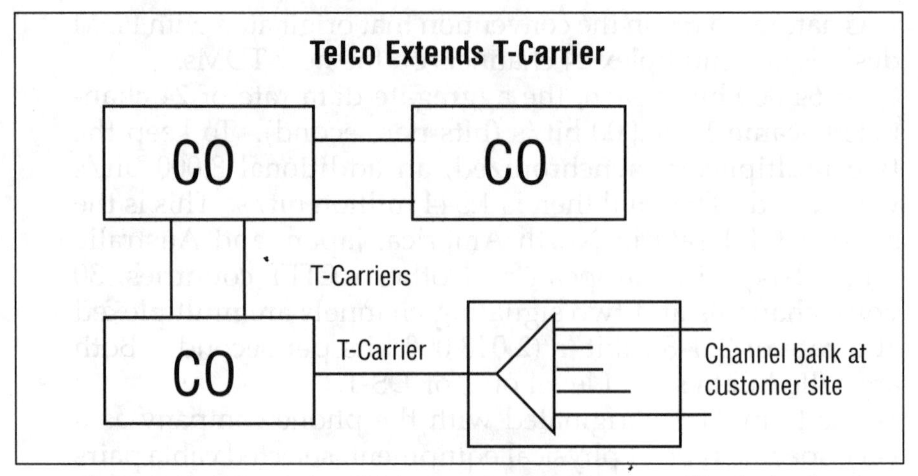

Figure 2-11. *Not only has T-1 become common between COs, it also has been used to extend multiple trunks to customer sites.*

Saving wires is an important consideration in densely built up areas, like New York City, where there is little or no room for additional wires in the sub-street conduits.

Likewise, the savings in wire pairs is important where the local loops are long. It these cases the telco must use heavier wire to keep resistance within limits. That is expensive. Long loops are common in rural towns and new suburban developments. The metal cabinet at the end of the block in those areas may contain a specialized channel bank, called a remote terminal, as easily as simple wire connections.

Through the spread of digital interoffice circuits and customer premises equipment, T-1 reached almost everywhere in the U.S. by 1980. But it was never available to end users except as a custom engineered installation at great cost. The only regular user up to 1982 was the federal government.

Phone companies historically had an aversion to selling bulk bandwidth, and the FCC considered it improper to discount the price of any service. Even though T-1 is less expensive to provide than the equivalent number of analog voice lines, the tradition of charging by bandwidth (as against cost) made T-1 uneconomical. The user who couldn't use 1.5 Mbit/s directly (as for inter- computer connections) had to add expensive multiplexing equipment that was unnecessary when buying separate circuits.

Starting in the 1980s, however, things changed. The FCC and telcos adopted the concept of cost-based pricing. Now it became possible to offer T-1 at less than the cost of separate lines. The saving was so great that demand for the first tariffed offering soon pushed the waiting time for new T-1 installations to over a year.

With the completion in the late 1980s of several fiber optic networks across the U.S., the availability of T-1 reduced waiting time to a few weeks in most metropolitan areas. Additional facilities constructed by local bypass companies, the teleport in many cities, mean that major office buildings have bandwidth available almost immediately.

The Plesiochronous Digital Hierarchies

Signal Level	Digital Bit Rate	Equivalent Voice Circuits	Carrier System	Usual Medium
North American				
DS-0	64 kbit/s	1	(none)	UTP Wire
DS-1	1.544 Mbit/s	24	T-1	"
DS-1C	3.152 Mbit/s	48	T-1C	Shielded
DS-2	6.312 Mbit/s	96	T-2	TP Cable
DS-3	44.736 Mbit/s	672	T-3	Microwave
DS-3C	89.472 Mbit/s	1344	T-3C	or
DS-4	274.176 Mbit/s	4032	T-4	Optical Fiber
European (CEPT)				
0	64 kbit/s	1	(none)	UTP
1	2.048 Mbit/s	30	E-1	UTP/coax
2	8.448 Mbit/s	120	E-2	"
3	34.368 Mbit/s	480	E-3	Optical Fiber
4	139.264 Mbit/s	1920	E-4	"
5	565.148 Mbit/s	4032	E-5	"
6	2,200.00 Mbit/s	30,720	E-6	"
Japanese				
1	1.544 Mbit/s	24	J-1	UTP
2	6.312 Mbit/s	96	J-2	Shielded TP
3	32.064 Mbit/s	480	J-3	Optical Fiber
4	97.728 Mbit/s	1440	J-4	"
5	397.2 Mbit/s	5,760	J-5	"

Table 2-1

International service on fiber under the oceans has spread digital connectivity to most major cities in the world. And when the newest VSAT technology put T-1 on affordable ground stations (see Chapter 5), T-1 truly became available everywhere.

Digital Voice

In practice, the economics of T-1 transmission are most attractive when combining voice and data, rather than when sending one type of information alone. To be mixed and handled efficiently on one network, however, the two forms of signals must be made compatible.

Mostly, on today's Public Switched Telephone Network (PSTN) — the normal switched long distance phone system with analog access lines we use every day — data sent by a phone subscribers is made like voice with the aid of modems (Fig. 3-1). In modern digital networks, analog voice signals are converted to digital bit streams and treated like data.

So, in one case, you make the data like voice for transmission; in the other you make the voice like data. Ultimately, everything ends up on digital transmission facilities.

Talking about T-1 transmission means we're talking about transmitting information in digital form. There are several ways to digitize voice. This chapter's review of digital voice covers only the most common techniques of digitization. The goal here is to describe where each is applied and to

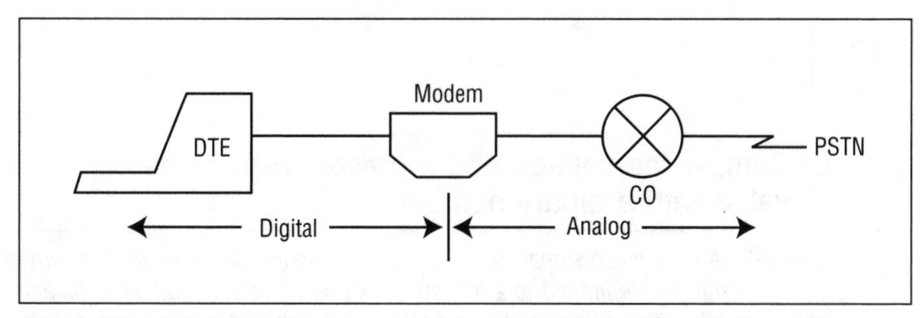

Figure 3-1. *Digital signals from a terminal or computer are converted to audible tones in a modem, to make them compatible with the analog PSTN.*

show how the selection of a digitizing method affects other aspects of networking. For a more complete discussion of digital voice, see the author's "Voice Over Frame Relay."

Analog versus Digital

Standard telephone signals have always been analog transmissions; they vary continuously, like the loudness in voice they represent. Analog signals can take on any value between wide limits (Fig. 3-2). Digital signals are allowed to take on only discrete values.

The problem of digital voice comes down to selecting the "best" way to represent a smoothly varying quantity. The constraint is on the limited number of bits that can be transmitted economically. "Best" often implies maintaining voice quality. But there are possible trade-offs among quality, bandwidth, cost, availability, reliability, and even pressure to adopt a fad.

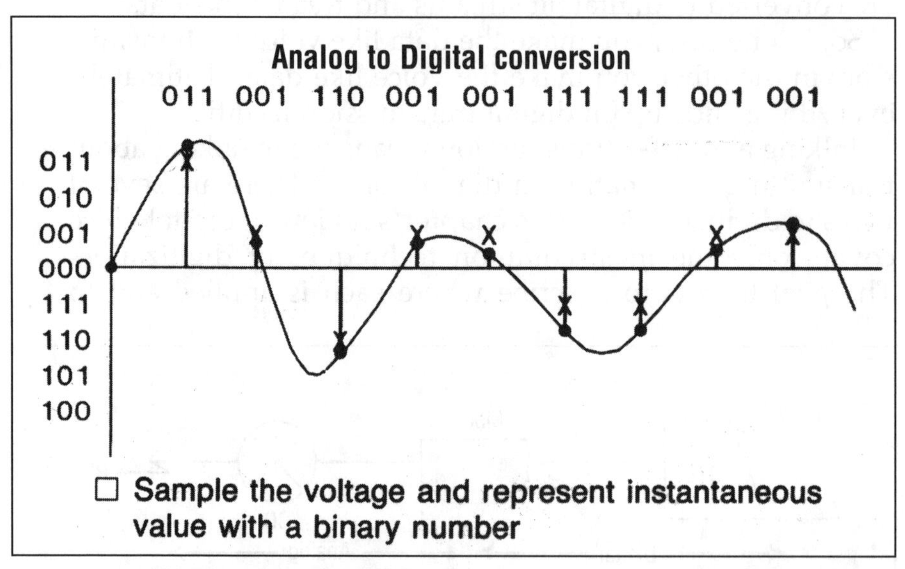

Figure 3-2. Analog voice signal (solid curve) takes on any value within wide limits. Digital signals are limited to a few steps (X points, corresponding to binary numbers at left). The difference between the actual value (dot) and closest digital representation (X) is quantizing noise.

Pulse Code Modulation

The existing worldwide standard for digital voice is called Pulse Code Modulation (PCM). Under good conditions, the quality of this method of voice transmission can be excellent. It is praised as "Toll Quality;" because it first appeared on interoffice and toll trunks between telephone company central offices. These have the strictest quality requirements.

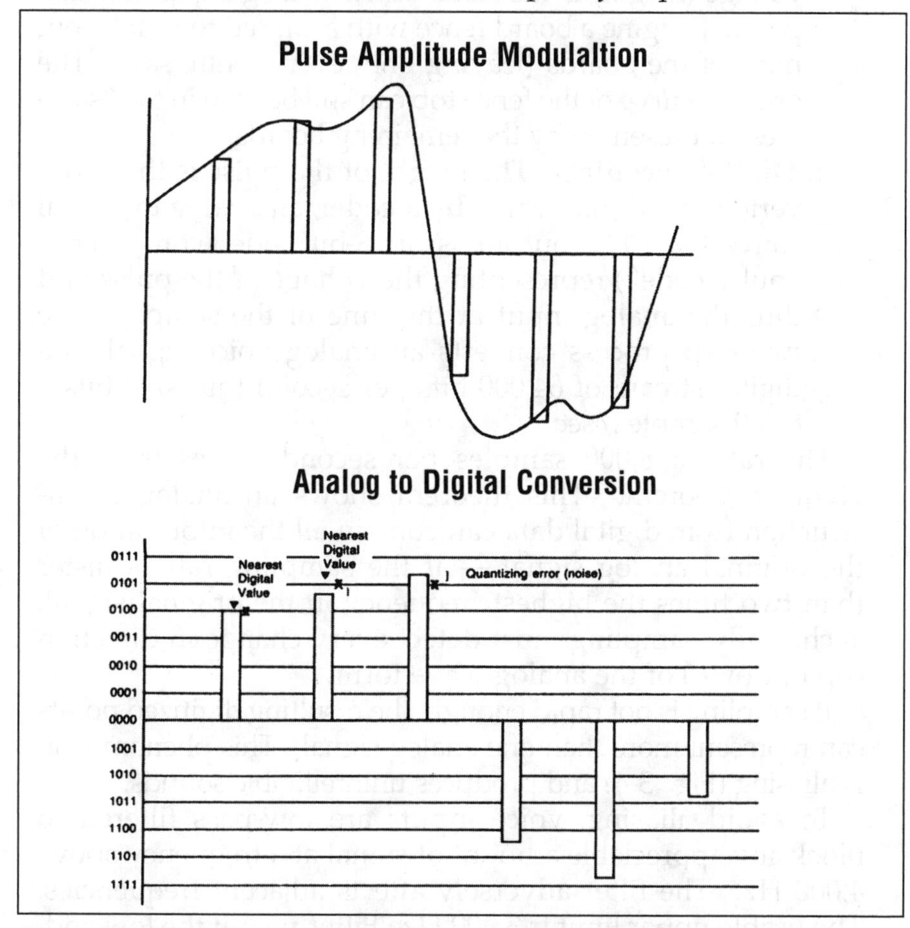

***Figure 3-3**. Samples of the analog input are taken at 1/8,000 second intervals. Sample values are converted to pulses (A), a process called Pulse Amplitude Modulation (PAM). Heights of pulses are measured in digital numbers (Analog to Digital conversion). The digital value is what is transmitted as a stream of binary bits. Received bits are converted back to numbers, then pulses. Pulses are filtered to recreate the analog wave.*

The PCM conversion between analog and digital can be done in one step, within a single integrated circuit chip, the codec (COder-DECoder). Traditionally, it is done in two steps (Fig. 3-3):

1. **Pulse amplitude modulation.** The incoming analog signal, representing the variations in a voice, is sampled or measured 8,000 times per second. The modulator uses the sample to send a very narrow square wave pulse whose voltage (height) is the same as the analog signal's at that point. Imagine a board fence with a curved top. Take out most of the boards (leaving only every tenth, say). The original curve of the fence top can still be seen in the "samples" represented by the remaining boards.
2. **Digital encoding.** The height of the pulse is then converted to a digital value by a coder, an analog to digital converter. The output is an 8-bit code word (hence "pulse code") representing the voltage of the pulse and thus the analog input at the time of the sample. The two-step process converts an analog voice signal to a digital stream of 64,000 bits per second (bit/s); 8 bits x 8,000 samples/sec.

The rate of 8,000 samples per second comes from the Nyquist theorem. This theorem shows an analog reconstruction from digital data can contain all the information of the original analog signal — if the sampling rate is faster than two times the highest frequency in the original signal. Technically, sampling must detect every change in direction (up or down) of the analog wave form.

If sampling is not rapid enough, the resulting digitized points can represent more than one analog signal. This phenomenon is aliasing (Fig. 3-4) and produces unintelligible sounds.

To avoid aliasing, voice inputs are low-pass filtered to block any appreciable amount of signal at a frequency above 4,000 Hz. The filter adversely affects adjacent frequencies. The usable upper limit is 3,300 Hz. Filtering out the low end, to block 60 Hz hum, puts the practical lower limit at 300 Hz.

The size of the sample, 8 bits, was determined after considerable experimentation, and a large amount of invention. The problem was to optimize the trade-off between bit rate and voice quality. The fact that computers were then settling on 8-bit characters favored the choice too.

The digital representation of a signal can take on only a small number of values, at discrete steps. An analog signal, by contrast, has almost infinite variability. Therefore, at the precise time of a sampling, an analog input is seldom exactly the same as a possible digital output step. The coder, however, must make a selection, and will pick the closest digital value. The difference (between the 'analog' dot and the 'digital' X) is digitizing distortion, or "quantizing noise." The human ear is very sensitive to quantizing noise. The distortion sounds bad. The quantizing process can be compared to someone measuring the height of the boards in a fence to the nearest foot.

Early listening tests showed that if many digital output values were very close together, the quantizing noise could be reduced to where it was not important. Unfortunately, the number of digital values required to cover the full volume range of a voice signal in such small steps is at least +/- 2,000. This is comparable to measuring the fence height in millimeters. To number that many steps requires 12 bits per sample. At 8,000 samples per second, that would be 96,000 bit/s.

Even 25 years ago, when T-1 was introduced, designers recognized the possibility of compressing voice. They simply gave less attention to the very loudest levels. That is,

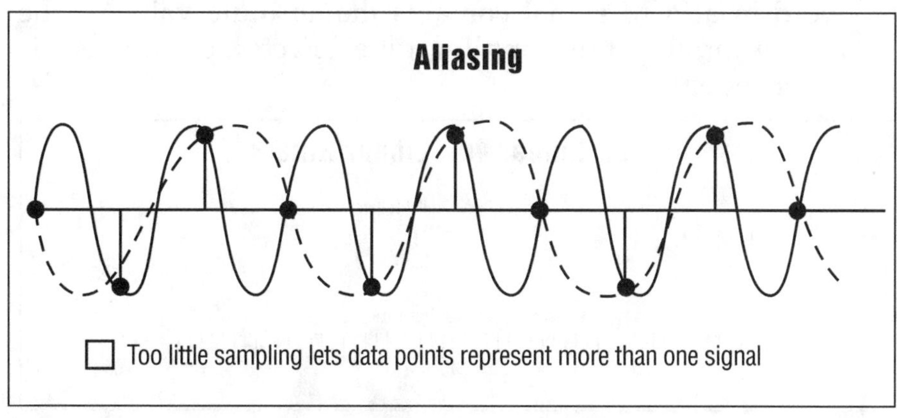

Aliasing
Too little sampling lets data points represent more than one signal

***Figure 3-4**. Too few samples of a wave allow the resulting digital points (dots) to represent more than one signal. Aliasing is avoided by filtering input signals to eliminate high frequencies. Filtered voice signals must have nothing higher than half the sampling rate of 8,000 times per second, or 4,000 Hz.*

concentrating the digital output steps in the low and normal volume range reduces the number of steps needed for "toll quality" to 256. In effect, the quantizing noise has been kept very low at low volume levels, but allowed to rise with loudness. The effect is masked by other distortions created by the microphone, receiver, and lines when the volume is high.

To use the fence analogy, a few inches may make a difference when it is small enough to step over. Once the fence is 10 ft tall, another foot hardly matters. To concentrate the measuring at the low end of a range requires a highly non-linear ruler (Fig. 3-5). To measure a fence with it, some graduations might be 1 mm apart; others, as much as 1 foot apart. Voice engineers designed a non- linear "voice ruler" with the "fineness" of an adequate linear rule near zero, and wider spacing at louder levels. This technique needs only 8 bits to measure pulse heights over the full volume range of a voice.

In other words, a non-linear voice encoder saves 33% of the digital bandwidth. The original signal is compressed for transmission, then expanded at the receiving end to the full range. The two-phase process is known as companding.

Thus PCM, today's standard, is itself a form of voice compression. The specific form of non-linearity is the "mu-law" algorithm in North America and Japan, "A-law" in the rest of the world. Note that each sample and each digital word in a PCM signal contains the absolute value of the analog input at the sampling time. Each measurement is independent.

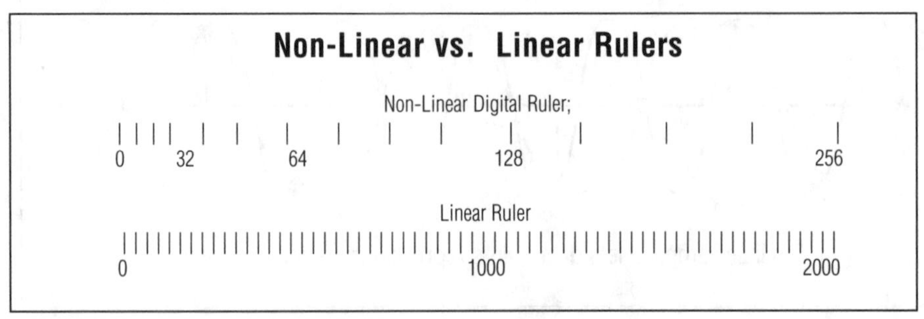

***Figure 3-5**. To concentrate on the low end means to be more exact with low signals than with loud passages. To do this the pulse height is measured in a non-linear way. For example, it can be compared to measuring a fence with a non-linear ruler. Companding makes the measurement more exact at low volume levels.*

Getting Your Voice Back: Digital To Analog Conversion

At the receiving end, the 64 kbit/s digital stream is divided into 8-bit words, which are decoded to values for analog output. Each received byte, every 1/8,000 sec, sets the height of a narrow analog output pulse. As closely as can be measured by the non-linear rule, the heights of these received pulses are the same as the pulses created by the sending PAM modulator. A low pass filter smooths the output into a continuously varying analog signal very similar to the input at the sending end. In effect the fence is being reconstructed by first erecting every tenth board, to measure, then filling in the spaces between to match.

The Voice Channel Bank

Millions of voice channels today are digitized and multiplexed in PCM voice channel banks (Fig. 3-6). The least expensive devices separate the two-step conversion process for economy, eliminating multiple coder-decoders.

1. **Pulse amplitude** modulation takes place at individual analog input ports. The pulse streams from 24 channels are interleaved on a single analog PAM bus. Each of the 24 signals appears on the bus in turn.
2. **Digital coding** is done by one codec, between the analog pulse bus and a high speed digital (T-1) line. Synchronization from a central control circuit ensures that each modulator sends its pulses at the correct times.

When receiving, a channel bank's codec converts the T-1 bit stream to pulses interleaved on the analog bus. Individual ports pick off their own pulses by watching a specific, repetitive time slot. The pulse amplitude modulated signal is then filtered at the port to recreate the original analog signal.Obviously, a device designed for voice and based on an analog PAM process must be modified to handle digital signals. More recent channel banks digitize voice channels individually or in pairs ("coder on the card"), then multiplex onto a digital bus.

Current channel banks are also much more sophisticated in handling data. They offer many options, in the form of hardware "plugs" or plug-in modules, to handle specific forms of data:

- DDS subrates (9600, 4800, 2400 bit/s)
- 56 kbit/s data ports (DDS)
- 64 kbit/s clear channels
- 64 kbit/s byte-synchronized in a 72 kbit/s signal (DDS-II)
- High-speed ports for N x 64 kbit/s (DSL).

These plugs occupy the same physical positions as voice modules (they plug into the chassis) and the same electrical position on the digital bus. In this way they can accommodate data channels without passing through an analog stage.

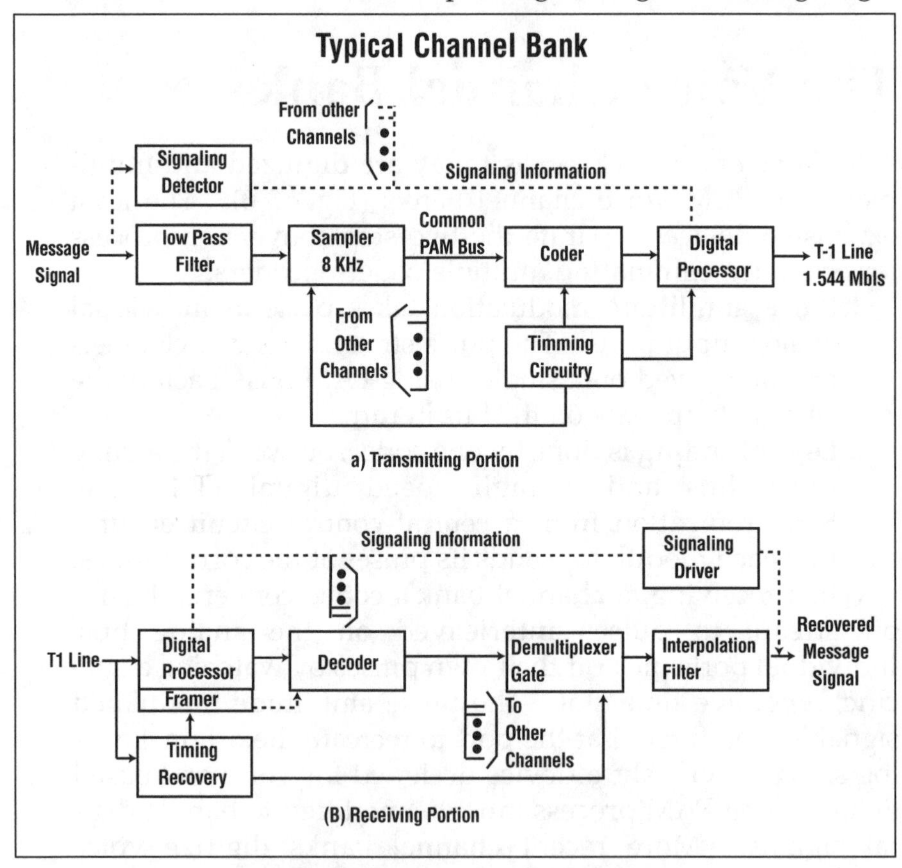

Figure 3-6. *Channel Bank block diagram shows separation of PAM and digital coding processes. Sending and receiving involve independent sections.*

Traditional channel banks offer more than 50 different plug-in modules, each specialized for a particular function. For example, the 2-wire and 4-wire E&M interfaces would be on different modules. There are, however, many devoted to applications with limited need in private networks.

The latest channel banks, first introduced in late 1987, employed more advanced semiconductor technology. New hardware (based on all the current buzz words) gives channel banks many of the features and functions of a T-1 multiplexer. For example, CMOS chips reduce the power consumption to a fraction of that consumed by traditional channel banks. Surface mount technology on thick film hybrids reduces the size to something unimaginable before 1985. And individual I/O modules made of custom chips can be programmed remotely to configure and manage the channel bank. This last feature drastically reduces the number of different I/O modules or "plugs" needed.

As channel banks advanced, as of early 1988 they became repackaged T-1 multiplexers. As T-1 muxes in disguise, they offered compatibility, on the T-1 line, with larger networking multiplexers. In other words, these new channel banks share the network management system while supporting many features and functions that go well beyond the capabilities of traditional channel banks:

- Soft configuration (from remote sites, too);
- Programmable modules (one can be set to perform the functions of many different "plugs");
- Remote diagnostics;
- Integrated CSU under the same management system;
- Local cross connections, between low speed ports;
- Protection switching to back up critical circuits.

(To quote Charley Dykas, "It's a mux.")

New designs incorporate ISDN technology. In 1990 there was no uniform standard for ISDN interfaces — all implementations were proprietary. Starting in the mid-1990s, they all started to migrate to a single standard in the US: National ISDN-1. Successive enhancements have been laid out through NI-3, and eventually all the US carriers will adopt it. In Europe there has been wider agreement among national authorities on a single

Euro-ISDN specification that simplifies building equipment and ordering lines. (Which may explain why ISDN is so much more successful in Europe than in the US.)

The Nature of Voice Signals

Most of the information in a voice is conveyed by frequencies around 1,000 Hz. The very low frequencies are more important in male voices. The highest frequencies, near 3,000 Hz add individuality and aid in recognition but do not carry a large percentage of the voice power.

Pulse Code Modulation

PCM sends a complete description of every sample. This completeness allows PCM to handle analog waves that jump from one extreme to the other while introducing a minimum amount of quantizing noise (Fig. 3-7). But this agility is called on only during the loudest passages of the highest frequencies.

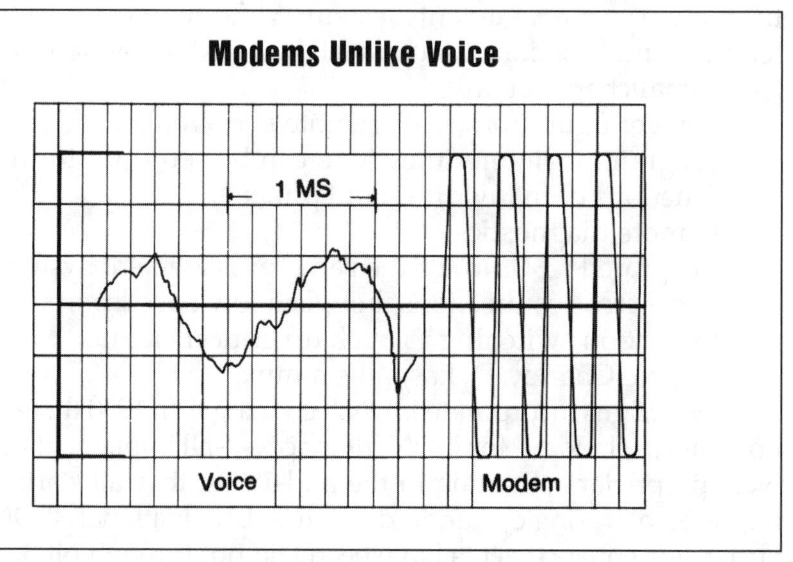

***Figure 3-7**. Human voices show the largest component frequency at or near 1,000 Hz. A modem signal, to take fullest advantage of the channel capacity, will more often use the combination of maximum frequency and full amplitude.*

Such an extreme situation might arise from a high speed modem signal, but normal voices do not change that rapidly. There is a strong correlation between adjacent samples in normal conversations. In other words, a PCM signal contains redundant information. The implication is that PCM uses more bandwidth than necessary.

Differential Pulse Code Modulation

Voice signals vary relatively slowly. The change between successive samples, the differential, is usually much less than the full volume range. This means that (for the same accuracy) fewer bits are needed to describe the change from one sample to the next (the differential) than to describe the full range.

The technique of Differential PCM (Fig. 3-8) can work with any number of bits. Generally DPCM employs 4 bits per sample. There is no free lunch, however. On those occasions when the input signal really does change drastically between samples, the DPCM technique is not able to follow the input precisely. The discrepancy introduces large amounts of quantizing noise and distortion. Consequently, 4-bit DPCM probably won't pass a modem signal of 4,800 bit/s. The proliferation of facsimile machines revived the concern for modem traffic in private T-1 networks. Where older forms of "digital" information were accommodated in

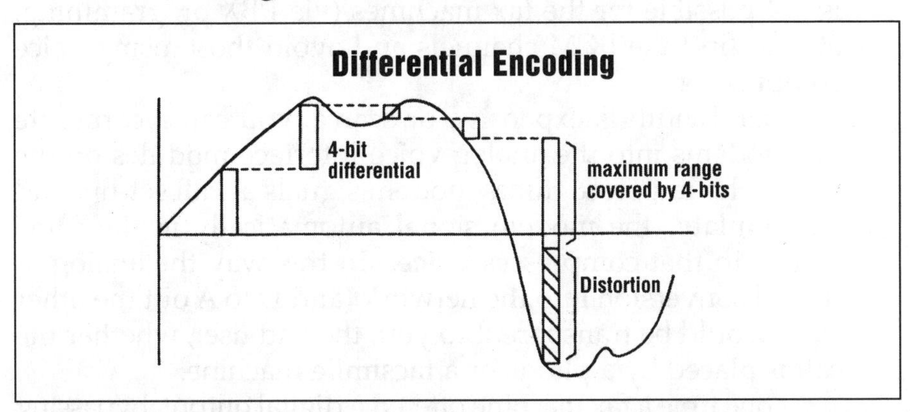

***Figure 3-8**. Differential PCM transmits only the change since the last sample. By contrast, PCM sends the absolute value of the analog input. Normal voice signals work well, as they change relatively slowly. However, sharp changes (high frequency and full volume) may exceed capacity and cause distortion.*

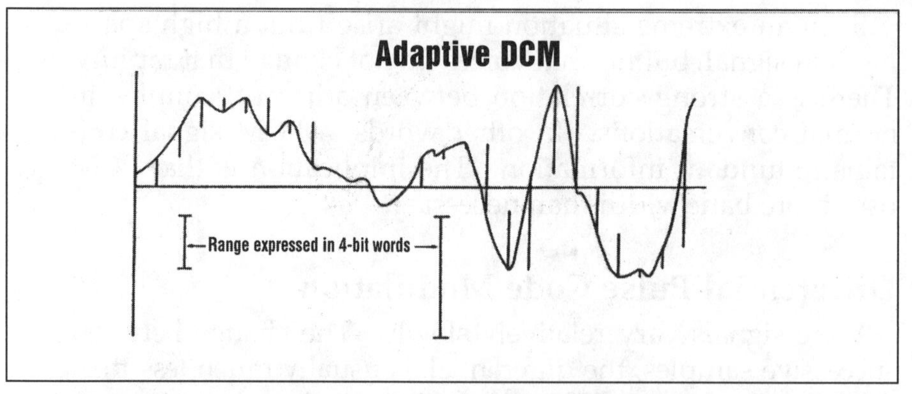

Figure 3-9. *The range or maximum size of the differential can be changed to accommodate the input signal. Loud passages at high frequency increase the differential. A running average of the most recent differential controls the range.*

digital form, the fax, with its internal V.29 modem, insists on working into analog equipment.

The problem involves voice compression. An ADPCM channel cannot carry a 9600 bit/s modem signal. Voice channels compressed to even lower digital bandwidth, like 16 or 8 kbit/s, may not be able to handle the fax fallback speeds of 4800 and 2400 bit/s. There are several workarounds available.

Certain voice channels can be run at the full PCM rate of 64 kbit/s, ensuring fax machines work at full speed. The administration of these channels may become a problem if it is not possible for the fax machines (via PBX programming, etc.) to find the PCM channels and avoid those using voice compression.

If bandwidth is expensive or scarce, you can incorporate fax modems into the analog voice interface modules on the mux. The mux recognizes modem signals at call set-up, and demodulates the modem signal automatically in the same DSP chip that compresses voice. In this way the analog to digital conversion into the network (and D to A out the other side) would be transparent to you, the end user, whether the call is placed by a phone or a facsimile machine.

At one time a fax machine offered a digital output, bypassing the internal modem when a digital transmission channel was available. There apparently was not enough consumer demand to make this a popular feature.

In the mid-1990s, frame relay vendors offered a way to carry only the necessary digital bandwidth for fax transmissions, and not the additional burden of an encoded analog signal. This method was first standardized as FRF.11 in 1997. The "network" side connections between two specific frame relay devices can be carried over T-1 networks, making mixed voice and fax transmission available without a frame relay switch or FR service.

Adaptive DPCM

DPCM can be improved without increasing the bandwidth. A sophisticated algorithm can assign different meanings to the 4-bit digital words to suit different conditions. Specifically, the volume range represented by the 4-bit word can be increased or decreased.

For example, if successive voice samples suddenly were very wide apart, the differential normally expressed in four bits could not match the actual change. The DPCM algorithm can be made to adapt in the situation of loud volume by increasing the range represented by four bits (Fig. 3-9).

In the extreme, 4 bits might cover the range normally represented by 8 bits.While reducing quantizing noise for large signals, this adaptation will increase noise for normal signals. So when the volume drops, the ADPCM system reduces the

Comparing Different Voice Digitization Methods

Encoding Technique	Bits per Sample	Sampling Frequency	Total Bit Rate	Interpretation of the Digital Data
PCM	8	8,000	64,000	Value of analog signal at this time (this sample)
ADPCM	4	8,000	32,000	Change in analog signal level since the last sample
CVSD	1	32,000	32,000	Does slope of the signal curve increase or decrease
VQL	3.5	6,66	32,000	How loud is this sample compared to the loudest "recent" sample
VQC	varies	8,000	8K, 16K	Which "vectors," or fixed patterns of 1s and 0s, does the PCM signal look like most recently
HCV	1, 2	8,000	8K, 16K	Similar to VQC with additional wave form coding and speech modeling

Table 3-1

volume range covered by the 4-bit signal. In this way the differential depends on the recent history of the analog input signal. And DPCM becomes "adaptive," or ADPCM.

In addition, ADPCM contains an adaptive predictor which tries to predict what the next sample is going to be. It is called "differential" because, instead of quantizing the speech directly, it quantizes the difference between each sample of speech and what the predictor thought it was going to be. There are several types of ADPCM, many standardized by the CCITT (now renamed ITU). a version of ADPCM that encodes speech at 32,000 bits per second has voice quality very close to that of PCM at 64,000 bit/s, but it cannot handle modem signals as well as PCM.

The standard ADPCM implementations send 4 bits per sample, but still sample 8,000 times per second. Thus the digital bandwidth needed is 32,000 bit/s. Bandwidth for signaling (dialing, off hook request for service, etc.) is a separate consideration.

In a reverse application of voice compression, Bellcore developed technology to double the analog bandwidth rather than halve the digital bandwidth. By doubling the sampling rate to 16,000 times per second, the voice response is extended to 7,500 Hz. This is a standard "program audio channel" that takes as many as three DS-0 time slots on some channel banks. It is used for radio and TV interviews and other "spoken word" applications. Oddly, wideband audio was the original idea behind ADPCM, but voice compression became the biggest selling point. (There are also "plugs" for channel banks that carry 15,000 Hz audio in six DS-0s.)

Sampling at the same rate as PCM, and being derived from it, ADPCM is more closely related to existing PCM equipment than other voice compression techniques. It is not fully compatible, of course. There is a need for a conversion device, a bit compression multiplexer (BCM or M44, a Mux for the 44 channels) (Fig. 3-10). The BCM goes between a T-1 line carrying ADPCM and two T-1 lines formatted for PCM. See also Chapter 7. A set of integrated circuits specifically designed to implement the BCM function has reduced the BCM to as little as one card in a T-1 multiplexer.

There are other devices, transcoders or similar, which compress voice channels by a factor of two. All of these devices work only with voice, not digital data signals. To compress data requires a completely different form of digital signal processing called data compression.

Note here that ADPCM does not send the value of the voice signal, only the change since the last sample. This situation implies that the receiving end must remember where the current level is at all times. Due to quantizing noise, this level may drift over time. An error on the transmission line can send it far off. To return the sender and receiver to the same levels, they are adjusted to zero each time the changes or differentials are zero for many samples. This happens when a speaker stops talking, which occurs frequently in normal conversation. To restore a normal operating range for continuous modem signals may take longer. As the large differentials from the sender attempt to increase a maximum memory point at the receiver, the error gets squeezed out by repeatedly clipping the analog signal.

CVSD: Continuously Variable Slope Delta Modulation

Imagine now a form of DPCM where the length of the digital word per sample is but one bit. With such a small word, many more samples can be taken and sent in the same digital bandwidth. But one-bit words cannot be used to measure loudness. One bit is either 1 or 0 — all or nothing.

Recall how PCM uses an 8-bit digital word to describe a voltage. ADPCM interprets a 4-bit word as a change in the

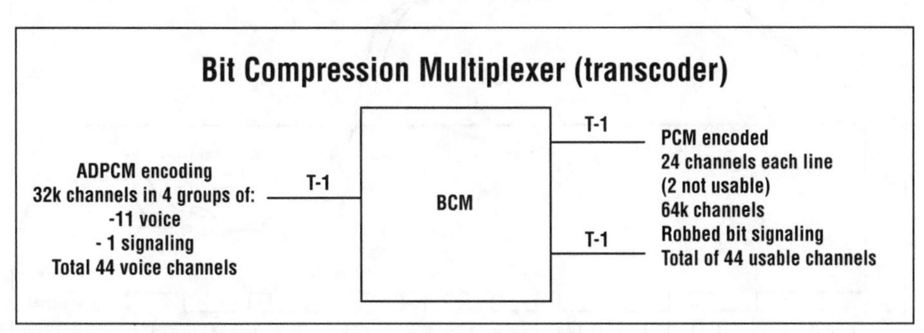

***Figure 3-10**. ADPCM is not compatible with central office switches. It must be expanded into two T-1 lines in the PCM format in order to route or switch individual voice channels at the DS-0 level. Because of the bandwidth dedicated to signaling on the ADPCM side, two channels in each PCM line are not available for use.*

voltage. CVSD looks at each individual bit and adjusts how fast the voltage changes. Rather than send the height of the analog signal curve, or the change in the height of the curve, CVSD sends information about changes in the slope of the curve. CVSD controls not the output voltage itself, but the rate of change in the output.

Slope here is the same as defined in geometry (vertical change divided by horizontal change) or building a roof (rise divided by run). In CVSD voice, the units are volts/seconds. The ranges of values encountered typically are a few volts and 1/32,000 second (32,000 bit/s).

At the sending end (Fig. 3-11), CVSD compares the input analog voltage with an internal "reference" voltage. If the input signal is greater than the reference, a "1" is sent and the slope of the reference curve is increased (bent up).

If the input is less that the reference, a "0" (zero) is sent and the slope of the reference reduced (curve is bent down). In other words, CVSD attempts to bring the reference to equal the input signal by "steering" the reference curve to follow the input.

The receiver end reconstructs the sender's reference voltage. The operation is exactly the same as at the sending end:

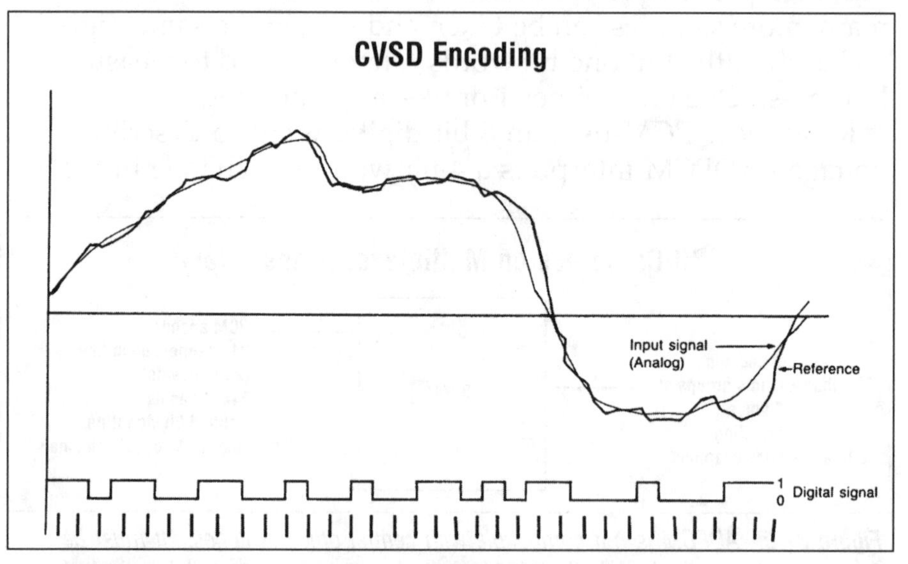

Figure 3-11 *CVSD coding works with a reference voltage. At each sample, the reference curve is "bent" up or down to follow the analog input. A "1" means increase the slope (bend up), a "0" means reduce the slope (bend it down).*

increase the slope when a "1" is received, and reduce the slope on a "0" (zero) signal. The result is an unfiltered reconstruction (practically identical to the reference) which approximates the input analog signal. Filtering smooths this to a replica of the input.

When the reconstruction is plotted, the result is a changing slope, thus the name: Continuously Variable Slope Delta Modulation.

The steeper the slope (up or down) the larger the output change between samples (Fig. 3-12). Therefore each time the change continues in the same direction the CVSD algorithm increases the size of the step taken between samples. Thus a series of 1s produces progressively larger rises in the output. This is a form of adaptation (as in ADPCM).

Figure 3-12 represents fairly the roughness of the reference curve. With 32,000 "bends" per second, there would be 32 corrections over a full cycle of a 1,000 Hz input wave.

CVSD was an early compression algorithm, mostly replaced by later developments. The usual rate for CVSD

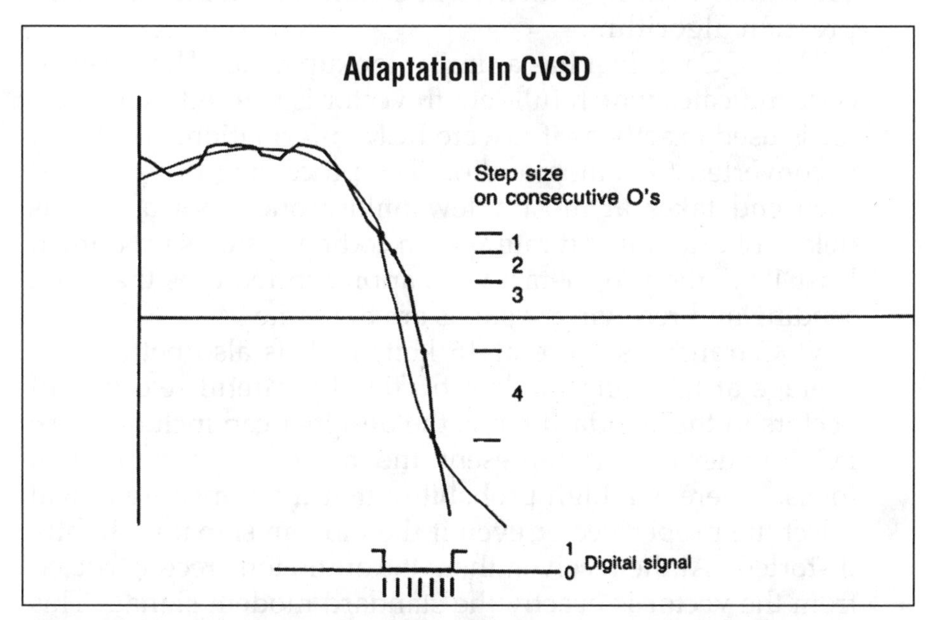

Figure 3-12 *Varying slope of CVSD reconstruction produces a change in step size between samples. In effect, CVSD is adaptive as it adjusts its response to the voice signal.*

was 32,000 samples per second, the same rate as ADPCM. Some implementations were set to run a CVSD voice channel at any standard data rate, typically from 9,600 to 64,000 bit/s.

With a good implementation of CVSD, voices are recognizable at 16,000 bit/s and still understandable at 9,600. Increasing the bit rate above 32,000 does not add much. At that speed CVSD produces a conversational quality.

VQC: Vector Quantizing Code;

VQC starts with PCM. Then, instead of transmitting individual samples, it looks at a series of samples, which in digital form is a string of 1s and 0s. A string of numbers (even 1s and 0s) is a vector. The vector that results from encoding the voice signal is compared to a set of vectors stored in the system, in a lookup table. The closest match is identified, and a shorter identification (a code) of that vector is sent to the far end. Hence the name.

VQC was an early form of Code Excited Linear Prediction (CELP) which has become the dominant compression algorithm.

The VQC receiver has a similar lookup table. The received code indicates which full-length vector is wanted. That vector is used exactly as if it were PCM information. That is, it is converted to analog audio. The processing of vectors at each end takes at most a few milliseconds, not a serious delay. Like all low bit rate speech coding systems up until at least 1997, the proprietaryVQC approach requires that both sending and receiving systems are the same.

VQC transmits voice at 16 kbit/s. It is also better than average at handling modem traffic. By careful selection of vectors in the standard table, the designer can include those PCM patterns that represent the most common modem tones. There is a high probability that the transmitters will select the proper vector even if the modem signal is slightly distorted. At the receiver, then, the audio tone reconstructed from the vector is exactly the standard modem signal. This modest amount of error correction allows VQC to handle group 3 facsimile signals (9600 bit/s) in a 32 kbit/s voice channel, and may work at 16 kbit/s.

High Capacity Voice (HCV)

Better voice quality at 8,000 bit/s requires a considerable amount of processing under a powerful algorithm. HCV expands on the principle of vector coding in VQC by adding other forms of wave form coding to model the vocal process (lungs, vocal cords, lips, etc.).

This processing intensity leads to an implementation on digital signal processor (DSP) chips. The processing takes some time, generally imposing a delay of up to 40 ms. When added to transmission delay, this is enough to require echo cancellation, so HCV includes EC for the tail circuits as part of the algorithm.

Because of the tight bandwidth limitation, transition signaling is preferred. By sending a signal only when the state changes (as

Modem Signals On Compressed Voice Channels

Voice Compression Technique	Voice Data Rate	Maximum Modem Speed	Perceived Voice Quality (scale of 5)
PCM	64,000	19,200 bit/s	4-5
	40,000	?	3
	32,000	?	2 (est.)
ADPCM	48,000	4,800	4
	32,000	4,800	4
	24,000	?	2.8
	16,000	?	1.7
ADM	48,000	4,800	4
	32,000	4,800	3.5
	24,000	?	3.1
VQC	32,000	9,600	Not rated
	16,000	4,800	Not rated
HCV	16,000	n.a.	Not rated
	8,000	n.a.	Not rated
CELP	8,000	n.a.	3.9

Table 3-2. *"Voice quality" is expressed as a Mean Opinion Score (MOS), a statistical analysis of subjective evaluations by a group of trained people who rate different systems from 1 (unsatisfactory) to 5 (excellent, "toll quality"). (Bell System Tech J, Nov 1978). Adaptive differential modulation is most like CVSD, though not identical. ? means "not evaluated"; n.a., not applicable.*

when someone hangs up) HCV avoids robbing bits that would add up to a significant portion of the 8,000 bit/s channel.

A Question of Voice Quality

There are at least two possible purposes for a digital voice channel:

1. To carry a human conversation.
2. To carry data modem signals.

Voice compression works most effectively on real voices. Modem signals are artificial replicas, using the same nominal bandwidth but in different ways. For example, to maximize the bits per second throughput, a modem will use all of the available frequency bandwidth and dynamic range (loudness) in a voice channel. Thus a modem will be far more demanding than the human voice.

To a human ear, a circuit may convey a spoken conversation with perfect clarity. The speaker's voice may be recognized easily. Yet a modem on that same channel will not be able to operate error-free at high speeds. The modem's stringent requirements for phase coherence and a need to send high frequencies at very loud levels with minimal distortion cannot be met.

Digital Speech Interpolation

Another characteristic of normal voice conversations is pauses. Usually one party listens while the other speaks; at times, both are silent. It is possible to use something like statistical multiplexing to eliminate these pauses from the T-1 line. DSI thus reduces further the average bandwidth needed by an individual voice path.

Because DSI relies on statistical probabilities, it works only when there are many voice channels — 72 channels permit a further 1.5:1 compression. For 2:1, about 96 voice channels must be contending for a single composite link. Newer implementations, usually called silence suppression, though similar to DSI, claim nearly 4:1 compression with far fewer active voice channels (see next section).

The penalty of DSI is some probability of "clipping" or loss of voice information. If all speakers talk at once when statistical multiplexing is in use, there will not be sufficient instantaneous bandwidth for everyone. The more channels, the less likely is clipping. The original DSI also required considerable overhead, typically 96,000 bit/s regardless of the number of channels. That overhead was needed to let the device at one end tell the other end which conversation is currently being transmitted. DSI formats are proprietary and not compatible with central office switches at the T-1 data link level.

DSI devices supported a single T-1 line — point to point only. They were used most often between PBX's for pure voice, as they offer no advantage for data channels, either as digital bit streams or modem signals. In the mid-1990s, some larger T-1 multiplexers and frame relay access devices started to support more than one T-1 by using compression and packetization.

Packetized Voice

A technique similar to DSI, packetized voice takes advantage of the pauses and quiet times in normal conversations to stop transmitting any information for a channel. Packets of voice information are sent only when there is a volume level above a threshold. A noisy background can fool the machine into sending additional, unnecessary packets, reducing the system capacity. Smarter servers therefore have dynamically set thresholds to minimize the impact of background noise.

In addition to the DSI function, packetized voice voice packetized servers apply additional compression techniques to reduce the average digital bandwidth per channel. A typical device puts eight voice channels on a single 56 kbit/s link.

Because they are based on DSI, packetized voice servers don't handle modem data as well as they do human conversations. Modems apply a constant carrier when transmitting, offering no quiet periods to save bandwidth. In the past there was often a requirement to keep modem signals (like fax calls) out of the voice server. An optional data port had to be time division multiplexed (TDM'd) onto the link, reducing the voice capacity significantly. Lately, packet voice servers incorporate fax modem features, making the voice channel transparent to voice and fax.

Like DSI, packets can clip if overloaded by many simultaneous talkers (or modems). By comparison, a straight TDM handles data as well as voice and is not subject to clipping.

Older packet voice devices couldn't separate individual voice channels for routing to different destinations. That is, packetized voice channels were point to point only. Later, channel switching was added, making it possible to network and switch packet-voice calls.

Sub-band Coding

In this method, the speech signal is divided into a number of different frequency bands using a special type of filter called a Quadrature Mirror Filter. The signal in each filter band is then quantized using a method similar to ADPCM. Since the energy in each filter band varies as the speech signal changes, the available bits are allocated to the different bands in an adaptive way so as to minimize the noise produced. The speech quality produced by sub-band coding is good at rates as low as 16,000 bit/s, but these coders can put lots of delay into the system. You normally don't notice the delays since it's smaller than the delay on satellite circuits. But the delay can affect the perceived voice quality and require echo cancellers.

VQL: Variable Quantizing Level

One example of this technique processes blocks of 40 PCM samples (40 bytes) taken 6667 times per second. The lower sampling rate, compared to PCM, reduces the top audio frequency to about 3000 Hz. After the 40 PCM samples are accumulated from one channel, the maximum amplitude (MA) is pick out of them. Being 8 bits, it can have 128 values, plus sign. This maximum value is then divided into 11 steps. If the MA is large, the 11 steps are also relatively large. But if the MA is modest, the 11 steps can be quite small. Their small size leads to high voice quality.Every PCM sample in the block is then compared to the MA and assigned a new code corresponding to the nearest of the 11 steps. Each sample now can be plus or minus 11 positions, or 22 possible values. This requires 5 bits per channel to encode. By pairing channels the number of bits is reduced to 4.5 per sample (22x22 = 484, or less than the 512 values which can be encoded with 9 bits). Each block of 40 samples also contains the 6 most significant bits of the MA, the A and B signaling bits, and a forward error correction code. A module with two voice ports presents a 64,000 bit/s stream to the mux/demux logic of the channel bank. All together, the block is 192 bits, or one T-1 frame.

LPC: Linear Predictive Coding

One other coding technique now available compresses a voice signal to as little as 1,200 bit/s. Linear predictive coding uses a far more complex algorithm, requires several times as much circuitry, and may cost 100 times as much as PCM.

Perceived quality is quite low. The voice of a familiar person may be recognizable, but barely so. The sound is intelligible, however, and is adequate for conveying instructions. Because of the high cost, LPC is found most often where the bandwidth is extremely expensive, as on land lines to remote areas, or across oceans.

Extensions to LPC produced several algorithms standardized globally by ITU at 8 and 16 kbit/s. Code Excited LPC (CELP) in several forms has become popular due to relatively low processing requirement for the low bandwidth used and high voice quality.

Telephone Signaling

When a caller picks up the phone (goes "off hook;") the request for service must be carried to the central office or PBX. Likewise a digitally multiplexed voice channel between PBX's must allow one side to "seize the line," carry dial pulses, respond with a busy indication, etc. Collectively these functions are called (or constitute) telephone signaling. As with any other information on a digital line, signaling over T-1 is done in bits. The presence of a specific signaling condition must be coded and multiplexed with the voice information.

PCM channel banks have a standard way to transmit signaling. Since the data stream is redundant, a small portion of it can be taken for signaling with no apparent effect on voice quality. In North America, the least significant bit in every sixth sample in each voice channel is devoted to robbed bit signaling. (See next section for details.) These bit positions are not available for voice information.

More important to users of voice and data networks, the least significant bit is not available for data either. The network can change the bit to meet signaling needs. Therefore the bit is completely unreliable. To avoid dealing with the signaling bit, the practice for data in North America is to ignore its position entirely. In a 64,000 bit/s voice channel one bit in eight is discarded by being forced to the "1" state. One result is 56,000 bit/s DDS. The practice also guarantees ones density of 12.5% even if the data is all zeros.

Newer transmission equipment, and switches that signal on an external channel rather than in-band, support "clear channel" service that allows full use of the DS-0 (see Index). That kind of transmission has been deployed by carriers in the 1990s as ISDN spread.

The Speed of the Voice Channel

Often the bit rates of voice channels are stated in rounded numbers. Seldom does the vendor specify whether the published figure includes signaling or not. For example, the 64,000 bit/s PCM channel includes 8,000/6 = 1,333 bit/s for robbed bit signaling. Does that make it 62,667 bit/s voice? (It is 56,000 bit/s data.)

One implementation of ADPCM robs one bit in four from every sample to handle signaling. In effect, this becomes 3-bit ADPCM. If the channel speed is 32,000 bit/s, would this be called 24,000 bit/s voice?

The ADPCM form chosen by AT&T uses common channel signaling in a "delta" channel format. That is, signaling messages for 11 voice channels occupy a dedicated 32,000 bit/s channel. Four such bundles fit on each ADPCM T-1 circuit for a total of 44 voice channels. Each usable voice channel requires (4 x 32,000)/44 = 2,909 bit/s for signaling. Does AT&T's ADPCM then become 32,000 + 2,909 = 34,909 bit/s voice? The point should be clear that the speed of the voice channel may not be what it seems. Fortunately, the discrepancies are not important in most private networks.

Compatibility

In the past, almost every vendor used a proprietary form of ADPCM. Some large vendors may impose de facto standards (like M44 ADPCM) but most are adopting the U.S. national (ANSI or T1.Y1) and international (CCITT or CEPT) standards. How then to provide interconnection until everyone is on a uniform standard?

To be compatible with the PSTN for switching individual voice channels, the T-1 data stream from a private T-1 network must be compatible with the D3/D4 channel bank format (the de facto standard in North America for signaling and encoding). This means that the bit stream on the T-1 circuit must:

- represent 8-bit data words of PCM,
- be for 24 channels presented in order from channel 1 to 24,
- come with robbed bit signaling, and
- have the proper framing bit sequence inserted every 193rd bit (between channel 24 and channel 1 of successive frames within superframes).

To access services provided in M44 format, the customer must have a bit compression multiplexer or transcoder (two T-1s of PCM in, one M44 out).

To use a T-1 circuit that is not switched by channel is much easier. It is necessary to conform only to D4 or ESF framing. That is, the user is free to organize the 192 data bits between adjacent framing bits in any way. Any carrier can switch almost any full T-1 circuit, regardless of the format (as long as it is framed). This means that customers can still use proprietary encoding and channelization techniques if they have the same equipment at both ends of a T-1 circuit.

In wholly private networks, therefore, the system designer has more flexibility. He can, for example, make the channels smaller or larger than 64 kbit/s. Almost every T-1 node vendor uses at least one proprietary format to frame the T-1 signal, particularly in connection with proprietary voice encoding. They do not respect the D3/D4 or D5 format, organized around DS-0s, in putting information onto the T-1 line.

One viable way to implement voice signaling is occasionally to add a signaling bit to the bits needed for the voice. For example, two 4-bit ADPCM samples could be associated with one additional bit for signaling (making a 9-bit byte). This gives each voice channel the equivalent of a half-bit per sample dedicated to signaling. In this process, voice quality is not reduced by bit robbing, though it does increase the bandwidth required and does not fit within the D4 standard.

Another reason most vendors need a non-standard format is to carry the supervisory channel that supports remote diagnostics and management. This channel may rob one bit per frame, or be a special subrate channel. Non-standard approaches, while valid for specific applications, have limitations when employed in hybrid (public and private) networking.

For example, a proprietary T-1 frame that dedicates one bit per frame for supervisory functions may cause problems with a DACS. A DACS can "false frame" on this bit position rather than the 193d or "F" bit. This trashes all data on the T-1.

There are vendors that offer flexible framing and remote management (even with a mix of proprietary and standard voice encoding methods), yet adhere strictly to the D4 and D5 standards. The trend is strongly toward standards support.

AT&T's ADPCM is their own format, yet has the status of a standard. It is called M44 for the number of multiplexed voice channels. M44 combines 11 voice channels with one signaling or "delta" channel. There are four bundles of six DS-0s each in a T-1, thus 44 channels. The delta channel carries all the signaling bits for the 11 voice channels in the same bundle. Therefore a bundle must be routed and switched as an entity to keep voice paths and signals together.

In Europe and countries that use the 2.048 Mbit/s CEPT format, all signaling information (off-hook, number dialing, busy signals between central offices, etc.) is carried in one channel (time slot 16). Bits are not robbed within the 30 voice channels.

Private networks have additional options by combining various approaches:

- Variable size bundles, putting 1 to 18 voice channels together with one signaling channel. This allows more flexible design of the logical topology.
- Signal processing, to allow individual voice channels, and their signaling, to be split out of a bundle at an intermediate node for separate routing.

Voice Applications

Up until the 1990s, PBX tie lines were the usual application for private digital voice. That is, trunks between analog or digital PBX's. Since that time, private and hybrid networks were deployed to replace leased lines between a PBX (digital or analog) and remote analog phones or key systems. With improvements in compression technology, facsimile calls can be carried as easily as voice.

Direct Digital Interfaces

A digital PBX may have an integral T-1 interface. In that case the voice circuits leave the network in the D3/D4 format on two-pair cable. That cable then plugs into the PBX, usually on an RJ-48 connector.

For T-1 interfaces to PBX's, the network nodes will not do voice encoding (that is, digitization). But the networking nodes can contribute Digital Access and Crossconnect System (DACS) functions, like grooming, and voice compression.

A multiplexed ADPCM interface on a PBX has never been generally available. M44 ADPCM, however, is a format available from certain carriers for access to their long distance networks. See Chapter 4, T-1 Circuits, for additional discussion of switching and networking.

Emulating Analog Ties

Trunks between analog PBX's most often terminate individually in the "E&M" interface (Fig. 3-13). E and M are signaling leads that accompany a voice path of 2 or 4 wires. A PBX seizes a tie trunk in one of several ways, typically by grounding the M lead (which appears as the E lead at the far end). Each PBX signals on its M lead and looks for signals on its E lead (remembered easily as Ear and Mouth, from the switch's point of view).

It is uneconomic to install physical wires over any great distance for analog E&M signaling. The phone company sometimes encodes the E&M signals onto the voice path using a single frequency (SF) tone. If present on the channel, the tone indicates "on-hook." Off-hook is absence of tone, so subscribers never hear it (except perhaps a brief chirp at the time a circuit is connected or disconnected).

The channel bank or bandwidth manager need only convey the signals generated by the PBX on the M leads and produce them on the corresponding E leads at the other PBX. How the T-1 nodes carry the E&M signaling information doesn't matter. If the same vendor's equipment is at both ends, it is usually proprietary signaling. All carriers are moving toward completely separate signaling paths, not involving the voice path network, in the form of common channel signaling like Signaling System 7 (SS7). Individual voice channels in a T-1 circuit also may be routed by a carrier (a form of switching) through a Digital Access and Cross-connect System. To do so requires that the T-1 use the D4 format. If the T-1 terminates on a switch there is almost always a need for D3/D4 robbed bit signaling, even when phones attached to the PBX use DTMF dialing. The switch could be that of a telephone carrier

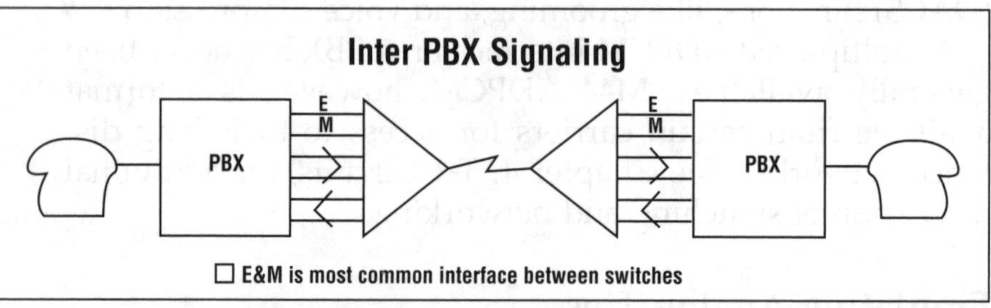

Figure 3-13. *Most common voice connection for a tie trunk between analog PBXs is the E&M interface. E and M are two leads that cross over to make the interface symmetrical. Each PBX signals on its M lead and looks for signals on its E lead (remembered easily as Ear and Mouth.*

as well as a corporate digital PBX. This D3/D4 interface frequently is required and promises to be a standard for many years yet — perhaps as long as the analog local loop.

OPX: Private Centrex;

Rather than put a complete PBX at every remote site, many corporations tie everything to their central PBX. The new digital machines have more than sufficient capacity if the "off premise extensions " can be delivered to all sites. The goal is to make every phone in the company act as if it were connected locally to the main PBX.

With a private T-1 network, they can be. This means emulating a phone in front of the PBX and emulating a PBX at the phone (Fig. 3-14). FXO (foreign exchange, office) is the interface on a network or T-1 node that "talks" to the PBX or "office" as in central office switch.

FXO is a 2-wire interface, usually loop start, which originally was intended to interface to a central office analog switch. It is defined by its set of functions:

- accept the 90 V a.c. ringing voltage (which indicates the PBX is calling a phone);
- draw loop current (accept battery) in response to ringing voltage to stop ringing output from the PBX and to indicate the called phone is off-hook (has answered);
- draw loop current (loop start) when the phone originates a call by going off hook;

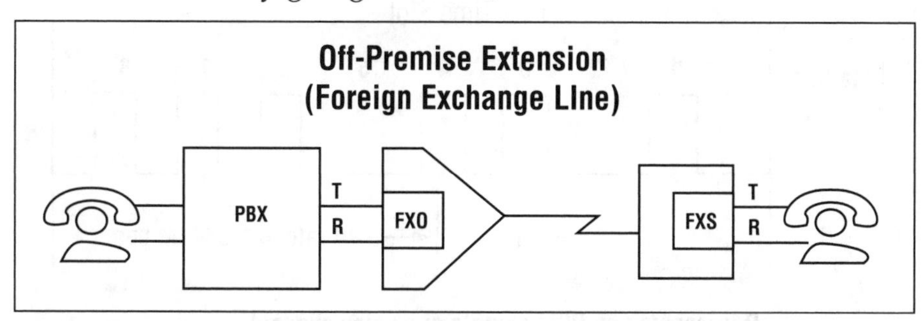

Figure 3-14. Tie lines between a PBX and remote phones involve Foreign Exchange, Office interfaces at the switch and Foreign Exchange, Subscriber at the phone. Together, FXS and FXO become an Off-Premise Extension, OPX. T and R refer to trip and ring, the two conductors of the twisted pair.

- deliver dialing information (DTMF or pulses) from the phone to the switch;
- establish the talk path for voice and dial tone.

FXS (foreign exchange, subscriber) forms the complementary interface at the other end of the OPX, facing the phone:

- generate a.c. ringing voltage into the telephone to announce incoming calls;
- give battery current to an off-hook phone to power it (the phone draws loop current to request attention);
- accept dial digits (DTMF or pulses);
- establish the talk path for voice and dial tone.

The configuration then is FXO at the central site and FXS at each remote site.

Obviously FXS and FXO are very different functions. Typically, different hardware supports them. The only exception that comes to mind is a trader turret on Wall Street. These monster key systems have a "manual ringdown" interface that both generates and accepts a.c. ringing voltage. They are designed for direct connection among themselves over relatively short 2-wire local loops . Outside of NY City, trader turrets operate like more conventional loop-start phones.

The private Centrex concept, then, involves many FXO ports on the network at the central site. While described here

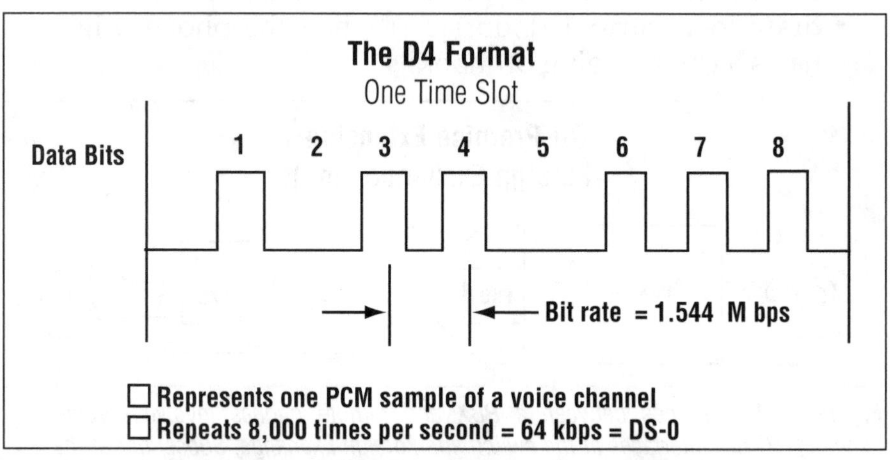

***Figure 3-15**. One time slot in the D4 format contains 8 bits. This digital word represents the instantaneous value or loudness of a voice signal. The bit rate is 1.544 million bit/s.*

as analog, there is no reason they cannot be digital, that is, T-1. In this case the node or network converts robbed bit signaling at the PBX to loop start interfaces at the remote phone sites. Some networking equipment will convert signaling between a remote FXS interface and an E&M interface on the PBX.

The central site equipment usually mounts in a rack. At the remote sites, the FXS terminations can be rack or wall mounted. New, smaller channel banks can mount on the wall, on the same plywood as the punch down blocks, etc.

D4 Framing: A Fundamental Concept of Digital Voice

Many concepts in digital voice are intimately connected with the way channel banks format bits on the T-1 line. An understanding of the D4 frame clarifies these ideas and permits a user to evaluate the various techniques for encoding voice digitally. The selection can impact switching of voice or data channels, and the design of major networks.

The Single Frame

A channel bank transmits on the T-1 line a series of bits at the rate of 1.544 million per second. They mean nothing unless the receiver knows how the bits are organized. A frame supplies organization.

The frame in a channel bank contains the value of one sample from each of the 24 voice input channels. Each value is represented by an 8-bit digital word. Each 8-bit word occupies a "time slot" in the frame (Fig. 3-15). The time slot is not the same as a bit time, but is eight bit times. The frames (containing a time slot for each channel) repeat at the sampling rate of the A to D converter in the PCM voice digitizer. That rate is 8,000 times per second or once every 125 microseconds. Thus one time slot per frame represents 8 x 8,000 = 64,000 bit/s. Throughout the world this is known as DS-0 or digital signal level zero.

The M24 Superframe

The standard D4 format (also called the M24 multiplexer format) is 24 time slots (Fig. 3-16). From the days of analog multiplexing, a "group" has been 12 voice channels. So the 24 channels in a D4 bank is two groups, also known as a "digroup."

The data in one T-1 frame is 8 bits x 24 channels = 192 bits long. The combination of 24 time slots (in North America and Japan) produces a composite speed of 1.536 million bits per second. In addition there is one framing bit, described below. The extra bit adds another 8,000 bit/s for a total of 1.544 Mbit/s. This is DS-1: digital signal level one.

Outside North America and Japan, in the CCITT countries, 32 DS-0s make a DS-1 of 2.048 million bit/s. There are no additional framing bits as one of the DS-0 channels is devoted to synchronization. See Appendix D.

The CCITT convention puts the functional equivalent of framing bits in one DS-0 channel time slot (the first, called zero). A second DS-0 (TS 16) carries only signaling. Thus the number of voice channels is reduced to 30. In many countries, the PTT will own the CSU and allow the customer only 1.92 Mbit/s for traffic. The PTT will insert the synchronization bytes.

In the D3 and D4 channel banks, the data bytes are sent in channel order, 1 through 24. (Earlier models, D1 and D2, used a pseudo-random pattern.) Thus all channel banks are "byte interleaved" multiplexers. It is possible to take only one bit from each channel, as is done in some T-1 "bit interleaved"

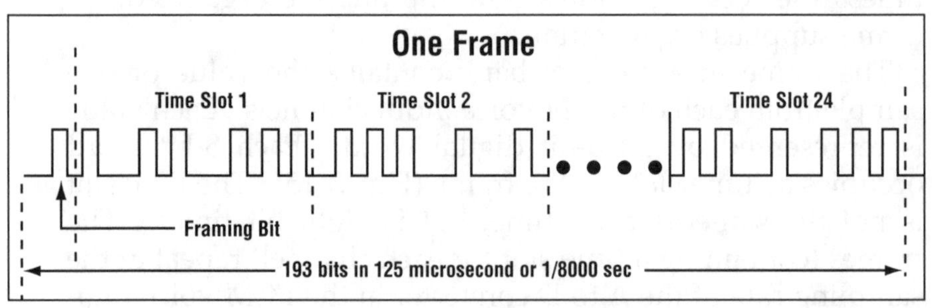

Figure 3-16. The D4 frame carries 24 time slots (of 8 bits each, or 192 bits) and a framing bit or F-bit to mark the start of the sequence, for a total of 193. The time slots correspond to the ports on a channel bank. Every time slot repeats 8,000 times per second, producing a throughput of 64 kbit/s in each slot. This is a DS-0 rate, or the standard voice channel.

multiplexers, but with various tradeoffs discussed elsewhere.

To decode and direct a byte, the receiver must recognize each one. The receiver must also be able to associate each time slot with a channel. Therefore the frame's beginning must be marked. Marking is done with a framing bit or "F bit" before every frame. That is, every 193rd bit on the T-1 line is added by the channel bank or data multiplexing function and is not part of the user's information.

The framing bit must be recognizable in a high speed bit stream. To find the 193rd bit the receiver looks for a defined sequence that repeats every 12 frames: 100011011100. Taken together, the 12 frames in the cycle constitute a "superframe" (Fig. 3-17). There are commercial chip sets that can find framing patterns within milliseconds. These circuits can bring two channel banks into synchronization so quickly that a loss of sync and recovery is barely audible.

A frame might have been marked by a sync character (rather than a bit) between frames. (The Bell T1DM data multiplexer for DDS effectively does this by putting a sync byte in time slot 24.) This character synchronization clearly marks the start of each frame. It would tell a channel bank which byte belonged to which voice channel ("mainframe synchronization"). But there would be no way to tell one frame from another within a superframe.

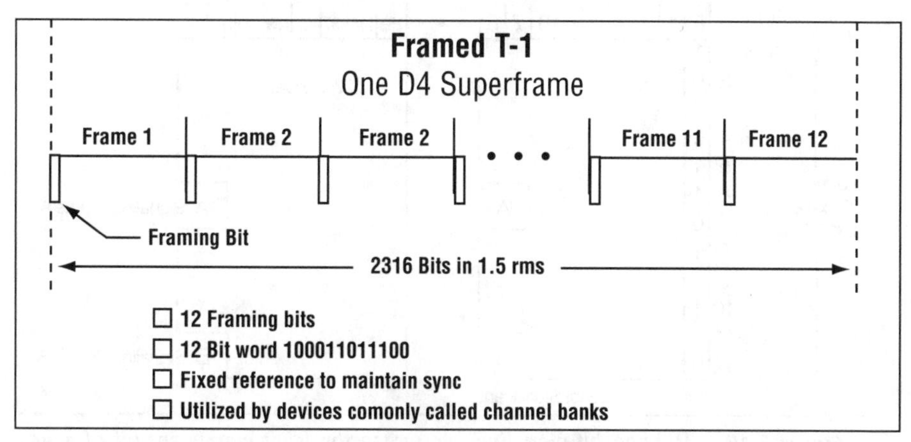

Figure 3-17. *A superframe is 12 D4 frames. An F-bit separates each frame from the next. The F-bit series is formatted in a fixed pattern to identify not only the start of the frame but also the start of the superframe.*

Robbed Bit Signaling

Channel bank designers chose to minimize the bandwidth spent on signaling. Rather than carry signaling with every time slot in every frame, they put signaling information in every sixth frame. The F-bit sequence, therefore, also identifies which frames have been modified to carry signaling.

In the 6th and 12th frames, the least significant bit of voice data in each time slot is overwritten with a signaling bit, leaving 7 bits for the voice code. This process is called bit robbing (Fig. 3-18). There are two robbed bits from each channel per superframe (A in the sixth frame, B in the twelfth). With two states (1 or 0) possible per bit, there are four signaling conditions.

When there is a tandem switch or DACS in the path of a T-1 circuit, the cross connection does not in general preserve frame or superframe alignment. As T-1 and E-1 framer chips receive a signal they extract the robbed signaling bits and pass them to the outbound framers. There each chip inserts the bits in a new superframe constructed for the outbound side.

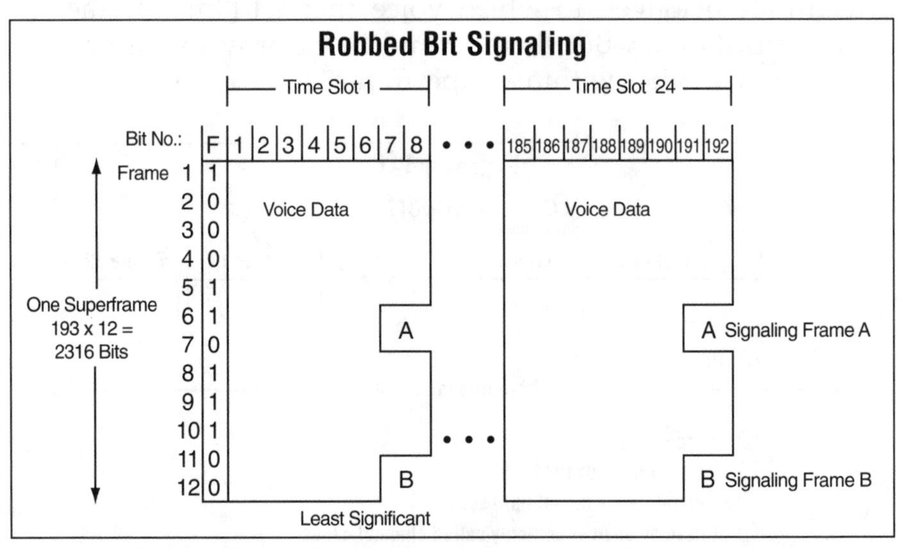

Figure 3-18. Robbed bit signaling overwrites the least significant bit of every time slot in the sixth and twelfth frames of a superframe. The state of the bit, 0 or 1, tells if the device at the other end is on-hook or off-hook. Signaling conveys dialing information and provides line supervision.

A byte that carries signaling coming into a switch likely will be placed in an outbound frame that does not carry signaling (5 chances in 6 or 83% chance). Each switch that passes a dialed call can overwrite the least significant bit (lsb) in a different time slot (byte). Eventually, all the lsb's may be clobbered by bit robbing. Only seven bits, or 56 kbit/s, is passed "clear" end to end by networks that use robbed bit signaling. In contrast, ISDN (with a separate path for signals) passes all DS-0s as clear 64K.

In most situations, though not all, both A and B are made the same: 0 for "on hook" and 1 for "off hook." These two supervisory states, and the sequence of transitions between them, can control a call setup or indicate disconnect, busy, dial pulses, etc.

For example, an idle channel is indicated by A and B in both directions being 0. When the caller goes off-hook, A and B sent from that end are set to 1. (AT&T's Megacom service inverts the signaling bits: 0 means off hook, 1 means on hook.) The called end might respond by sending 1s in the signaling positions. These 1's might be continuous or for a few hundred milliseconds (a "wink"), depending on the type of line or trunk. Either response indicates "ready to receive dialed digits."

Dialing today is most often by tones. Many digital trunks still mimic rotary dial pulses by alternating the signaling bits between 0 and 1. Signaling bits represent the state of the M lead on the E&M interface of the sending party. When the called party answers, that side of the line returns continuous 1's. When either side hangs up, their signaling bits revert to 0s meaning on-hook.

ESF: Extended Superframe

When digital voice was introduced, four signaling states were enough, but new services could use more. When only voice was transmitted, occasionally changing one bit meant nothing. Bit robbing was a very creative solution. Now to change even a few bits is disastrous. Terminal data and computer communications make bit error rates critical.

To respond to these new needs, the extended superframe, ESF, was tariffed by AT&T (Fig. 3-19) and eventually adopted by all other carriers as well as standardized by ANSI. (Formerly, it was called Fe, "F sub E," for extended framing.)

"Extended" means the number of frames in the F-bit pattern is doubled from 12 to 24. This makes room for two additional signaling bits, C and D, in frames 18 and 24 within each superframe. Until ESF was fully implemented, the most common practice was to repeat signaling bits: C equals A and D equals B. As the necessary central office equipment was installed, users were offered new features based on a reorganization of the 8,000 bit/s stream of framing bits.

Channel banks and multiplexers of 20 years ago may have needed all 8,000 per second of the framing bits to hold synchronization. This is no longer the case; 2,000

Comparison of Framing Standards

	SF (1970)	ESF (1983+)
Frames per Superframe	12	24
Bit rate per function:		
Terminal framing (24 channels)	4000	2000
Mark signaling frames (every 6th)	4000	included
Block error check (CRC-6)		2000
Multipurpose data link (or FDL),		4000
Uses of the Facilities Data Link		
Bellcore 1985 Proposal *(not adopted)*:		
(far end block error) (1 P-bit)		333
Alarm channel (1 A-bit)		333
Local operating channel (4 L-bits)		1333
End-to-end (express) (6 E-bits)		2000
1989 Proposal to ANSI:		
Message Channel (6 M bits)		2000
Overhead for ZBTSI (6 Z bits)		2000
(ZBTSI supports Clear Channel 64 kbit/s)		

***Table 3-3**. Superframe (SF) has been used by channel banks for 20 years. AT&T started to deploy extended superframe (ESF) without defining the multipurpose data link. Bellcore proposed specific uses for the 4000 bit/s data channel, all for internal telco operations. Later, when deploying ESF in quantity, AT&T defined messages on the FDL for gathering error statistics from the CPE.*

bit/s is sufficient. With ESF, only every fourth F bit (in frames 4, 8, ...24) is used for synchronization, in a 6-bit framing pattern sequence:

FPS = ...001011...001011....

This leaves 6,000 bit/s for new functions:

1. 2,000 bit/s for continuous error checking.
2. 4,000 bit/s for network information.

A cyclic redundancy checksum is calculated from the data (but not the F bits) in each extended superframe. The resulting 6 check bits (CB1-CB6) are inserted into specific F-bit positions in the next following ESF (before frames 2, 6, 10, 14, 18, and 22). This means that at every point where the bit stream can be monitored, the quality of the transmission medium can be measured. This will be true even if the content of the data is encrypted. All the tester needs to do is monitor the flow in one direction, calculate the CRC from any superframe, then compare those six bits with the CRC spread over the following superframe.

Originally, the data channel (now known as the facilities data link, FDL) was intended for supervisory and diagnostic functions, status messages, and monitoring of T-1 circuits — by the carrier.

Facilities Data Link

In mid-1985, AT&T released the ESF format for use without defining in any way the 4,000 bit/s channel, the FDL. Two years later, this channel was preempted for network

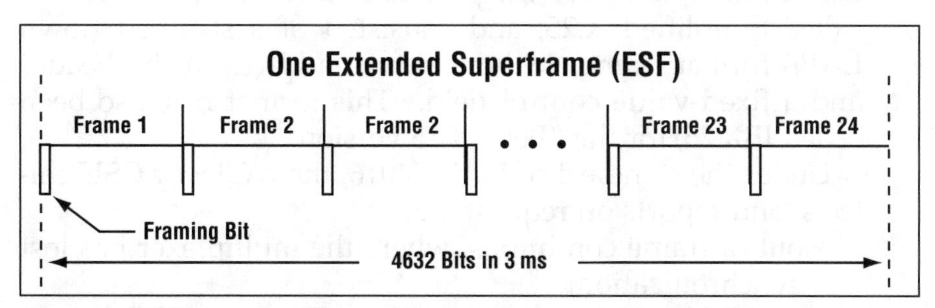

Figure 3-19. The extended superframe (ESF), at 24 frames, doubles the length of the older 12-frame superframe. The F-bit series becomes 24 bits long. It still identifies every sixth frame for robbed bit signaling, but the additional bits support new features.

use, primarily diagnostics involving loop backs and statistics. The standards are spelled out in the requirements for network channel terminating equipment (NCTE) (Pub. 54016) and CSUs, and have been added to by the application of "zero-byte time slot interchange." ZBTSI is a way to preserve ones density that did not catch on and never became widely available.

If a T-1 passes through a DACS, the DACS regenerates the ESF for each output port. This terminates the arriving FDL. If the carrier performs any diagnostics, they will interrupt the FDL. However, in a campus environment, or when using locally attached intelligent channel banks to feed voice into a T-1 node, the FDL is still available to the end user. By putting the supervisory channel into the FDL, all 24 DS-0 time slots remain available for user traffic. Yet the channel bank comes under the control of the central network management system. (See the chapter on network management.)

At the time of divestiture CSUs were ruled by the FCC to be Customer Premise Equipment (CPE). Thus the regulated carrier cannot provide them. This situation indirectly makes ESF more attractive, as ESF monitoring offers a way to isolate faults between lines and hardware.

Under the original ESF format defined by AT&T (Tech Pub 54016), the data channel (FDL, or facilities data link) is used by the carrier to interrogate the CPE, whatever (and whose ever) it is, about the condition of the line. What the carrier expected in response, also over the FDL, is a report of statistics on line quality for the past 24 hours. The protocol was called Simplified X.25, and consisted of a stripped down LAPB format frame, without source address in the header and a fixed-value control field. This format has also been called IBX.25, for the "Bell" X.25 version.

Under the standard of Pub. 54016, the NCTE or CSU collects (and reports on request) about:

- out of frame conditions, where the multiplexer has lost synchronization;
- errored seconds: seconds with one or more CRC errors;
- bursty seconds: 2-319 CRC errors;
- severely errored seconds: 320 or more CRC errors; and

• failed seconds, also called unavailable seconds (UAS).

Each report covers the preceding 96 intervals of 15 minutes each, the last full day, or a separate command elicits a 1-hour report over four intervals.

The AT&T definition for "severely errored" represents a very high error rate indeed, about one bit in a thousand (10^{-3}). With 24 frames in a superframe, and 8000 frames per second, there are about 333 CRC checks per second. "Severely errored" means that 96% of all frames during that second have at least one error. While some voice traffic might be understandable, expect data in SESs to be pure trash.

However, the tariff says you get a rebate only for failed seconds. FSs are defined as the time starting after 10 consecutive severely errored seconds, and ending when there have been 10 consecutive seconds that are not severely errored.

Contract with your carrier for better performance if you transmit data. In light of improvements in equipment during the 1990s, and the almost universal deployment of optical fiber for long distance lines, these standards for errors are considerably out of date. To assure their customers of reliable data transmission, many carriers now offer quality commitments based on bit error rate (BER). A rate of one errored bit in 10^6 is often a starting point; 1 in 10^{11} is possible on some circuits.

Another common quality measure is availability, or the percentage of time the line is in working order (meets the BER specification). This has improved because fiber optic networks have become self-healing and can reroute a "broken" T-1 in milliseconds. The availability rates that are now practical make it hard to express as a percentage: too many 9s after 99.99+%. Instead, the inverse is specified: outage time. This number can be less than 30 minutes per year.

In 1989 ANSI adopted a new ANSI standard, T1.403-1989 (from the T1E1 committee of the Exchange Carriers Standards Association). A major change from 54016 requires the NCTE to report on CRC and other errors each second, without waiting for a poll. Each second, CPE sends a report for the last four seconds to the network. Overlapping the reports allows for several reports to be lost without depriving the central office of any information. The benefit (from the ANSI committee's point of view) is that it is not necessary to break into a T-1 circuit to measure how well it is doing — all that's needed is a tap or monitor connection. The test equipment can determine if the framing is proper, and if so, what the error rates are. By contrast, 54016 requires the tester to inject a command message to request a report on error history.

There is no longer a requirement to save the statistics for more than 4 seconds. However, most CSUs continue to save the statistics according to 54016 so that you (the customers) have a record to examine locally. See Fig. 4-2 for more information in connection with CSUs.

In addition to a requirement to monitor the CRC errors, the ANSI approach suggests three optional measurements:

- controlled slip, the loss or repetition of an entire 192 bit frame, which may not be obvious to the CSU and hence not meaningful in some cases;
- framing error, which is any error in a framing bit; and
- bipolar violations, where successive pulses are the same polarity.

All ANSI reports are formatted as 112-bit message frames (LAP-D) as described in CCITT recommendation Q.921, the same basic format used for messages in 54016. Each performance report message (PRM) contains information on the current and previous three seconds, to cover short error bursts. T1.403 refers to the FDL as the "embedded operations channel", EOC.

The new standard is good and bad news:

- every carrier supports it;
- T1.403 can coexist with 54016;
- it is non-intrusive, as the monitor can simply "listen in" on the T-1 line — there is no need to break the circuit or insert messages which might interfere with user data;
- any T-1 terminating device (like a DACS) will terminate the FDL too — the customer will probably not be able to test end to end on the FDL;
- there is no provision for customer use of the FDL for things like network management;
- users cannot set thresholds for error rates.

Into early 1990, practically all ESF installations were operating according to 54016. It was the de facto standard. By that time, however, all carriers (including AT&T) had announced support for the ANSI standard. Though neither equipment nor lines to support T1.403 were widely available in 1990, migration to this standard was effectively complete by the mid-1990s. ANSI revised T1.403 in 1995.

Certain commands in the FDL, for loopbacks, may be either in the message format (Q.931) or a repeating bit pattern. For example, the FDL is filled with repeating 001 patterns for 5 s to command a digital loopback in the CSU.

False Framing

There are two conditions where framing may be lost due to the nature of the traffic on the T-1. It is possible for the bit patterns in the payload to mimic the synchronization pattern in the F bits. The receiver synchronizes on bits other than the F bits, resulting in "false framing."

This may happen if:

- A pure 1000 Hz tone is transmitted over a link that uses SF (12 frames). Since ESF has become the framing method of choice, this condition is becoming rare.

Now you know why the official test tone; for telephone networks was changed from 1000 Hz to 1004 Hz. The "digital milliwatt" is a bit pattern that describes a 1004 Hz tone that will convert to an analog signal at dBm0, 1 mW into 600 ohms.

- Frame relay traffic at one time was reported to cause mis-framing on older T-1 lines. The alleged cause was a pattern of bits in the frame headers. This problem disappeared about 1994, without explanation.

Proprietary Framing

The practical significance of the framing requirement depends on the networking situation and the user's applications. To meet the minimum requirement every 193rd bit must follow a D4 or ESF pattern (Fig. 3-20). The user who does only that is permitted to connect any T-1 terminal

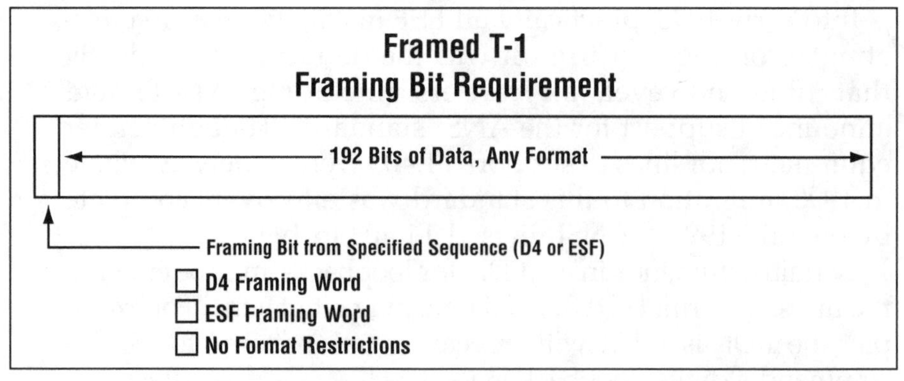

***Figure 3-20**. The requirement for "framing" on the T-1 line demands the proper sequence of F-bits be inserted in every 193rd position. There is no requirement on the content of the 192 data bits other than the ones density and maximum zeros rules.*

device and route DS-1 circuits through any carrier facility. This is true regardless of the information format sent by the user. It is common for a data multiplexer to create its own frames, entirely different from the 24-channel D4 frame, within the 1.536 Mbit/s stream. The data multiplexer or bandwidth manager can add its own characters for synchronization, in addition to and ignoring the F bits, to create frames longer or shorter than 193 bits. A separate CPE function, which may be software or hardware or both, inserts the

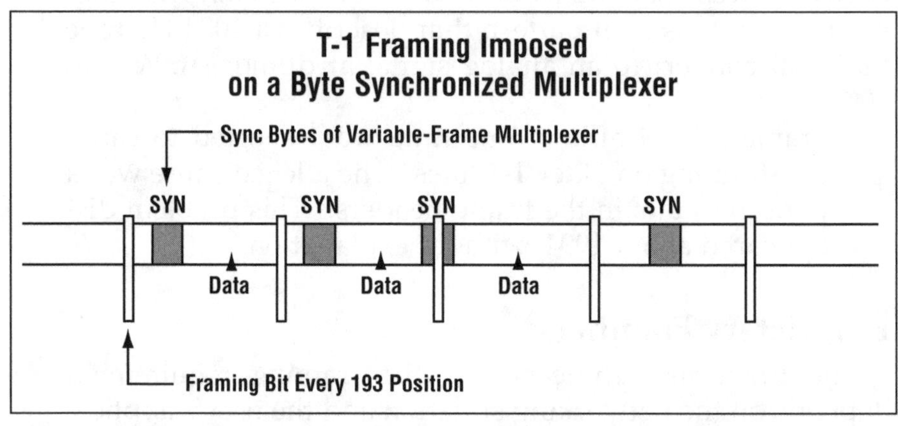

***Figure 3-21**. D4 framing (F-bits in the 193rd position) may be imposed on a proprietary frame based on byte synchronization. Insertion of F-bits at the sending end, and their removal at the receiver, is done in hardware. D4 framing is transparent to this type of multiplexer function.*

proper framing bits (F bits) to ensure compatibility with the network. At the receiving end, the F bits are removed first. To the CPE, it appears that it has a clear channel at a rate of 1.536 Mbit/s, framed in any way it wants. This is true even if a D4 framing bit falls in the middle of a sync byte (Fig. 3-21). Consequently, the framing requirement need have no affect on the switching, data bypass, drop-and-insert, or voice compression or other proprietary capabilities of customer-owned equipment.

To participate in the special carrier services that fan out or switch DS-0 channels at central offices, naturally the customer premises equipment (CPE) must create those channels. That is, the CPE must be fully D4 compatible at the DS-0 level.

T-1 Circuits

▶ To date, most T-1 or DS-1 circuits have been dedicated leased lines. Even the switched T-1 services of the major carriers require a fixed connection (local leased line) for access from the customer's site to the serving office.

Terrestrial

The standard for performance is the land line, what used to be dedicated copper pairs and has migrated to optical fiber. The original "T-1 facility" is still extensively used (Fig. 4-1). It is a run of selected copper pairs, one pair for each direction, in a standard cable bundle. Care must be taken to pick wires that are of consistent gauge, free of branch circuits (taps), and don't have loading coils installed.

Integral to T-1 are digital regenerators inserted at 1-mile intervals (the distance between manholes in large cities). This form of T-1 is used over distances of fewer than 50 miles; more than 50 repeaters in a line introduce too much distortion and jitter. Note that there are no terminal devices shown — the "T-span" is independent of what it carries.

For very short runs, phone companies are now offering "dark" T-1 facilities. That is, the line may contain no repeaters. If there are repeaters they may be powered by a constant current source provided by the customer. This is necessary because the circuit runs directly between sites operated by the customer. The cable does not pass through a central office that could power the repeaters.

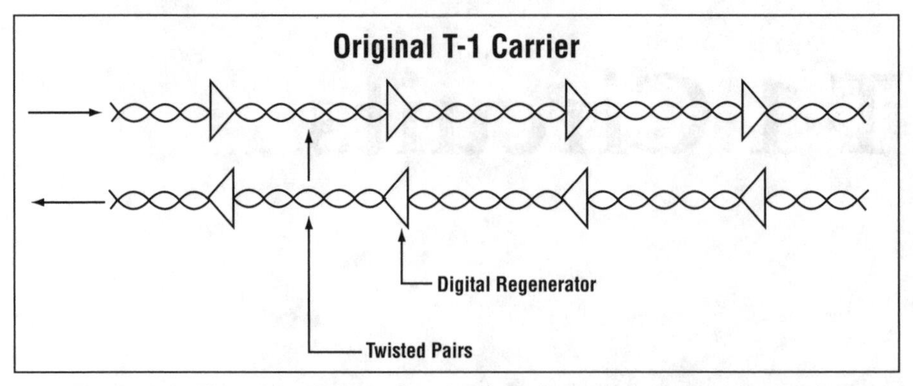

***Figure 4-1**. The original T-1 is a "carrier" based on existing pairs of twisted copper wires. Digital repeaters at 1-mile intervals regenerate the digital signal up to 50 times.*

On long-haul legs, T-1 circuits were usually carried by microwave relay facilities until 1990, when about 75% of the nation's long distance voice traffic was on microwave. Error rates were reasonably low, about one bit in a million.

After 1990 microwave systems were replaced by fiber. Error rates dropped to 1 bit in a billion or better. By the turn of the century we will communicate mostly on glass. Not only is fiber very "clean" (i.e. fewer errors than on microwave links), it has kept its promise to be very cheap, and has helped long distance charges continue to decline.

Terrestrial routing can be fairly direct. Due to the large number of existing cables, the land route is the shortest physical signal path between most pairs of points (beyond line of sight). The shortest path results in the shortest time delay for information to travel from one end to the other. Even with repeated multiplexing and other forms of handling at multiple carrier offices, the transit time for a T-1 circuit across the US is typically 40-50 ms (milliseconds). It can be as little as 10 ms for a hundred mile leg. (Voice circuits can have much higher delay.) Terrestrial delay is also fairly constant, changing only when the route is altered.

For the highest reliability, diverse routing of redundant T-1's is available to prevent the infamous "random backhoe" (a farmer digging his field, but more likely a suburban home builder) from disrupting all of a firm's communications. Between central offices full redundancy is normally provided by the carrier. Optical fiber normally runs between SONET switches, which operate on fiber rings that allow automatic protection switching to restore circuits on broken fibers within a few tens of milliseconds. Because the carrier provides redundancy, there is little point (and great expense) in providing leased backup circuits for T-1's.

Unfortunately, a catastrophe is most likely to happen in the local loop — the "weakest link" — near the users' offices. In most locations, the local loop is the hardest leg to route diversely. But if reliability is a prime concern, here is where to add leased lines for diversity.

On The Line

Regardless of the vendor, a dedicated T-1 line will appear at the customer's end as two wire pairs: for data send and receive. The physical connector used to be a DB-15 (15-pin) type, but the current standard is the RJ-48C. Fig. 4-2 depicts these connectors and one end of a T-1 line, with equipment attached. We will refer to this figure often in this chapter.

Some circuits supplied to the government are unframed; that is, the full 1.544 Mbit/s bandwidth is used for data (often encrypted). These interfaces have additional conductors for loopback control and testing. All are terminated in a single connector, which must be a DB-15 for that many conductors.

The data leads of the terrestrial T-1 line operate with a bipolar signal (Fig. 4-3) that sends positive and negative voltages. Alternate mark inversion reverses the polarity of alternate 1's, or marks. That is, logical zeros are zero voltage, while logical ones alternate between plus and minus 3 volts. Between two successive marks, or ones, the voltage returns to and pauses at zero. This helps maintain a zero reference at the receiving end.

Keeping Time

The T-1 interface has no separate clock signal. Timing information to keep a T-1 terminal synchronized with the network must be derived from the received data signal at each node. To

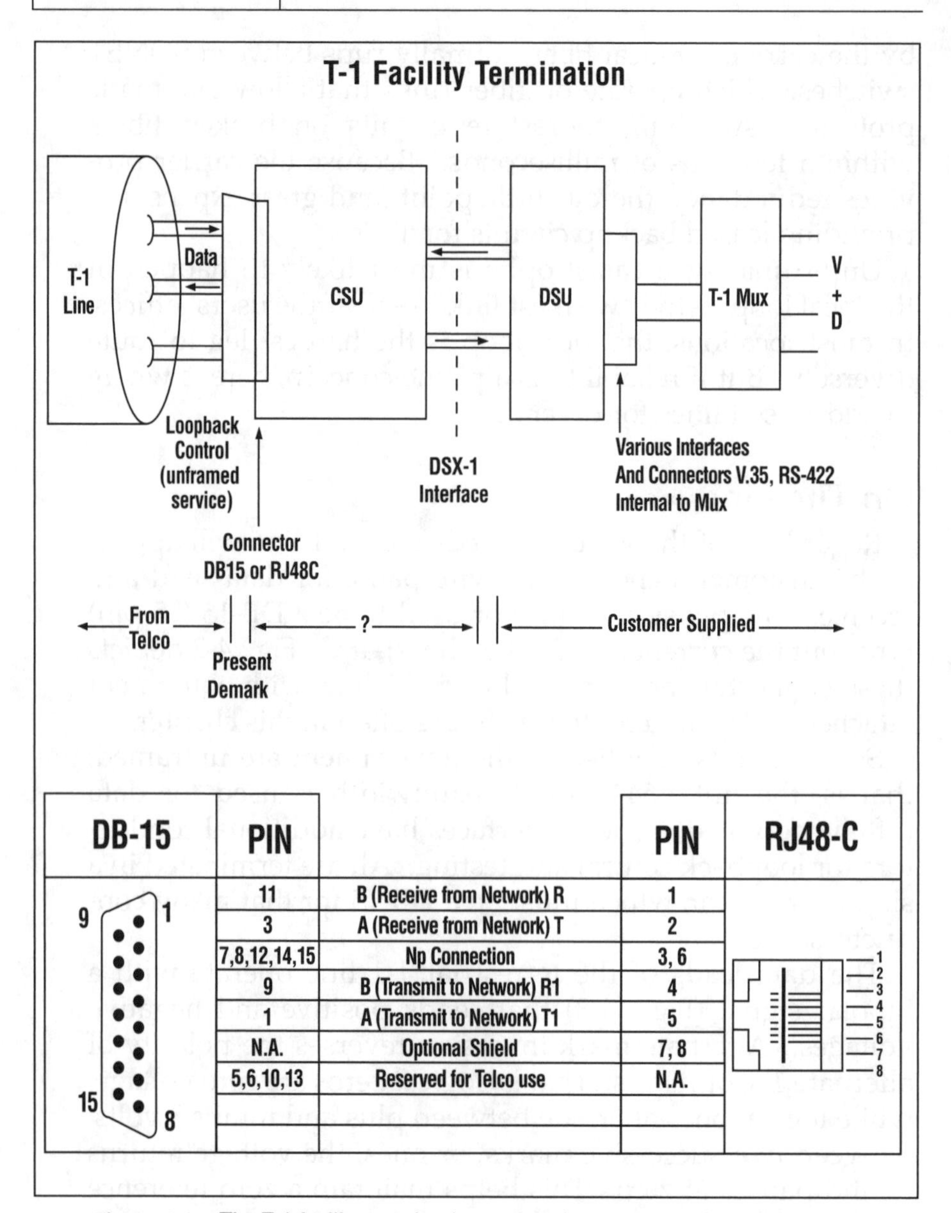

DB-15 PIN		RJ48-C PIN
11	B (Receive from Network) R	1
3	A (Receive from Network) T	2
7,8,12,14,15	Np Connection	3, 6
9	B (Transmit to Network) R1	4
1	A (Transmit to Network) T1	5
N.A.	Optional Shield	7, 8
5,6,10,13	Reserved for Telco use	N.A.

***Figure 4-2**. The T-1 facility termination enters the users premises as two wire pairs for data. There may be additional wires for loopback control and testing of unframed lines (government only). The channel service unit (CSU) provides loopbacks to the network and is the last regeneration point for incoming signals. "DSX-1" as an interface is well defined in terms of electrical and mechanical features. A digital service unit (DSU) matches various terminal interfaces to the DSX-1 standard and usually is part of the user's terminal equipment: multiplexer, bandwidth manager, PBX, computer, etc.*

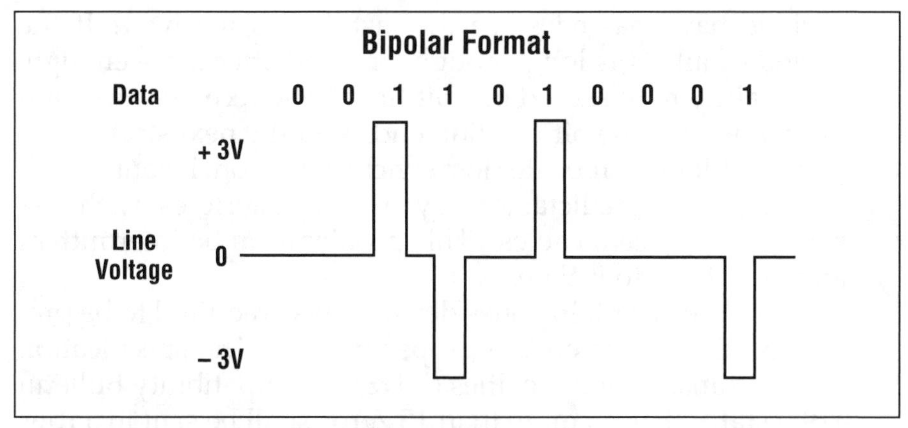

Figure 4-3. Bipolar signal format at the DSX-1 interface (and on the T-1 line) "inverts alternate marks" or ones (AMI). Spaces or zeros are zero voltage. Voltage returns to zero (RZ) between successive marks. Both 15-pin and RJ-48C connectors are found installed.

make clock recovery possible, there must be information to recover it from. Successive zeros produce a steady, zero output voltage — no information. Only 1's provide timing information.

At the regenerators and perhaps the receiving end, the clock used to be recovered in part by a tuned resonant circuit, later by a phase locked loop circuit driven by the ones pulses. Without sufficient ones, the circuit no longer produces reliable timing output. The phase of the output pulses may vary with time, and be either ahead of or behind the master clock, a condition known as jitter. Through multiple repeaters the jitter can accumulate. The maximum jitter mentioned in the standards is 5000 degrees of phase: almost 14 bit times.

Switches in the public network do receive a definite clock signal (see below). But T-1 networks that do not run through a switch, or a DACS, need to supply a frequency source very near to 1.544 Mbit/s.

Ones Density

Repeaters and terminals (multiplexers) must track the data bit pattern through rapidly changing phase shifts of many bit times. They cannot track on 0's, only 1's. A long string of 0's increases the uncertainty in timing recovery. How many bit periods have passed since the last 1 was received? If the period of all 0's is long enough, and the jitter large enough, uncertainty may exceed one bit time. The receiver then cannot tell where one bit position ends and the next starts.

If an older line repeater lost synchronization, it could oscillate in an unpredictable way, which interferes with T-1 signals in adjacent cables. This problem can be intermittent and very hard to isolate.

Therefore, a certain "ones density" was required to be present on the line to ensure proper timing. The specification for T-1 transmission facilities (AT&T "Compatibility Bulletin 119") states that no more than 15 zeros shall be sent in a row, AND there shall be at least N ones in every 8(N+1) bits for a minimum average ones density of 12.5%. This number of ones is sufficient to keep any type of T-1 circuit synchronized. Ones density is particularly important to the standard line repeaters used in common T-carrier spans. They are far from central offices where they could receive accurate clocking.

In 1987 the FCC recognized that the newer equipment which now dominates the public network is not as sensitive to strings of zeroes. Nor are new repeaters subject to oscillation. Therefore the FCC relaxed the ones density requirement, as far as "harm to the network" is concerned, to 80 zeroes in a row. However, the average density must still be at least 12.5%. This requirement was included in a EIA/TIA standard. Regardless, many central offices set their equipment to raise an alarm after 15 zeroes. Carriers naturally want to avoid alarms and so still want users to maintain the old one's density.

Because of the variations among specific circuits, some lines may operate satisfactorily despite substandard one's density. Out-of-spec performance cannot be assured, however.

The signal format on the T-1 line is designed for optimum use of cable pairs. The alternating pulse format halves the effective frequency of the signal, to about 3/4 MHz. This brings the frequency component carrying the most energy to half the bit rate. Bipolar is not compatible with conventional terminal or computer interfaces, however. Some conversion is required. Historically network channel terminating equipment (NCTE) performed the conversion in two steps

CSU: Channel Service Unit

The first piece of equipment connected directly to the line is the CSU or channel service unit (Fig. 4-4).

Traditionally (before divestiture in the US and always outside the US) the CSU has been considered part of the network. A CSU contains the last signal regenerator on the line before the data terminal equipment (DTE), and a mechanism to put the line into loopback for testing from the central office. A third section of the CSU monitors the signal to detect violations of the bipolar, 15 zeros, and 1's density rules — and to look for loss of signal. Violations typically produce warnings, via lights.

During the 1980s, CSUs acquired several additional functions. To summarize them:

Loopback

Originally, in 1983, the loopback was switched in by a relay controlled with a separate wire pair. These days, the loopback command is "in band," on the T-1 line. There are three types defined, all of them for the carrier's use only:

Line Loop Back (LLB)

On D4 or ESF framed lines operating to 54016, loopback of the full T-1 is latched in by a bit sequence of 10000 (binary) repeated at the T-1 rate for 5 seconds. Release of LLB results from 100 repeated for 5 sec. Under T1.403, the pattern 00001110 11111111 in the FDL actuates the LLB, while 00111000 11111111 releases it.

Payload Loop Back (PLB)

The extended superframe format includes the facilities data link, buried in the framing bits. A CO uses this channel to send a simplified X.25 message to the CSU commanding a loopback. Under T1.403, the pattern 00010100 11111111 in the FDL actuates the PLB, while 00110010 11111111 releases it.

DS-0 Loop Back

Though not usually handled by the CSU, there may be test commands present on the T-1 for loop backs on individual DS-0's or subrate channels.

To regain control of the local loop, particularly the ability to set up a loop-back, carriers invented an additional device to mark the juncture of outside wiring (the local loop) and inside wiring (owned by the customer). The "Smart Jack" generally goes in the wiring vault of a larger building, or outside a smaller building. The FCC accepts that the smart jack is not CPE, and hence can be owned and supplied by the carrier.

Keep Alive

Another important function of the CSU is to generate a "keep alive" signal when the attached terminal equipment

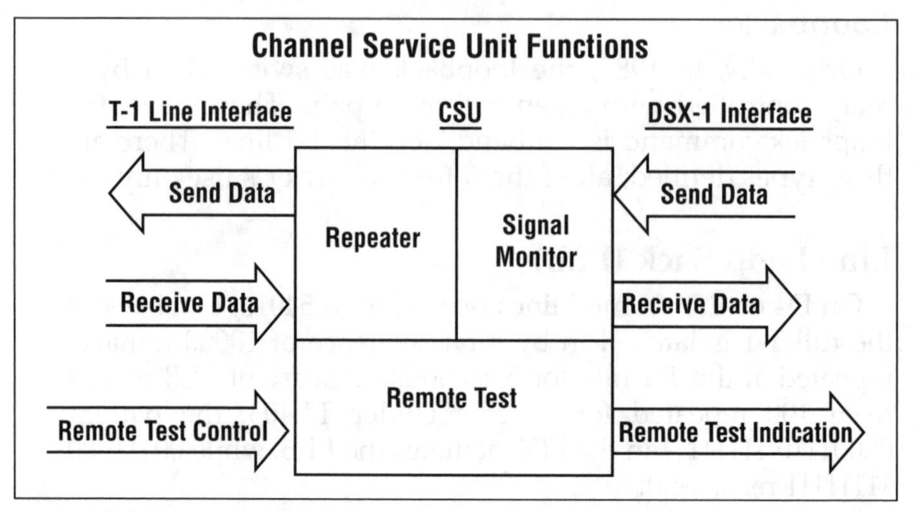

Figure 4-4*. The channel service unit (CSU) regenerates the wave forms received from the network and presents the user with a clean signal at the DSX-1 interface. It also regenerates sent data. The remote test functions include loopback for testing from the network side. Signal monitor warns of bipolar violations and low ones density. CSUs are being integrated into multiplexers.*

fails to deliver a stream of valid data (which may be HDLC flag characters, or idle) or is disconnected. AT&T used to request one of three types, depending on the equipment they used to supply the service:

1. Unframed, all 1's
2. D4 (or ESF) frame containing all 1's in data positions
3. Loopback of signal received from the network.

A later version of 62411 (December 1988) specifies only the all 1's type. This is also known as a Blue alarm or Alarm Indication Signal (AIS). In other words, when the T-1 multiplexer fails or is disconnected, the CSU must send continuous unframed 1's to the network. While an alarm, all 1's does not set off bells and lights in the central office the way loss of signal does. The same all 1's is usually required on ESF as well as SF lines, though some carriers may request other types of keep alive signals.

Since the customer must now provide the CSU as well as the DSU, some vendors are offering combined devices. These designs can save space, but when not built into the DTE, they raise the question of where the CSU draws its power.

When the CSU belonged to the network, the carrier supplied DC power through the signal leads. Send and receive data pairs are balanced and isolated from ground by transformers. Each pair can therefore be used as one power lead. AT&T originally thought it important to keep the CSU functioning during a local power failure at the customer site (to avoid maintenance alarms). A working CSU would allow test personnel to distinguish that type of power failure from a network problem.

Now that they no longer are responsible for the CSU, carriers are reluctant to power it. Generally they refuse to power the DSU. With optical fibers , there is no way to provide this power. So the FCC now allows carriers the option of powering the CSU or not, for both wire and fiber cables. Policy came down to no power almost universally for new installations.

The result is that end users will be required to provide uninterruptable power for the CSU.

Powering CSUs locally may not be much of a burden when CSUs are integrated into the terminal or multiplexer. And integration is a definite trend. Smaller CSU circuits are a reality now, with at least two sources of "CSUs on a chip" (and even 4 CSUs per chip) that vendors can build into multiplexers. The first internal CSU in a T-1 multiplexer was delivered in early 1988.

Integral CSUs are not only easy to power, they are less expensive due to reduced size and parts count. Therefore they should also be more reliable. When added directly to a T-1 interface card, they also:

1. come under centralized network management;
2. offer CSU redundancy when T-1 cards are redundant.

ESF Statistics

The latest job commonly assigned to the CSU is to calculate and collect the ESF error statistics described toward the end of Chapter 3. There are two ways this is done.

AT&T started to deploy equipment that assumed the CSU would hold the latest 24-hour history of errors, in 15-minute intervals. Then, on a command from the central office over the facilities data link (FDL), the CSU sends the error history (for 1 hour or 24 hours) to the central office, also over the FDL (Fig. 4-5). The Line Monitor Unit (LMU) in the CO inserts the commands and picks off the statistics.

Note that this arrangement requires two things:

1. the CSU sends no information until it is polled by the LMU, which must inject the request into the FDL.
2. that means the LMU terminates the FDL, that is, the LMU must break into the line physically;

The second version is the ANSI standard, adopted by the local exchange carriers (who supply most of the local loops connected to your CSUs). AT&T has since adopted the ANSI standard too.

In the ANSI version, the CSU sends a report to the CO every second, describing the quality of the transmission during the previous 4 seconds. Only the CO is required to collect the statistics, but most CSUs continue to save them in the old format for user access. The physical connection in the CO is only a tap, there is no need to inject messages into the FDL. The LMU still takes the reports off the FDL.

The catch with the ANSI standard, from the network operator's viewpoint, is that the CSU that meets the standard no

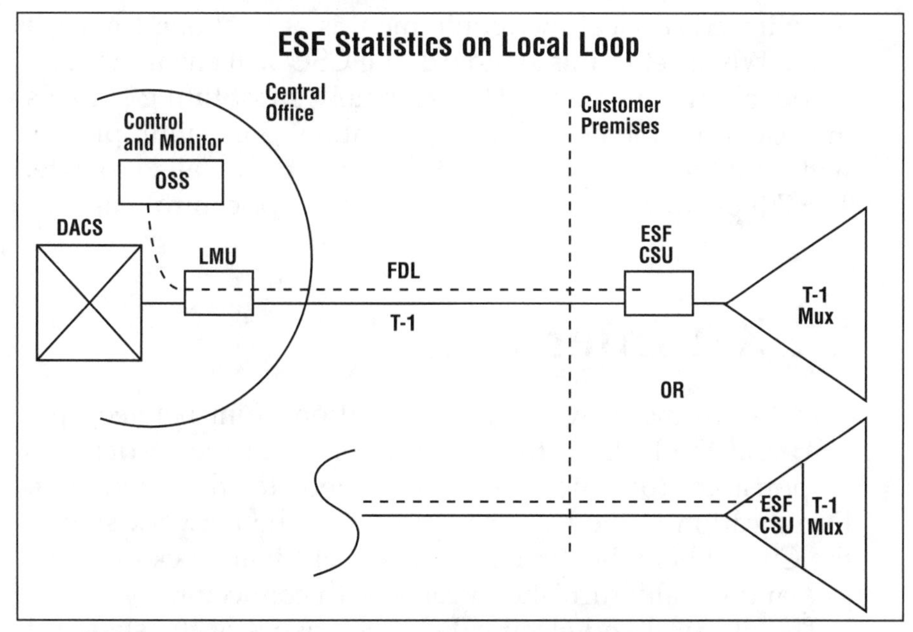

Figure 4-5. *Line Monitor Units (LMUs) in the central office issue commands to the CSUs over the facilities data link (FDL), or merely listen for periodic reports. ESF quality statistics are returned from the CSU to the LMU, also over the FDL.*

longer necessarily records statistics —only the carrier knows. For full management capability, the network designer will specify that the nodes, if not the CSUs, will make ESF quality statistics available to the customer as well as the carrier.

1's Density Enforcement

To meet FCC and AT&T standards for 1's density, some CSUs can overwrite a zero bit with a one when necessary to avoid transmitting long strings of zeroes. In addition to transparent mode, there are two options:

"Part 68": The FCC allows up to 80 zeroes in a row as long as the average density of 1s is at least 12.5% over a short interval. That many zeroes are often found in data streams from video codecs and data encryption devices. Some carrier equipment will raise a maintenance alarm on finding as few as 16 consecutive zeroes.

62411: The AT&T limit in general is 15 zeroes in a row, with a 12.5% average density. However, for access to channelized or switched services the requirement is at least one 1 in each byte. When set to this standard, the CSU will ensure that no carrier alarms are raised. Unfortunately, substituting 1s for 0s may cause problems with the application. Some multiplexers will "jam bit 7," or place a 1 in the 7th bit of an all-zero byte. The 7th position avoids altering signaling or control bits.

DSX-1 Interface

DSX-1 formerly was the demarcation point between the CSU and the DSU — that is, between what the carrier was responsible for and what belonged to the customer. Demarcation in the U.S. has moved to the network side of the CSU, where the carrier installs a terminal block or smart jack and usually furnishes a cable with connector.

The physical and electrical characteristics of the standard T-1 cross-connect are set by the DSX-1 (Fig 4-3). DSX-1 defines details like the maximum and minimum pulse width, the voltage limits of the ones pulses, maximum jitter, etc. Also included in the standard are the size and shape of the connector and pin functions. It is quite stringent.

The DSX-1 specification arose in central offices. There each arriving T-1 signal is freshly regenerated by an Office Channel Unit. The OCU corresponds to the CSU, being the first/last device on the local loop, but on the CO end. Once regenerated, the DSX-1 signal will carry 655 feet to a cross connect bay, then an additional 655 feet to another digital device (repeater, multiplexer, etc.). The tight specifications ensure interoperability among devices from many different manufacturers.

Part of the T-1 interface spec (Tech Pub 62411) covers the allowable deviation in bit rate. It used to be +150 parts per million (ppm), but has been narrowed over the years, most recently to +50 ppm or 75 bit/s. This means that the basic transport facilities of a carrier network will operate at any bit rate from 1,543,925 to 1,544,075 bit/s.

Who determines the exact bit rate? Either the user or the carrier. It depends on the type of network and the nature of the service provided by the carrier. However, somebody must decide what the master clock source will be and design the network to use it.

When two multiplexers are not synchronized, they suffer from frequent error bursts caused by frame slips. When the receiving mux runs slower than the sender, the receiver's buffer will overflow. If the slip is controlled, one full T-1 frame (193 bits) will be discarded, losing information. If the sender is slower, the receiver's buffer will "underflow" which causes one frame to be repeated. Slips are audible on voice circuits as clicks. On data circuits the error causes a data frame to be discarded and retransmitted to correct the error.

User Clocking

It is possible to lease only basic, point-to-point T-1 facilities that don't contain sources of precise timing (that is, no DACS). The carrier traditionally provisions these circuits on T-carrier systems in the local loop and M13 multiplexers (see Chapter 7) between COs. DSX-1 cross connections are set up manually in wiring bays. There are OCUs and perhaps additional repeaters within the CO. All of these devices synchronize themselves to whatever signal they carry, as long as it is within 50 ppm of 1.544 Mbit/s.

Even though the DS-3 speed of the M13 multiplexer is controlled by the Telco master clock, the T-1 ports are "asynchronous." The nature of M13 multiplexing allows for some variation in bit rate for each of the 28 inputs at T-1 speed. The M13 "bit stuffs" to accommodate the variation. This means the user's multiplexers need not be synchronized exactly to the network — again, as long as they fall within the range tolerated by the M13s. In these cases it is the responsibility of the user to establish the precise clock rate for the T-1 network and see that it is within 32 bit/s of "true" T-1.

The master clock source could be an oscillator inside one of the nodes. Or the node's internal oscillator could be locked to some known external source. Possible sources are the carrier itself, stable oscillators, and timing sources tied to the Global Positioning System (GPS) satellites or LORAN marine navigational aids (near the coasts only). See Chapter 5 for more information on clock sources.

The customer's T-1 multiplexer (and it must be only one) that supplies the 'master' clock may be configured for "internal clock" or "free running." This mux uses its clock rate to control the outbound transmission rate on all of the T-1 lines attached to it.

The exact rate of the "T-1 signal" is the frequency of this clock. It may be anywhere in the range 1.544 Mbit/s +/-32 bit/s. This speed range (stratum 4) is smaller than the range that the network can accept. This relationship ensures the network transmits the signal successfully.

Other multiplexers in a private network synchronize to the clock master. Nodes connected to the clock master have that T-1 port configured for "loop synchronization" or "loop timing." In that mode a T-1 line interface in the mux extracts the bit-rate signal from received information — the rate of the clock in the master node that's sending the signal. This clock then drives all the transmission rates on all T-1 ports of that mux. Note that by using the received clock rate to send back to the master node, the transmission stays synchronized.

Nodes not attached directly to the clock master can derive their clocking from a T-1 sent by a node attached to the master node. In this way the rate of the master clock is propagated throughout the network.

Note that only one T-1 interface per multiplexer can be configured to receive clock. All other T-1 ports use this derived clock, which becomes the node's internal clock.

Which port should be set for loop timing? The one with the "best" clock signal. Best in this context means the one with the most direct derivation from the master clock.

Sometimes it is relatively easy to decide which is the better clock source. In larger networks the answer is not always easy. The nature of the transmission line may affect the quality (stability) of the clock signal. For example, a T-carrier line with many repeaters will introduce more jitter than a single span of optical fiber.

When a network changes configuration any node may find that its port with the best clock signal may change. Multiplexers smartened up considerably since the mid-1980s, in terms of finding a reliable clock. At one time Dorgan's dilemma (Fig. 5-17) would throw a network out of synchronization. Later they could handle it by knowing more about clock quality on each line. Less expensive atomic clocks, GPS receivers, and precise timing from central offices offer generalized solutions.

Carrier Clocking

All the digital nodes in a central office (switches, M13 multiplexers, DACSs, etc.) share a station clock. From the days when digital switches were introduced up to the 1990s, there was a single precise timing source, the basic system reference frequency (BSRF; originally the Bell SRF). It provided master clock for all of the Bell System in the U.S. and other carriers as well. It was based on a cesium atomic clock, the primary time reference. Carriers have since migrated to independent local clocking sources for each CO, most often based on the Global Positioning System.

A GPS receiver in a central office can generate a clock to within 10^{-11} of the nominal rate. This is close enough to having an identical clock that the expense of distributing clocking from one source no longer makes sense.

The cost of cesium oscillators has come down considerably. It is economically practical to make such a clock part of a large switch, like the 4ESS, and even a T-1 mulitplexer.

The station clock ensures that all T-1 lines and DS-0 channels behave well when cross connected within a digital switch, either a 4ESS, a 5ESS or a DACS. Locking all switches in the network to a switches single clock ensures that they all send and receive bits at the same rate. There are no bits lost because one switch sends them faster than the next switch can receive them.

Temperature changes (which vary the length of cables) and other influences cause very minor variations in bit rate. Buffer memory or elastic stores hold extra bits temporarily until they can be processed. If the rate changes too much, the buffer overflows and there is a "frame slip." The buffer is cleared and operation continues, but data (193 bits in 1 frame) are lost or repeated.

Having one clock one for all switches reduces the amount of buffering needed to minimize the periodic errors due to frame slips. Small buffers also introduce minimal transit delay.

Some carrier services always pass through these digital CO switches: ACCUNET T1.5 with CCR is based on DACS, MEGACOM connects directly to a 4ESS switch, etc.

Even facilities ordered as plain transport service may be routed through a DACS. The carrier will always inform the user if this is true. Because DACS equipment automates provisioning, its use will spread until every T-1 circuit passes through at least one.

With a CO switch or DACS in the circuit, the carrier determines the bit rate precisely. The user has no option but to synchronize to the network. If not synchronized, then the bits can build up, causing frame slips as often as every few seconds. PCM voice may remain intelligible, though noisy, at this error rate. Data circuits would be nearly useless.

At the customer office, the CSU locks its clock to the signal from the switch or DACS, and thus to the local CO's station clock. The CSU passes that rate to the multiplexer at the DSX-1 interface. CPE will lock onto the incoming bit rate when configured for "loop timing." In this state, the multiplexer will transmit bits at the same rate, matching the speed of the CO.

Even one DACS in the facilities of a private network will establish the bit rate for the entire network. And it's a lot cheaper than buying your own cesium clock.

DSU: Digital Service Unit

The digital service unit (Fig. 4-6) takes an active role in shaping the T-1 signal being sent. That is, the DSU converts synchronous interfaces like RS-422 to bipolar. The DSU enforces rules about bit density and bipolar format. For example, the DSU could stuff the least significant bit in each byte with a one. Or 1s density may be left to the CSU.

This is exactly what is done to maintain the ones density in 56 Kbit/s DDS circuits. The least significant bit is always sent as 1 and ignored on receipt to avoid changes made by robbed bit signaling. Thus even if the live data stream is all zeros it meets minimum 1's density requirements: only 7 zeros are put in each time slot.

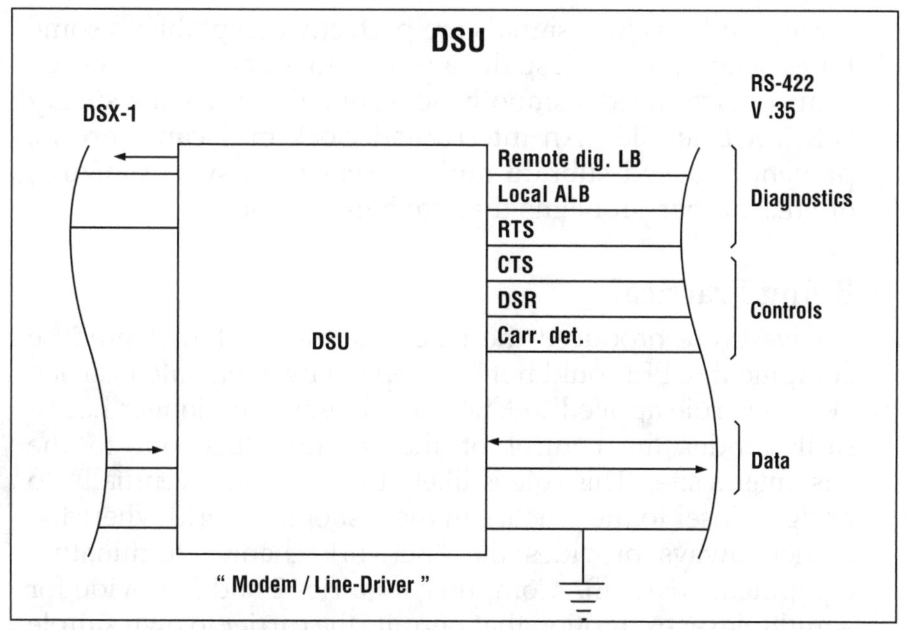

Figure 4-6. *A data service unit (DSU) prepares the customer's data to meet the format requirements of the DSX-1 interface, for example by suppressing zeros with special coding. The DSU also provides the terminal with local and remote loopbacks for testing. The DSU is usually built into the terminal equipment or multiplexer, and will eventually disappear.*

The DSU reacts to bipolar violations of certain types when they are seen on the incoming signal line. Some bipolar violations are interpreted as commands from the network to go into (and out of) loopback modes.

More and more, vendors are building the DSU and CSU into terminal equipment like multiplexers. At the same time, a parallel trend in T-1 CSUs is to bring in multiplexer functions. A CSU can deliver not just a DSX-1 to a multiplexer (or PBX), but also split off some number of DS-0s into as many as six serial data ports.

The time slots devoted to a serial port are filled with idle pattern (all 1s) in the DSX signal.

The serial ports on a CSU are synchronous, usually V.35 with clock signals on leads in the connector. You might think the clock rate on a 56 kbit/s port would be 56 K. Perhaps, but not necessarily. In some CSUs the clock delivered to a 56K serial port is 1.544 Mbit/s, but it is stopped at intervals so the average is 56K.

Stop-and-go clock signals are perfectly acceptable to some types of equipment. Usually a router doesn't care. However, some devices need a smooth clock, one that runs at a steady 56K, for example. An interrupted clock may cause errors, prevent attached subrate multiplexers from synchronizing, or create other problems that are hard to isolate.

Being Practical

Divestiture produced the FCC rule that customer premise equipment (CPE) could not be supplied by a regulated carrier. With this rule applied to CSUs, the line was no longer necessarily under the control of the central office out to the customer's site. This rule is likely to be relaxed eventually, to bring it closer to the practice in the rest of the world where the carrier always provides the "network channel terminating equipment" (NCTE). Computer Inquiries 2 and 3 provide for a multiplexer exemption that permits the carrier to own simple CPE. This exemption has been used by Nynex to deploy a digital overlay network to provide 56K and T-1 access services.

If the customer does not provide the loopback functions of the CSU, there can be finger pointing during a line failure. A carrier-compatible CSU helps reduce time to repair. The user should also be able to initiate loopbacks — local analog (line loop back) and remote digital (payload loop back) — within the DSU or inside the multiplexer. When used in conjunction with test equipment or the capabilities of the terminal, these diagnostics help isolate faults.

In part it is the lack of certainty about the CSU/DSU that makes Extended Superframes desirable. The low-speed data channel in the framing bits allows the central office or long haul carrier to interrogate the customer's equipment regardless of what part of that equipment answers the inquiry.

It could be the CSU or DSU that keeps statistics and reports them to the CO. Or these functions could be part of the multiplexer or channel bank.

If you have the older D4 framing from D4 to ESF, and want to move to ESF, what do you replace: the CSU or the mux? You could install ESF CSUs, for as much as $2500 each. They offer remote diagnostics and some vendors provide a network management system just for their CSUs. The newer multiplexers and channel banks with integral ESF CSUs may be more economical in the long run. For example, a channel bank complete with integral ESF CSU can cost only two or three times the price of an ESF CSU.

Clear Channels

The ability to send any data pattern, including an indefinite period of all zeros, defines a clear channel.

B8ZS

A bipolar violation underlies one method to keep the ones density up when transmitting consecutive zeros on a T-1 carrier. The method substitutes a known pattern of ones, with bipolar violations, for a group of zeros. A bipolar violation occurs when two 1's pulses in a row have the same polarity.

It is not possible to signal control functions with normal bipolar patterns (alternate mark inversion, AMI), because all of them

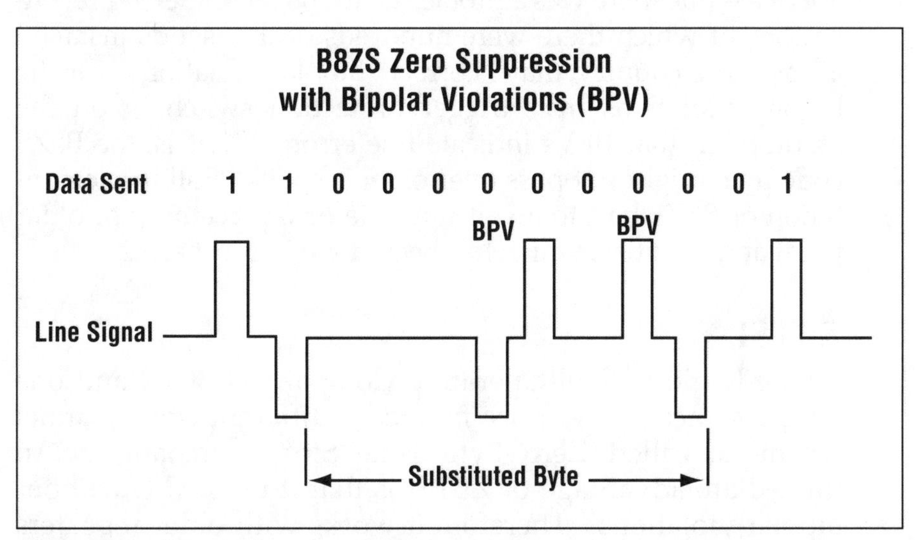

Figure 4-7. *Zero suppression via bipolar violations substitutes a known pattern for eight consecutive zeros. B8ZS maintains ones density while remaining transparent to data.*

can be valid data. The technique of "binary 8-zeros suppression," or B8ZS, violates the AMI bipolar pattern by sending consecutive pulses of the same polarity (Fig. 4-7). The violations distinguish a byte substituted for all zeros from a normal byte of data containing legal ones. The device that does the substitution typically has been a CSU, but could be the multiplexer.

If the last mark (1 pulse) is plus, the B8ZS feature sends 0 0 0 + - 0 - + instead of 00000000. After a -1, it would be 0 0 0 - + 0 + -. Two pulses of the same polarity in a row alert the receiver that this sequence is not normal data. If the receiver sees the known pattern, it restores the 8 zeros. B8ZS or something similar is needed to transmit 64 kbit/s clear data channels with absolute assurance of maintaining 1's density to AT&T standards.

This B8ZS pattern was selected to preserve the balance of plus and minus voltages, to prevent DC components on the wire. B8ZS is limited to "T spans," typically local loops that access a public network, and is not carried end to end through higher order multiplexers (like M13). An M13 converts all pulses to a single polarity, and so loses track of the bipolar violations.

For a while there was a problem with B8ZS. Older T-1 regenerators, of which there were hundreds of thousands installed all over the country, may "correct" bipolar violations. Or the bipolar violations may trigger protection switching on the assumption that BPVs indicate line errors. That is, the B8ZS code byte might not pass over older T-spans. Full implementation of B8ZS had to await upgrade or replacement of older plant and facilities, which has been done in most areas.

ZBTSI

One Regional Bell Operating Company (RBOC) and one long distance carrier once favored a different clear channel technique called Zero Byte Time Slot Interchange. An immediate advantage of ZBTSI is that it uses AMI without bipolar violations. Therefore it works with older repeaters that won't pass BPVs as well as newer spans configured for B8ZS. It also is carried end to end through M13 and higher level multiplexers.

ZBTSI involves considerable processing to ensure sufficient 1's density. There is also additional delay (greater than 0.5 ms or four frame times per switch). These factors, and the loss of CRC readability (see below) discouraged use of ZBTSI.

Framing Bit Patterns

Extended Super Frame						Super Frame	
Framing Bits[1]				Frame Number	Signaling bits	Multiframe Alignment	Terminal Framing
Z	C	M	F				
1				1			1[2]
	2			2		0	
		3		3			0
			(0)	4		0	
5				5			1[2]
	6			6	A	1	
		7		7			0
			(0)	8		1	
9				9			1[2]
	10			10		1	
		11		11			0
			(1)	12	B	0	
13				13			
	14			14			
		15		15			
			(0)	16			
17				17			
	18			18	C		
		19		19			
			(1)	20			
21				21			
	22			22			
		23		23			
			(1)	24	D		

1 The values of the F bits are fixed in the ESF format; Z, C, and M bits are variable.
2 These D4 framing bits are changed to "0" when the following group of four frames contains ZBTSI codes.

Table 4-1. *The F bits in the Extended Superframe perform the same functions as the terminal alignment and frame alignment bits in the superframe. That is, they allow the receiving multiplexer to identify where the DS-0's are, and to locate every sixth frame that contains signaling information. The Z bits are used with ZBTSI encoding for clear channel capability. The C bits are for a CRC-6 error check. The M bits form a message channel of 2000 bit/s when configured for ZBTSI. Together, the Z and M bits constitute the facilities data link.*

To clarify the process, we look in a little more detail at the framing bit patterns for both SF and ESF framing (Table 4-1).

The F bits at frames 1, 5, and 9 in the super frame (which are normally 1's) mark off blocks of four frames. In the same way, the Z bits in the ESF pattern mark off the same size blocks of 96 time slots. The Z bits occupy half of the facilities data link (FDL) described in Chapter 3.

The ZBTSI encoder works with four frames at a time, applying two stages of processing to prevent long strings of zeroes. First it scrambles a block of 96 time slots to get 768 bits of homogenized information. This step alone reduces to less than 2% the probability of finding more than 14 zeroes in a row.

If there are no longer strings of zeroes, the scrambled block is sent as is, except that the 96th octet is sent first. The receiving decoder simply unscrambles the bits back to 96 time slots and

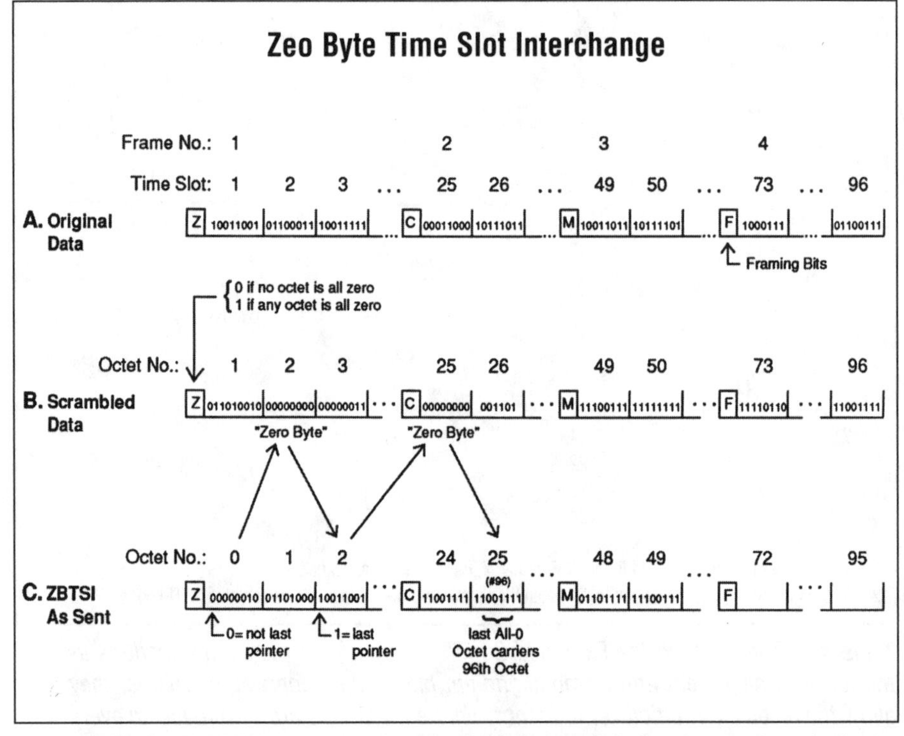

Figure 4-8. *ZBTSI scrambles original data (A) to minimize the number of all-zero octets (B). As sent (C), a "1" for a Z bit means the first octet points to the first all-zero octet (even itself). Once pointed to, an octet can carry a pointer to the next all-zero octet and still be received as all 0's. The last all-zero position carries octet 96.*

moves octet 96 to the end of the block. The framing bits are not scrambled, just the payload. Therefore the CRC check sum in the C bits is not changed to match the scrambled data. Hence every frame's CRC will be in error. This feature may be used to identify lines with ZBTSI coding. The Z bits remain all 1's.

If there are 15 or more 0's after scrambling, the ZBTSI encoder continues processing:

a. The last or 96th "octet" of the 768 bits is held aside. ("Octet" implies a meaningless mix of bits, not a coherent "sample" or "byte.")

b. The leading F or Z bit is set to 0, indicating the block following will contain interchanged time slots.

c. The first octet sent in the frame (number 0 in Fig. 4-8) is not user data but a pointer to the first octet that is all 0's. The last seven bits are the number of that octet.

d. If this is the last pointer octet in the block, the first bit is set to 1, otherwise it is zero.

Note that if the first octet itself is the only one of all 0's, the pointer would be 0000000, but preceded by a 1 (last pointer). Thus it would no longer be all 0's.

e. If there are more all-0 octets, the second points to the third, the third to the fourth, etc. Pointing puts 1's in them, eliminating strings of 0's. The last pointer is the one before the last octet of 0's.

f. Therefore the last all-0 octet is identified and yet can carry octet 96, which is inserted from memory. If number 96 is itself all 0's, it can be thrown away.

Criteria in the proposed ANSI standard allow the processing to ignore all-zero octets that do not cause a 1s density violation. For example, if the octets on either side contain enough 1s, an all-zero octet will not create a string of 15 or more 0s. The algorithm need apply ZBTSI only to a Violating All-Zero Octet (VAZO).

Note that the two clear channel techniques are incompatible with each other. B8ZS introduces BPVs which are seen by a ZBTSI decoder as errors. The ZBTSI encoding makes all the CRC values wrong. And by taking half the FDL for the Z bits, ZBTSI creates a speed for the message channel (2 kbit/s) that doesn't match the D4 or B8ZS FDL (4 kbit/s). Therefore great care must be taken when installing software-configurable equipment to make sure that the clear channel provisions are uniform.

Satellite

The satellite services can be attractive for remote locations, off the main network. Over very long distances, and especially when broadcasting information one-way, they are very serious contenders. Some satellite carriers price their T-1 circuits far below terrestrial lines over long distances.

In many ways the satellite circuits resemble the terrestrial ones. They run the same bit rate, have the same signal format, and often share the other specifications of the DSX-1 standard. Other interfaces may be available, including V.35 and RS-422. Satellites will support D4 and ESF framing and signaling, but other than AT&T's offerings, they do not necessarily require either now. In the future, ESF may be adopted by all satellite companies.

The greatest single difference between satellite and terrestrial circuits is the time delay. The closest an earth station can ever be to a satellite is 22,300 miles (Fig. 4-9). And that's only if the earth station is on the equator directly under the "bird." All U.S. earth stations are farther away, leading to a two-way transit time, even at the speed of light (186,000 miles/sec.) of

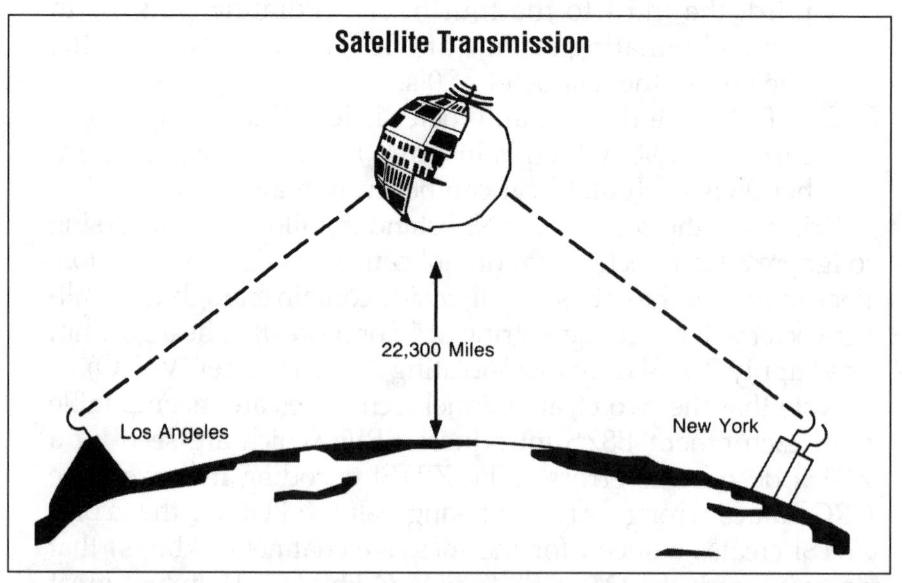

Figure 4-9. *Transit delay is the biggest difference introduced by satellite circuits. Not only is satellite longer, but it is subject to variation.*

1/4 second between earth stations (up and back). The round trip, from a grounded user's point of view, is 1/2 sec.

A half second can play havoc with communications. That much delay lets both parties to a normal phone conversation start talking without realizing the other also has started. These "voice collisions" can be annoying, and reduce "conversational throughput."

For data traffic, the communications protocol must account for the time very carefully. For example, if an SNA circuit were carried as part of a T-1 path via satellite, protocol options would have to be set to accommodate the long time between the transmission of a block of characters and the acknowledgement (ACK) of that block by the receiver.

The T-1 bandwidths commonly available via satellite make the problem worse. For an example, take a 64,000 bit/s subchannel sending 2,000 character blocks to a CAD/CAM station. In the half second before the first block can be acknowledged, the originator could send as many as 31 additional blocks. The SNA standard practice is to halt after 7 blocks and wait for the ACK. As a block is acknowledged, another can be sent.

In this situation, waiting for ACKs reduces throughput drastically. But it can be worse. Any additional delay, like processing time at the far end, could lead the protocol at the sending end to "time out." This may lead to a reset or outage, unless anticipated and accommodated.

Setting the block size very large would reduce the number of blocks outstanding, but increase the possibility of errored blocks. The chance of an uncorrectable error is proportional to the length of the block. Even one hard error causes the entire block to be sent again. Large blocks can encounter errors so often that they reduce throughput and cause noticeable delays.

Note that the subchannel was 64,000 bit/s, not the 56,000 bit/s found on DDS and other terrestrial services. The reason is that satellites can offer "clear channels." Because they do not switch, they do not use robbed bit signaling. They need not avoid the eighth bit position because there is no fear it will be changed arbitrarily for telephone operations.

The digital T-1 signal actually modulates a microwave radio beam. There are no T-1 repeaters on a satellite. So they can also tolerate long strings of zeroes.

The Wobblies

Wobble is the slow movement of the satellite in the sky. This motion varies the distance between it's antenna and the earth station. Changing distance means changing transit time. The ground station sees a slow shift in phase of the received signal. Over longer periods (many hours to weeks) the phase can shift by tens of bit times. Consequently, satellite equipment usually incorporates a buffer, or elastic store, to soak up the variations.

Slow "jitter," fade from atmospheric conditions, flying objects, and animals in the transmission path all contribute to the error rate via satellite. Intrinsically, that rate is much higher than for terrestrial microwave. However, satellite carriers, having bandwidth in abundance, use some of it to reduce the effective error rate through forward error correction, FEC.

Many protocols include a cyclic redundancy check with each data block to permit the receiver to test for transmission errors. Most common are 16 bit numbers. These are enough to detect almost all errors in blocks of 4000 characters. Correction is through a retransmission of the block containing the error.

By making the CRC very much larger, it can contain enough information to permit the receiver to correct most transmission errors, not just recognize them. "Large enough" can be as large as the information itself, but is usually less. Having extremely wideband transponders, satellites can spare the bandwidth to support FEC and thus get overall error rates down to the one-in-a-million range of land lines.

Who Does What

A leased T-1 line routinely involves three carriers: a local operating telephone company at each end and the long-haul carrier (land line or satellite) between. In the most general case, each is independent and must coordinate installations with the other two. Exactly how many line segments and how many companies depends on the LATA factor.

The U.S. is divided into just under 200 Local Access and Transport Areas . These may cover a metropolitan area, or an entire state. The local telephone company can carry traffic only within a LATA. That is the rule of the Divestiture Agreement, which broke up the Bell System and laid down ground rules for what the Bell operating companies would get, and what AT&T would get.

Calls that cross a LATA boundary must be carried by an inter-LATA or long distance carrier, also known as an interexchange carrier (IXC). This requirement holds even if the same operating telco controls the LATAs at both end points (Fig. 4-10). This means Nynex (NY Telephone) can't carry a call or support a data circuit from New York City to Buffalo, though it occupies and serves both of those LATAs.

In addition to the local loop (or terminating channel, TC), there probably will be an intra-LATA channel needed (Fig. 4-11), at either or both ends, An ILC connects the first central office with the inter-LATA carrier's "point of service" or "point of presence." This segment also has been called the LATA distribution channel, LDC. AT&T usually has one to a few serving offices (POS or POP) for T-1 service in each LATA. T-1 is not offered at every central office. Several rural LATAs have no AT&T POS at all. This means there can be three vendors and five segments to a "simple" T-1 line.

In the AT&T tariffs approved on 27 April 85 (FCC 9 and 11),

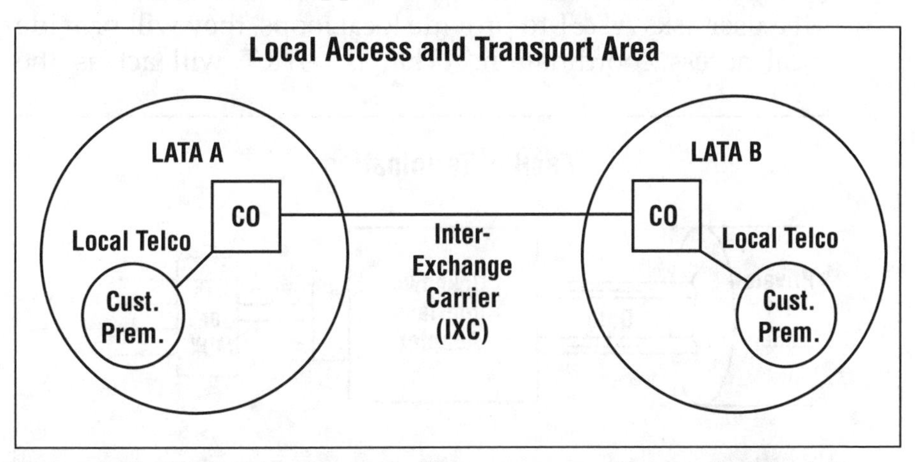

Figure 4-10. *The local telephone operating company provides circuits (leased or switched) within a local access and transport area (LATA). Service between LATAs must be furnished by an interexchange carrier (IXC) or long distance company. These restrictions apply even if the same telco operates both LATAs.*

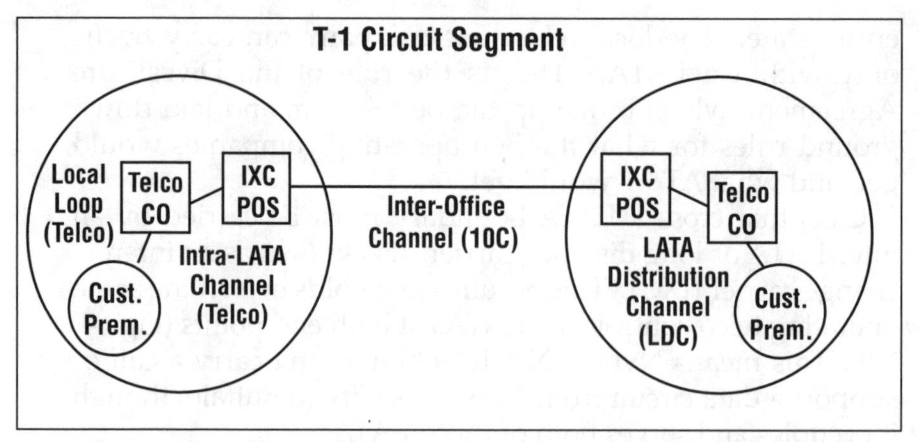

Figure 4-11. A "simple" T-1 circuit will have up to five segments if the interexchange carrier's point of service (POS) is not in the central office that serves the customer's premises.

the cost of leased lines was unbundled into multiple parts: central office connections, inter-office channels, local loops, and other services. Tariff #10 lists the various services available at each central office, but no pricing. As the local loop, including the intra-LATA channel, is now priced separately, the customer is free to accept it from AT&T or to provide it by any other legal means. The regulated telco usually provides it, but short haul broadcasting is a logical candidate.

If a user asks AT&T to provide local loops, they will provide "local access coordination." That is, AT&T will act as the

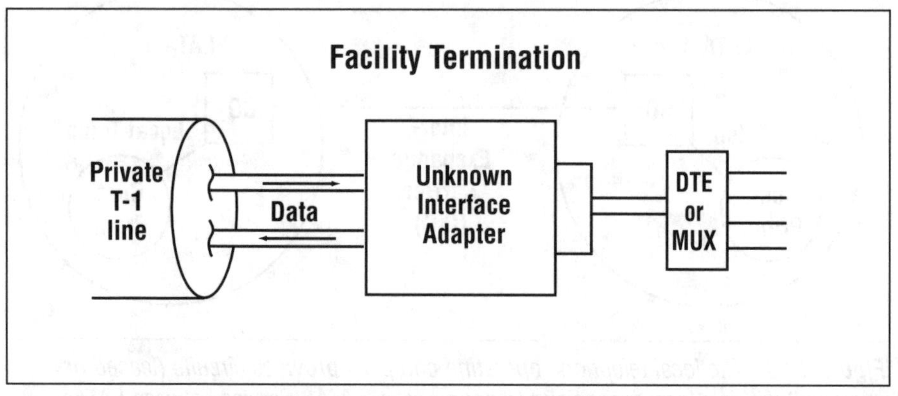

Figure 4-12. Private T-1 facilities have no standard interface. Each vendor must be consulted for specifics.

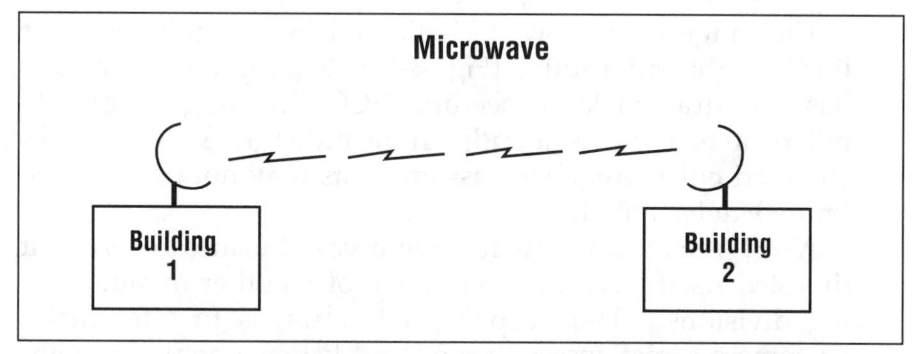

Figure 4-13. *Private microwave can be quite inexpensive to fairly expensive, depending on size. It requires line of sight paths. Its biggest drawbacks are the requirement for FCC licensing and a shortage of frequency allocations in some areas.*

user's agent to arrange for T-1 access circuits from customer premises to AT&T's point of presence. A caution is in order: the user is responsible for each segment individually and pays from the time each is installed. On occasion, AT&T has arranged for local service before the long-haul circuit was available. The customer then pays for those unusable circuits until AT&T finishes its installation.

Other T-1 Transmission Media

Now that AT&T has recognized and accepted the option for a user to provide access, the bypass technologies become of great interest as T-1 transmission media. The following brief review is intended merely to alert readers to the possibilities of alternate or private media. Rapid changes continue in all these technologies.

Private media have a wide variety of interfaces, with and without framing (Fig. 4-12). There is no general rule. Vendors must be consulted on specific products.

Short-Haul Microwave

Low end microwave systems offer fairly inexpensive hardware costs. From the bottom, bandwidth starts at T-1. More than one vendor offers two to four T-1's on a single pair of transceivers.

The range of microwave is limited by a requirement for line-of-sight path routing (Fig. 4-13). In addition, microwave has the drawback of needing FCC licensing of private microwave: there is a radio transmitter at each end. In crowded cities, frequency assignments may not be available immediately, if at all.

AT&T retained its microwave access business even as it divested itself (second wave, 1996) of its other manufacturing divisions. The purpose, it is clear, is to offer major customers direct access to the long distance network, without using facilities of the local exchange carrier.

VSAT

Very Small Aperture Satellite stations for private T-1 networks were shown in early 1988. Private earth stations for T-1 need dish antennas less than 4 meters in diameter. That means a new low in the cost of the ground equipment. Ultra small aperture dishes (USAT) for 56 kbit/s service can be as small as 0.5 meter in diameter.

With the large amount of optical fiber put into the ground from the late 1980s into the present, satellite transponders became available at relatively low cost. Some were converted to private networking, for voice and data backup, for example. Put it all together, and the very large user who can tolerate delay gets full control of T-1 (and greater) bandwidth at a reasonable cost.

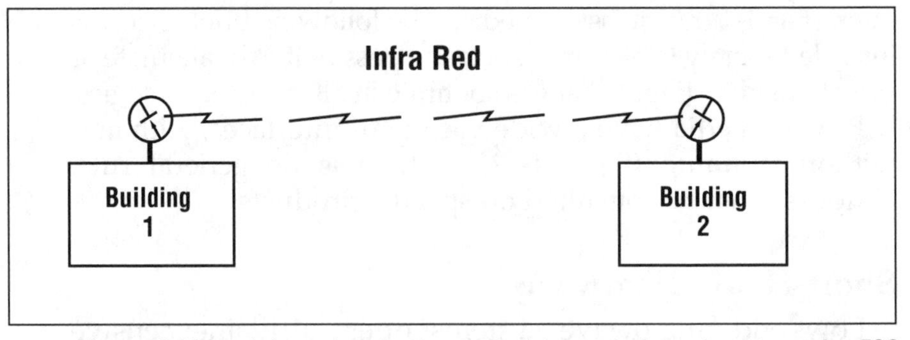

Figure 4-14. *Infra-red and other optical transmissions do not require FCC licensing and can be inexpensive. But they are affected by snow, birds, etc., which limits them to about 3 miles.*

Infra-Red

Recent improvements in infra-red transceivers have increased their range from under 3 to about 5 miles. Rain, birds, and so on can affect transmission. For reliability in storms, the distance may have to be reduced. For light, as for microwave, routing must be line-of-sight (Fig. 4-14).

On the plus side, no FCC transmitter license is required. Bandwidth can be quite large, up to T-3 at about 45 Mbit/s. And the equipment is very small and not expensive. Therefore a private link is often built up of two redundant transceivers aimed over slightly different paths. A separation of even a few feet makes simultaneous signal failure very rare.

Optical Fiber and Coax

These two media have many similarities. Both have very good immunity to noise, and very large bandwidth, depending on the form of line driver (Fig. 4-15). Both come in the form of relatively expensive cables (while the glass fiber itself has come down sharply in price, the cable assembly has not declined as much). Cables may be had with steel reinforcement for aerial suspension, or jacketed for direct burial in earth.

One of the large cost elements is installation. Both forms of cable need careful placement. A sharp bend in a coax creates a local change in impedance, introducing a reflection (echo) that disrupts signals. Optical fiber bent too sharply loses light that should be kept inside, greatly diminishing the signal strength. A kink in either kind of cable can block signals completely.

And of course, laying a cable requires a right of way. Fiber optic systems now support distances above 100 miles without repeaters; coax can be driven about 40 miles; distances are increasing constantly. A clear route that long can be hard to find. Particularly in cities and incorporated municipalities, the obvious rights of way may be under franchise to utilities. For greater distances there is usually a cost in reduced bandwidth: the product of BW x DIST is roughly a constant for a particular cable.

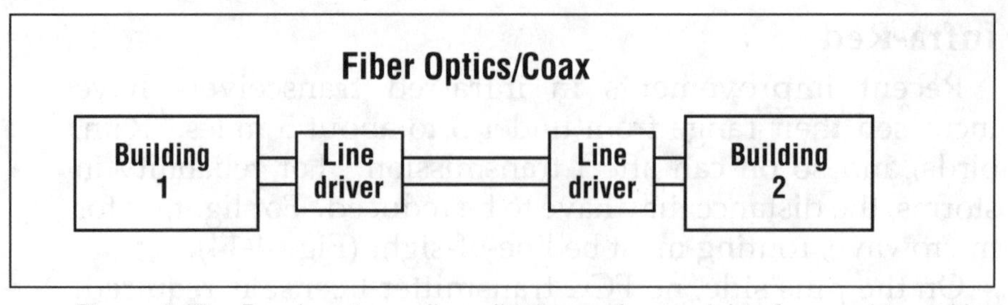

Figure 4-15. For short to medium distances where a right of way exists, cables of optical fiber or coax offer large bandwidth and good immunity to noise. Materials and installation may be expensive.

When installed without repeaters the fiber is referred to as "dark." While usually owned by the user, dark fiber may also be leased from a carrier — maybe. The demand for bandwidth has risen so fast that owners of fiber want to retain control so they can enjoy the benefits of new technologies that increase the capacity of each fiber. Faster electronics and laser transmitters have been deployed for years. Wave division multiplexing promises to multiply the capacity of a fiber by as much as 16 times. Whoever has control of the dark fiber can reap the income for that bandwidth.

Twisted Pair

Common telephone wiring was the original T-1 medium, and still has many applications at this rate. If it is used in the local loop between the customer premises and the CO, why not elsewhere?

A spare pair of twisted pairs (four wires) can extend the DSX-1 interface several hundred feet. With proper line drivers, or CSUs, a T-1 line can operate over 6000 ft of twisted pair. This medium is one of the easiest to install, and boasts low material costs.

Twisted pair installed for phones can be found throughout most buildings. However, older wiring may have loading coils, unused branches, taps, extra terminal blocks, and other features that will add excessive inductance or capacitance and make the line unsuitable for T-1. Twisted pair for T-1 must be engineered carefully for both local loop and inside distribution.

Digital Subscriber Line

Copper pairs can carry a signal frequency beyond 1 MHz. The normal T-1 CSU doesn't begin to take full advantage of category 3 unshielded twisted pairs (UTP), which is normal house wiring. Starting in the mid-1990s, most new cable installations used category 5 UTP, designed for 100 Mbit/s LAN traffic as well as POTS voice. Cat 5 wire has even broader bandwidth capability.

The fact is, the capacity of the existing wire plant is underutilized.

At the higher frequencies UTP acts like a broadcasting antenna. It sends part of the signal energy into space, weakening the pulses as they travel along the wire. The higher frequency components weaken faster. The circuits compensate for this frequency response.

Hardware developments like DSP chips and sophisticated signal processing software can push much more signal over UTP. Carriers have taken advantage of technology to replace the old "T carrier" systems.

There are many products that can send a T-1 or E-1 signal over the entire length of a local loop (18,000 ft of 26 gauge wire) with no repeaters in the outside plant. Recall from an earlier section that T-carrier requires a repeater every 6000 feet. Installing and maintaining repeaters is a large expense and a potential source of outage. Finding a fault can be difficult.

Digital subscriber loop technology replaces the CSU and OCU with new line drivers. DSL is a way to make the local loop run faster. The central office must have DSL equipment to match any DSL device a customer installs.

Note that all comments about DSL have tied it to the local loop. That is the only part of a circuit where DSL exists. Beyond the loop, into the carrier's network or a customer's terminal equipment, information travels on other media. For example, you might distribute data on a LAN. A carrier will use an existing backbone network to carry traffic between central offices. That transmission technology could be SONET, frame relay, ATM or something else — but it probably won't be DSL.

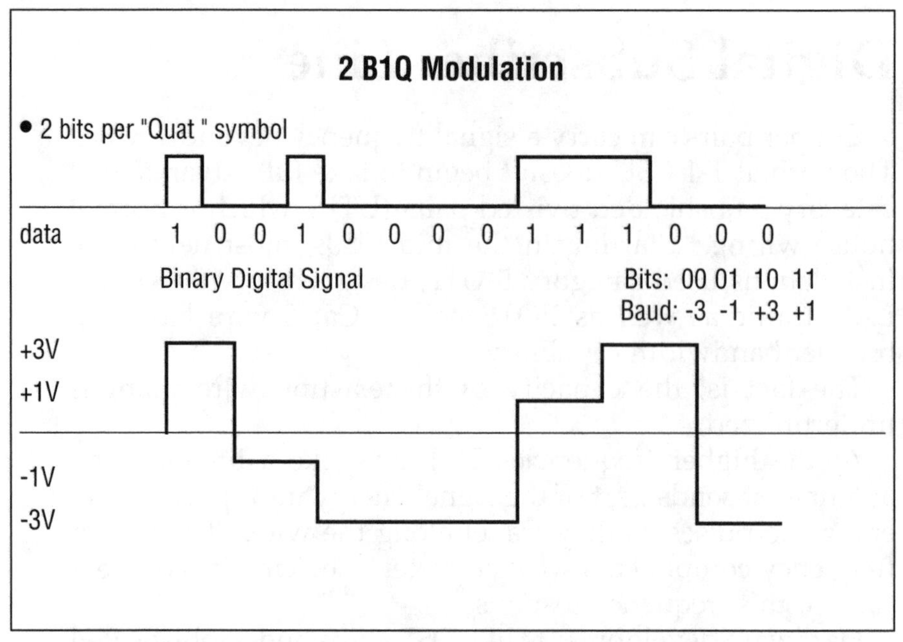

Figure 4-16. *The oldest "DSL" technology is the same as ISDN basic rate interface: 2B1Q. Four voltage levels allow each pulse (symbol) to convey two bits of information.*

Each carrier makes its own technology decisions, so as DSL-based services roll out don't assume that the CPE technology will be the same in all parts of the country or the world.

Base Technologies

There are any number of modulation methods that will encode a digital signal for transmission on UTP. Ethernet uses one; Token Ring LANs, another.

Listed here are the four that have been commercialized as of 1997 for local loop access between customer premises and the central office. Expect others to follow. When a better method is designed, it is likely some company will decide to offer a product based on it.

Your job is to understand enough to avoid the hype that comes with the many products based on these foundations.

2B1Q From ISDN

As long ago as the late 1960s, ITU (then CCITT) defined a new local loop technology for the basic rate interface (BRI) to the Integrated Services Digital Network (ISDN). BRI differed from the AMI used with T-1. In AMI, each pulse has the same voltage (of alternating polarity) and thus can represent only one bit. BRI was defined with four possible pulse voltages, +3, +1, -1, and -3. Each pulse, called a quaternary (quat, for short), can represent two bits (Fig. 4-16). Hence the name 2B1Q (2 bits 1 quat).

The bandwidth of the BRI is two 64K user channels, one 16K signaling channel, and 16K of overhead and framing: 160 kbit/s. Because each pulse conveys two bits, the pulse rate is halved: 80 kbit/s. By randomizing the bits before sending them, the pulses are also randomized, plus and minus. This reduces the effective frequency of the transmitted signal by almost another factor of two. So even in its earliest days, when electronics were not as powerful as today, BRI could reach over 18,000 ft of UTP.

2B1Q sends full duplex — both directions at the same time — over one pair. Each end applies echo cancellation

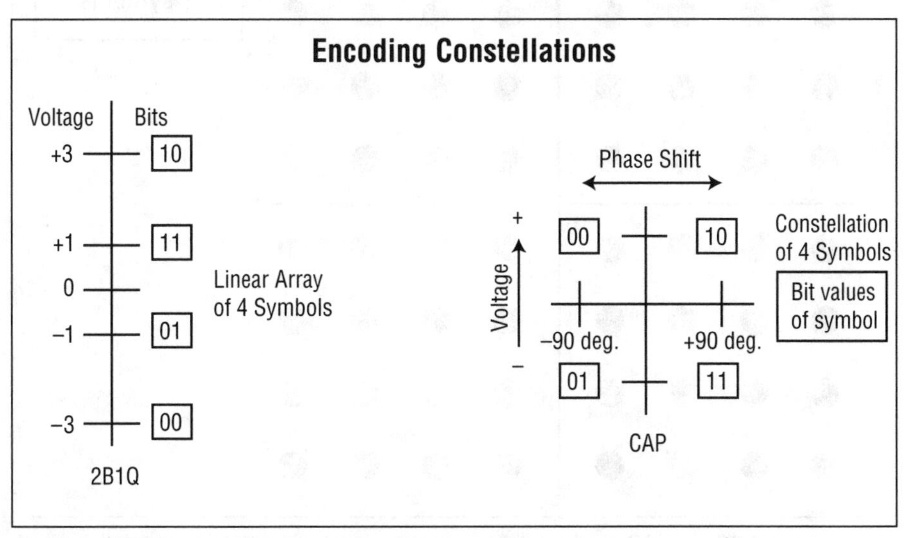

Figure 4-17. *A linear or one-dimensional array, as in 2B1Q, can't convey as many bits per symbol as a multi-dimensional array or constellation, as in CAP. Each CAP symbol can represent up to 15 bits.*

to hide its own transmission and reveal the received signal pulses.

Now crank up the pulse rate. You will find that the very same 2B1Q signal can carry a full E-1 over a usable distance (though not the 18,000 ft of a full length "standard" local loop). Multiple pairs are commonly employed to reach the full distance for T-1 (two pairs) and E-1 (three pairs).

Carrierless Amplitude/Phase (CAP)

CAP has been around almost as long as 2B1Q. It's a modem: a special modem, but still a modem.

Special means that it is restricted to the local loop: it cannot dial up over the PSTN. The CAP signal cannot be encoded as PCM in a DS-0. A matching CAP device in the CO terminates the signal and converts to DSX-1 or another standard interface.

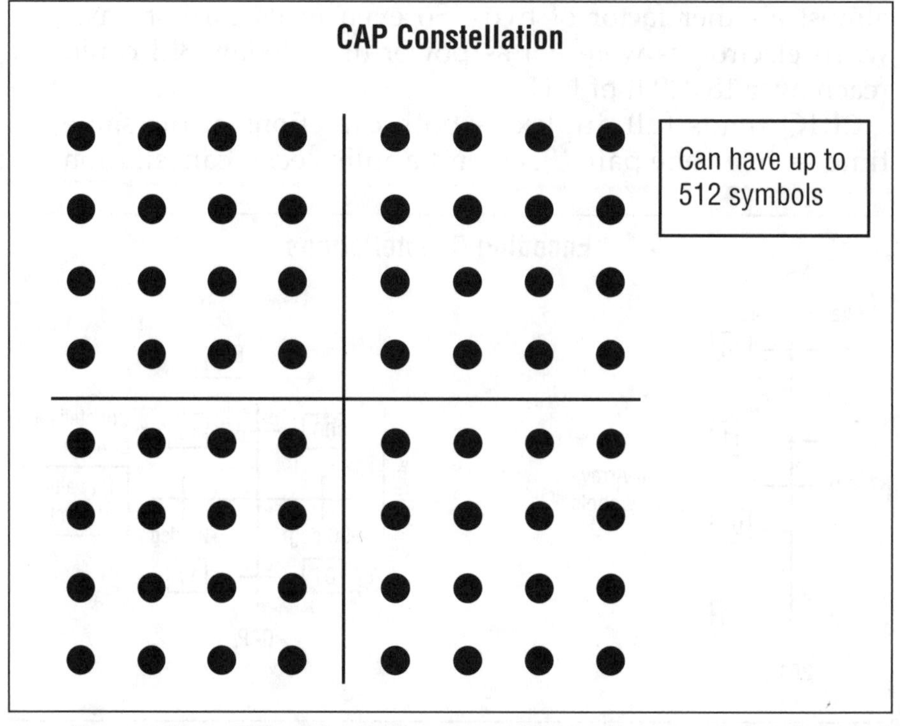

Figure 4-18. *CAP modulation on a good circuit will create a 'constellation' of up to 512 different symbols. This ability allows multi-megabit speeds because of the large number of bits per symbol.*

In a sense, CAP is an extension to 2B1Q. The BRI signal can be represented as points on a line (Fig. 4-17). CAP adds a dimension, phase shift, that spreads the line into an array, also called a constellation. CAP may have 4, 64, or more points in the constellation (Fig. 4-18). Each voltage and phase combination in a pulse (more properly called a symbol) can represent as many as 15 bits. This means the frequency of the symbols can be reduced, lowering the frequencies used on the local loop. Lower frequencies mean less loss, which translates into greater reach and/or higher bandwidth.

Discrete MultiTone (DMT)

DMT is not just a modem, it is 250 modems.

The frequency spectrum of UTP loop wire is divided into "bins" each 4 kHz wide. That is about the bandwidth of a normal voice channel. Since a voice channel will carry a modem signal, it follows that the UTP wire will carry 250 modem signals.

The channels are dramatically different, because they are centered on widely separated frequencies. Recall that the higher frequency signals fade more with distance. Thus the bins at the low end of the spectrum have higher inherent capacity than the bins at the high end (Fig. 4-19).

DMT not only accounts for frequency response, but also looks (automatically) for interference in each frequency bin. Certain bins will be swamped by local radio stations (we are talking radio frequency here). Motors, lighting, and other things may reduce the quality of certain bins.

The DMT "modems" assess each bin, then assign it a modulation constellation that it can handle. Since this evaluation is on the actual line, the process will discover the maximum capacity of that line. In most cases, the carrier will limit the bandwidth to what the customer orders. There could be an option to increase it later.

But there is only one UPT. How can there be 250 modems operating on the same line? The answer relies on fairly sophisticated math. It's not new math: Fourier, the French mathematician who did the original work, died in 1830.

You have probably had a chance to see his work in a frequency scan or spectrum analysis. The chart that results plots the strength of the signal against its frequency, just like Fig. 4-19. The sum of that information describes a single-value input, averaged over the time of the analysis.

Recall that at any given moment, there can be only one voltage on the interface wire that delivers the RF signal, but that voltage varies rapidly over time. The spectrum analyzer examines that single voltage for a certain length of time to tease out the frequency components.

This is the Fourier transform: a manipulation from the time domain (one voltage that varies with time) to the frequency domain (a stable list of the amplitudes for each frequency component). The sending-end DMT device reverses the process of spectrum analysis by applying an inverse Fourier transform; (IFT). The time interval of each transform is 1/2000 second.

Each DMT bin is a narrow section of the frequency spectrum. The information encoded into that modem signal is one point on a constellation, that is, a voltage (and phase). Taken together, the values in the 250 bins are a spectrum analysis of the DMT signal that must be sent. That is, the 250 points describe a Fourier transform of the desired signal. To get the signal, the sender applies an IFT. The resulting waveform takes half a millisecond to send.

To implement the IFT requires a powerful processor. The processor optimized for math transforms is the digital signal processor (DSP). It was about 1995 that DSPs were first able to tackle this process in real time.

Time Shift Keying (TSK)

The signal on the line is almost a sine wave. This nearly pure tone acts as a carrier signal which is easy to generate and simple to detect and lock onto at the receiving end. The subtlety comes from the way information is encoded: a small shift in the quarter periods that holds the wave peaks in a constant phase but shifts the point at which the voltage goes through zero. The zero-crossing point may occur early or late. Signal processing in the receiver extracts the bits encoded into each half wave.

This technology is in the early stages of commercialization.

Product Categories (Applications)

Any of the four technologies described (or others not marketed at this time) can support a wide range of services and product families. Many more products are offered than there are technologies so, despite the different names, you know they are similar "inside the case."

What follows are current Categories of xDSL products (where "x" stands for any of the present or future modifiers). More could be added at any time.

ADSL: Asymmetrical

What's asymmetrical is the bandwidth. That is, the transmission speed in one direction is much greater than in the other direction. Common pairs of speeds mentioned are 6 Mbit/s and 640 kbit/s; 3M/384K, and T-1/64K. The higher speed is assumed to be from the CO toward the customer.

An "ideal" application is video on demand, where 6M is enough bandwidth for two "VCR quality" video channels. At least that was the idea until it didn't sell. Cable TV delivers more

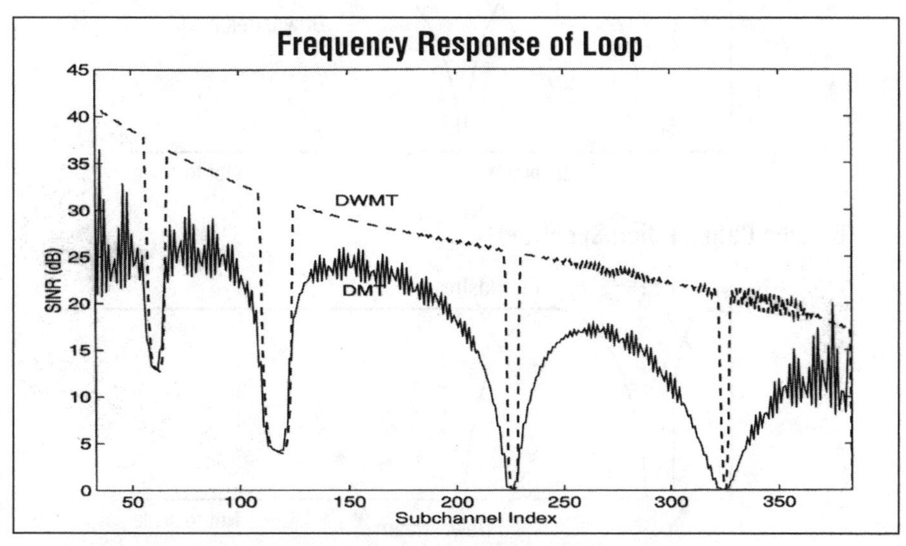

***Figure 4-19**. The top dashed line approximates the frequency response of a twisted pair local loop. Discrete multitone (DMT) can't use it all when some channels or 'bins' are suppressed. Discrete wavelet multitone (DWMT), an alternative to the inverse Fourier transform, takes better advantage of the bandwidth.*

simultaneous channels and costs less to install.

The second "killer application" was Internet access, specifically surfing the World Wide Web. The idea is that users seldom upload much to the network, and want really fast delivery of the next screen. ADSL could make even multimedia presentations "pop" on the screen. At least they could if the WWWeb server sending the information could deliver it fast enough, which almost none can.

Commercial customers and businesses, the groups most able to pay for big bandwidth, seem to prefer Internet connections and other CO access that run at the same speed in both directions.

This acronym has been implemented primarily with DMT. Some vendors split the frequency band into up-stream and down-stream segments (Fig. 4-20A). Others apply echo cancellation so the better bins at the lower frequencies can be used in both directions (Fig. 4-20B).

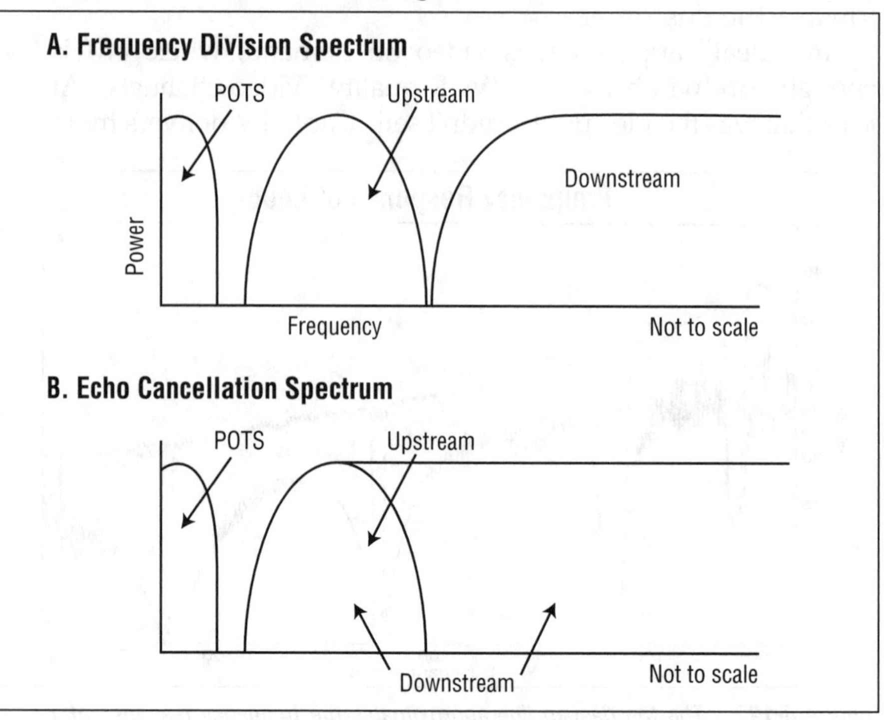

Figure 4-20. ADSL may separate the frequencies of the signals sent in each direction (A), avoiding the expense of echo cancellation. EC lets both direction use the more effective lower frequency bins simultaneously.

HDSL: High-speed

The workhorse of DSL, HDSL is widely deployed as a way to deliver T-1 and E-1 service on access lines. T-1 needs only a single pair for short loops, and can reach to the full 18,000 ft on two pair. When on two pair, each set of wires carries a full-duplex signal, using echo cancellation, at half the T-1 rate. By comparison, a T-carrier uses one pair for each direction at the full 1.5 Mbit/s. An E-1 will need from one to three pairs, depending on loop length.

Both 2B1Q and CAP are used.

IDSL: ISDN (Basic Rate)

IDSL is exactly the ISDN basic rate interface (BRI): 2B1Q running at 80,000 symbols per second to carry 160,000 bit/s. The channelization is the same, too, but only the pair of 64K channels are used. The signaling channel (D channel) is simplified (with dummy addresses) as are the embedded operations channels in the overhead bits.

The phone companies find IDSL easy to accept because it is familiar from 20 years exposure to BRI. CPE vendors like the idea because any device with a BRI is a candidate for a relatively easy migration to IDSL: some software changes are all that is necessary.

Several makers of central office equipment jumped into the DSL market with the IDSL format because the ISDN chips needed are readily available and easy to work with.

MDSL: Medium-speed

Medium in this case means fractional T-1 or Nx64 for N from 2 to about 12. Any of the technologies will do it, but 2B1Q and CAP seem favored because they are simple to put on a single wire pair.

RADSL: Rate-Adaptive

If the need is to wring the absolute maximum bandwidth from a given local loop, put ADSL on the job. Rate-adaptive means the equipment will analyze the line and adjust the transmission rate to the maximum speed that preserves the required error rate (higher speeds make the signal fade more and harder to analyze, leading to more errors).

Most DMT implementations can do rate adaptation. These vendors often promote this ability. CAP works differently (changing the modem constellation) but achieves about the same result. There are arguments over how fine a polish each technology puts on the job of maximizing line speed, but the differences so far are small in practice.

SDSL: Symmetrical

If the speed is the same in both directions, you have SDSL. Vendors using this handle have used mostly DMT, at FT-1 speeds and above. There is no reason another technology won't work too.

VDSL: Very-high-speed

If you want to go really fast, like 52 Mbit/s, you need an optical fiber or — if the distance is short, up to 1000 ft — you can use VDSL.

The laws of physics say that the product of bandwidth and line length is roughly constant. At 52 Mbit/s the signal fades out of sight fairly fast.

The available distance is fine for a LAN. It has also been described as the distribution link between a fiber that ends in a roadside cabinet and the customer's terminal equipment.

Specialized Circuits

More than one carrier will be offering circuits with switching as a value added feature. The first two of these new offerings both switch DS-0 channels, but in different ways. To establish more clearly just what these switched digital services offer, we must look at the layout inside a telco central office.

Access and Cross-Connect Systems

Voice service started with POTS, plain old telephone service. In pure analog voice there was a need to switch incoming local loops. Not only was it necessary to form a circuit, it was also necessary to provide a convenient electrical tap for "access" to the line for testing and diagnostics (Fig. 4-21).

Originally the cross-connects were done manually, on large frames that carried thousands of wires — much of that continues, in wire centers . Access to leased lines was on the wiring frame, or through a patch panel where each line had a jack. At this point a repairman could get on any line to monitor it or run tests.

As Central Offices became larger and more automated the cross connect progressed by improving the connector technology — from soldering to wire wrap to punch block to connectorized wiring. Access was made more automated by using the capabilities built into switches (crossbar and later). This was still analog cross-connect and access. Voice channels on a T-1 were demultiplexed before switching. Leased analog lines for voice and data transmission fit the same pat-

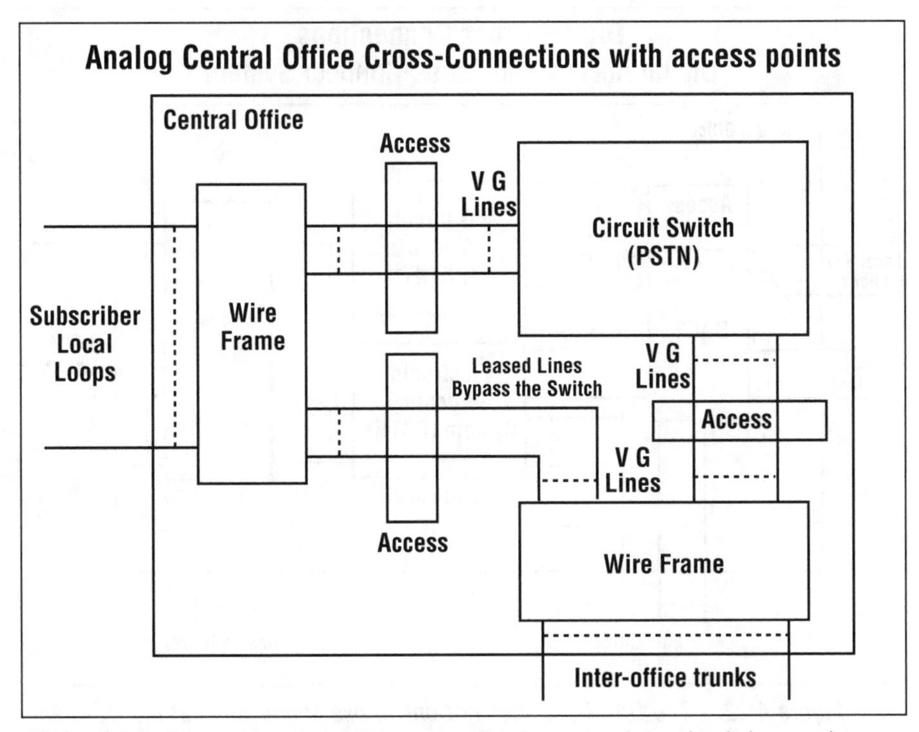

Figure 4-21. Cross connections among subscriber loops, trunks, and switches require access points where technicians can insert test equipment or monitor line signals.

tern, but they do not pass through the switch.

The practice of making voice cross-connects at the channel level continued into the digital era. When first offered, DDS channels were distributed manually, the same as voice channels. The main change was from analog to digital form, at the DS-0 level. But it was still done on wiring frames or by plug ended jumpers on DSX-0 cross connect frames.

With the advent of pure digital switches, another option appeared. It was no longer necessary to demultiplex digital signals from leased T-1 lines to the DS-0 level for cross connection and access. They could be manipulated in T-1 form electronically (Fig. 4-22). But because leased lines change infrequently, and there is no need to collect billing information for a line dedicated to one customer, the digital cross connect device could be much simpler than a PSTN switch. The functions, however, are identical: operate on T-1 bit streams to move bytes among time slots and among T-1 lines.

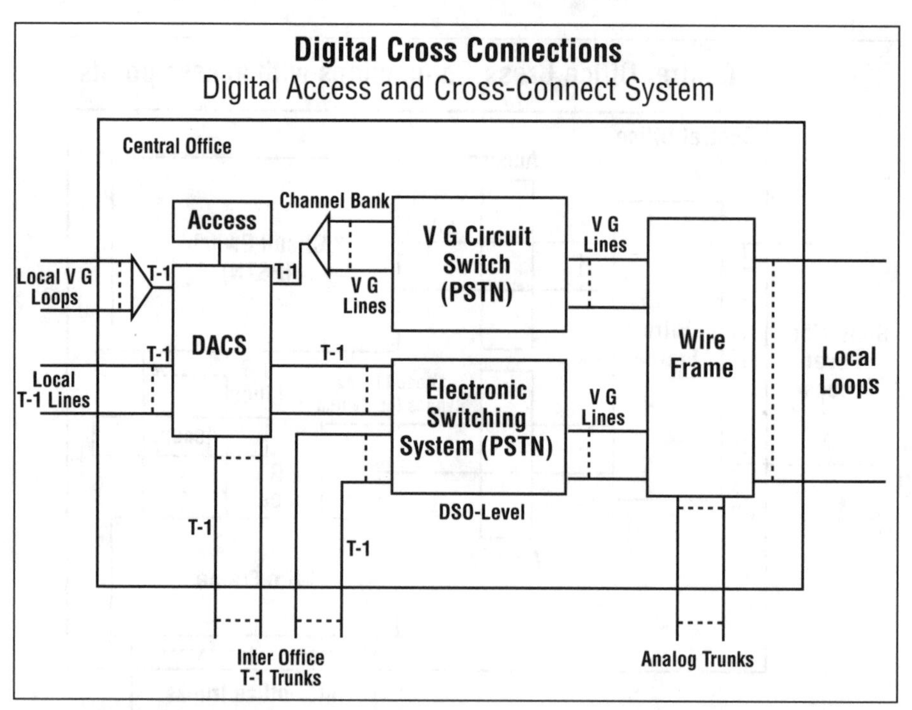

Figure 4-22. The DACS provides not only cross connections at the DS-0 and DS-1 levels, it also gives access to individual channels, either DS-0 or DS-1, for testing and monitoring.

Briefly, a Digital Access and Cross-connect System (DACS) switches by temporarily storing the 8 bits from an inbound time slot in a buffer. Some time later those bits are inserted into any other time slot of any line of the DACS, even the originating one. There is a delay. The length depends on the relative positions of the incoming and outgoing time slots. By routing all the time slots from a given T-1 line to another, in the same order, the entire T-1 circuit is switched.

To the telco, a DACS provides access. This means it provides a way to monitor traffic on individual DS-0 channels or break into them to conduct tests and perform diagnostics. To gain access to a circuit for monitoring, the DACS simply sends the data bytes to the access line as well as the destination line. In the same way, test signals from the access line may be inserted into the channel under test, which could be a DS-1 or a DS-0.

So the DACS evolved from switching technology, but in simpler form and smaller size. In fact, within carriers a DACS is sometimes referred to as a "slow switch." DACSs offer 128 or more T-1 lines. Some vendors claim thousands of T-1 ports. At least one T-1 per DACS must be dedicated to test equipment access. That is, a typical DACS can support 63 "incoming" and 64 "outgoing" T-1 trunks, though all are full duplex in the D4 format. That is, a DACS "terminates" T-1's.

The purpose of a DACS is to offer DS-0 connectivity among many T-1 circuits. If using PCM voice, for example, a DACS could collect channels from several local feeder lines to maximize the fill of a long distance T-1 circuit devoted to voice tie lines.

Fractional T-1/T-3

Now that carriers have DACS functionality in their central offices, they no longer have to hard-wire each channel on a cross connect panel between channel banks. They have an easy way to deal with individual DS-0's. In other words, because of advancing technology, the DS-0 has become a practical product.

But how do you sell a few DS-0's? For the long distance carrier, there is no problem. Simply allow many customers to share the public network backbone by grooming their DS-0's into T-1, T-3, or faster digital transmission lines. This procedure is essentially the same as leasing individual voice grade lines over digital facilities. However, in the digital network, any bundle of DS-0s can take on the characteristics of a single fast channel, a Fractional T-1 (FT-1).

The local exchange carrier faces a different situation. The local loop can't always be shared among many customers, and one customer may want only a quarter of the T-1 bandwidth. There are at least two ways to deliver the FT-1 service at the customer premises demarcation point:

T-1 Interface, Partially Activated. You install a normal T-1 multiplexer, one that recognizes D4 format. The carrier tells you which time slots are turned on (cross connected to anything), and you put your voice and data in those time slots.

When you need more bandwidth, the carrier can turn up more DS-0's without installing new hardware or visiting your premises. It's a matter of setting up new cross connections in the DACS. Not only is it easier, it promises to be much faster than normal circuit provisioning.

In the U.S., where the carrier cannot own the multiplexer, this was the only practical way to deliver Fractional T-1 until DSL technology.

Super-Rate Data Port. In the CEPT countries, notably France and Germany, the PTT will own the first device attached to the line on the customer premises. For a full E-1 service, that device is a CSU equivalent. But for fractional E-1 service the PTTs have chosen to deliver exactly what the customer ordered. The interface is a synchronous, full duplex standard (like X.21 or RS-449) running at 256 kbit/s, 756 kbit/s, etc.

Therefore a networking multiplexer must support these super-rate aggregates directly. The main concern is framing. The PTT passes the full user bandwidth as a single channel, end to end, so the mux must supply framing within this clear channel. In most cases, hardware vendors furnish a special card or interface module to deal with these services.

The PTT's approach avoids any question about the order of DS-0's as delivered. In the U.S., it is possible to route each DS-0 individually, possibly over different routes, and to change routing on some of the channels in a Fractional T-1 service. Differences in propagation delay, or changing the order of time slots during transmission could impact your applications, particularly superrate data channels (Nx64K). If any of your logical channels exceed 64 kbit/s, obtain assurances from the carriers for FT-1 that all DS-0's will be routed over the same path and kept in the same order within a contiguous bundle.

Fractional DS-3 service is a logical extension of the FT-1 concept, but based on a DACS with DS-3 ports. At least one carrier supports both a partially filled DS-3 and multiple T-1 local loops (see also Fig. 4-23).

The DS-3 interface may be a traditional M13 format. The newer "C-bit parity" framing format is preferred.

Multiple T-1s require inverse multiplexers at the customer premises on both ends. In the backbone the T-1s are carried as individual channels, but within the same DS-3. The service is sold as one channel with a speed of 4.63, 6.18, 7.72, or 10.81 Mbit/s.

By using the commonly available T-1 circuits, FT-3 services extend faster channels beyond the few hundred cities where full DS-3 access is available.

Customer Controlled Reconfiguration

Up until 1985 the carrier always programmed the DACS. It was simply a faster and easier way to set up leased digital circuits. Later, carriers let the customer participate in reconfiguration.

Customer Controlled Reconfiguration (CCR) was AT&T's name for customer access to the computer that controls all of its DACS nationally (Fig. 4-24). The idea is to automate line provisioning (a telco word for installation) and let the customer do his own. By contrast, the first DACS installations, in central offices in the 1960s, required the operator to address each DACS individually.

The carrier allows a user to enter an alternate network map into the central computer. Filtering the input, the computer produces a command list to reconfigure the network of DACS. Filtering means the changes are limited to what that

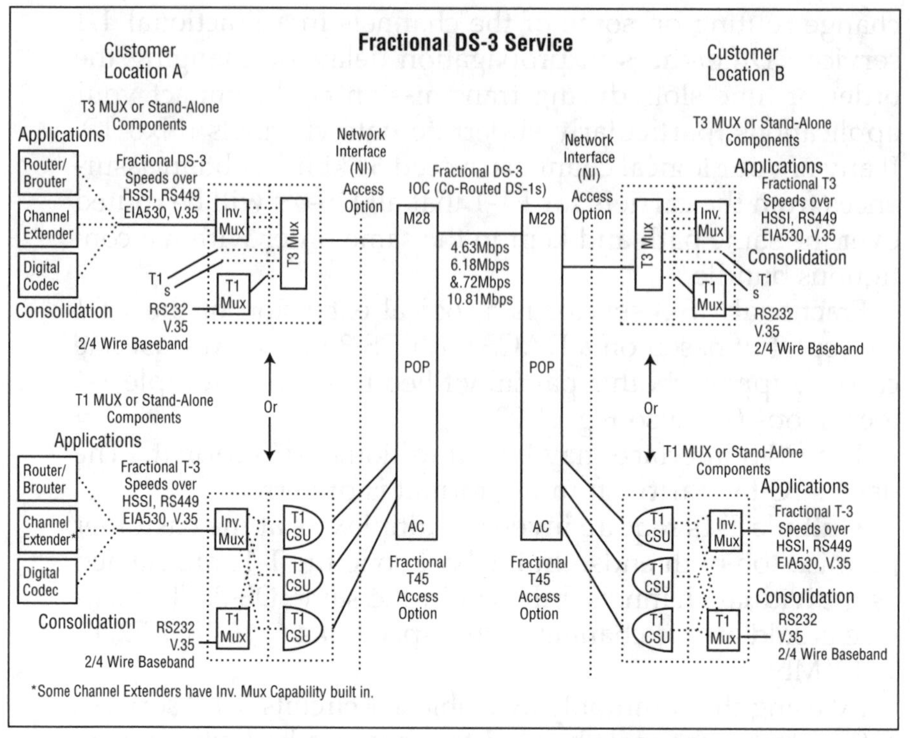

Figure 4-23. *Fractional T-3, like FT-1, may consists of a partially filled line or several slower lines joined by an inverse multiplexer.*

customer has purchased or authorized. In late 1988 the system was not completely automated. The control computer now printed the command list for people who manually reconfigured each DACS. This process produced a relatively long and variable delay in changing maps. Those operators were the final filter to prevent unauthorized instructions from affecting the national network.

With CCR, the corporate network operator can route DS-0 channels among T-1 access circuits (that is, the local loops or terminating channels). Access must be leased. Basic demand for digital traffic would be met by interoffice channels that also are leased full time. But it should be possible to handle peak loads through ACCUNET RESERVE, a comparable T-1 service offered on demand, switched DS-0 channels when available, or other service that is available on demand and billed by usage. Eventually, the goal is to migrate to ISDN where any service

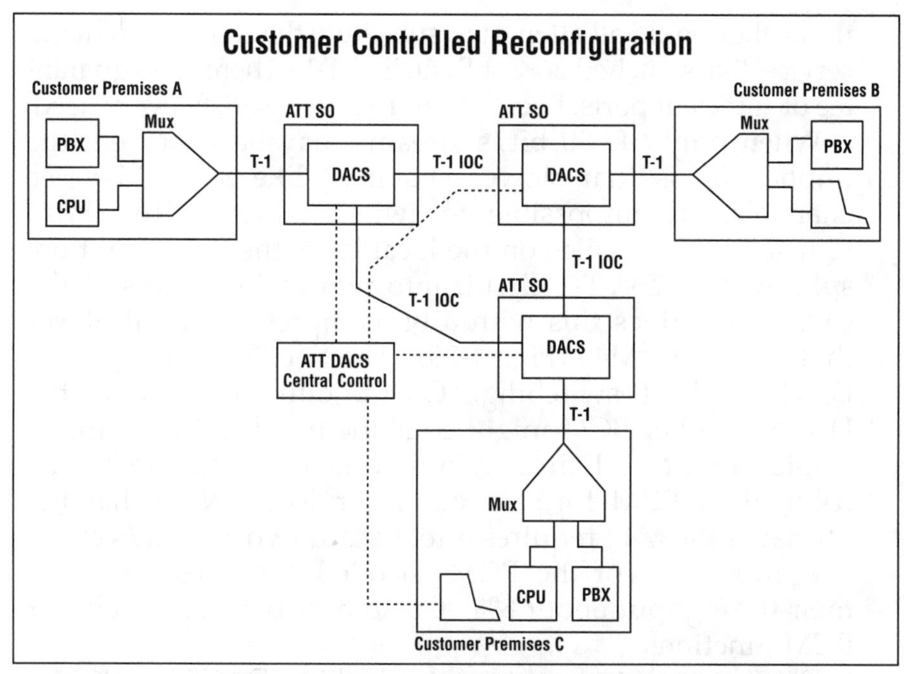

Figure 4-24. Customer controlled reconfiguration (CCR) allows the user to reconfigure a network based on shared facilities. The customer puts alternate network maps into a central control computer. The computer implements the new map by issuing instructions to all DACS nationwide.

would be available over the same T-1 access loop.

DACS Control Limitations

In reality, the DACS is still under the control of the carrier. There is no guarantee on timing. As it can, the single computer reconfigures all the DACS in the country. Exactly when it will get to a specific channel is not predictable.

For less than critical applications, most customers probably will opt for dial-up access to the DACS computer. In case of a disaster, many DACS customers will be attempting to give instruction to (put new maps into) the one computer. Leased lines are available to those who do not want to risk getting a busy signal.

The DACS control computer also arbitrates between customers if their commands conflict, as when competing for the same bandwidth. The same bandwidth? A DACS doesn't create bandwidth, only switches it. If a fault interrupts many users, they could all attempt to substitute the same on-demand service, like switched 56K or Switched T-1. There is no guarantee of sufficient ports, lines, or bandwidth to satisfy every need.

When only 64,000 bit/s streams may be switched, the implication is that slower channels, like ADPCM voice channels, are impossible to switch individually. With reduced bit rate voice on the local loop, there is an option: split an ADPCM T-1 circuit into two PCM circuits at the CO. AT&T does this with a bit compression multiplexer that it calls the M44 (Fig. 4-25). The two T-1's then go to a DACS for DS-0 switching. On the outbound side of the DACS, another BCM might combine two PCM lines into a single ADPCM T-1 circuit. Individual channels can be distributed in PCM form to other services. Note that the format of the M44 requires it to discard two 64 kbit/s channels from each of the PCM encoded T-1 lines, reducing their throughput about 8%. Customers pay separately for BCM functions.

Data channels too must be switched in a DACS in 64K segments. This is true whether they carry a 64K clear channel or only a 1200 bit/s modem signal. Sub-rate multiplexing can put multiple data channels in a 64K time slot. In most installed DACS, all subrate channels in a DS-0 must be switched or routed together. In 1987 some ability to switch

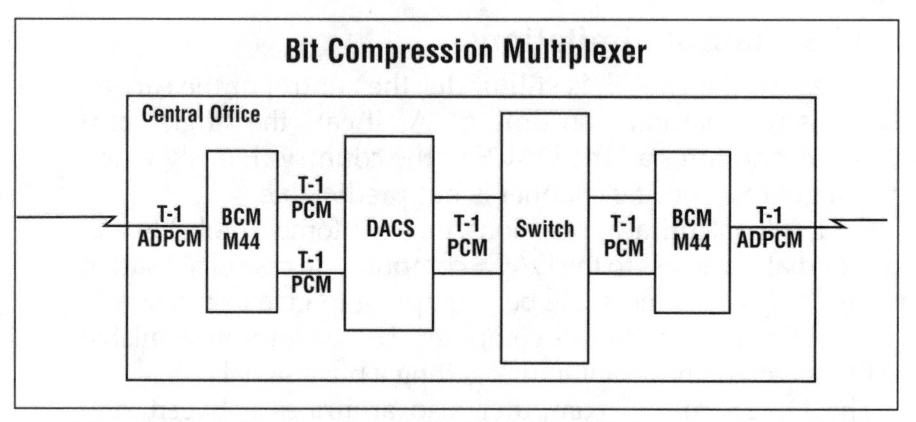

Figure 4-25. *A bit compression multiplexer like AT&T's M44 converts 64K PCM to 32K ADPCM. Compression increases the number of voice channels carried on a T-1 line to 44 (4 channels are needed for signaling). Data channels of 56 Kbit/s cannot be compressed, nor can the ADPCM channel carry high speed modem signals.*

subrate channels independently was introduced into the public network. Bandwidths larger than DS-0 may be switched only as multiples of 64 kbit/s.

Because the DACS is simple it must be programmed externally to alter a channel's routing. In-band signals cannot control a DACS connection. That is, a terminal user cannot switch a connection to another computer as is possible with switching statistical multiplexers or a data PBX. The network operator must intervene by instructing the controller computer to change the DACS.

There is one aspect of DACS performance to be wary of — the "transparent" T-1 connection. This setup pipelines a full T-1 through the DACS with minimal delay. While both input and output T-1 bit streams are framed every 193rd bit, they are not necessarily framed in the same way. That is, a superframe that enters a DACS as [ABCDEFG...XYZ], where each letter is a frame, may leave as [DEFG...XYZABC]. The slip is in the framing bits. If the multiplexing devices are framing on the 193rd bits, they may stay in frame sync but will lose track of the sixth and twelfth frames. This loss could garble data or hide the signaling bits. In normal voice cross-connections, the DACS extracts the signaling bits and re-inserts them in the proper outbound frames.

Byte-interleaved multiplexers that establish their own framing, in addition to F-bits, can use the transparent DACS connection. They prefer this type of cross connection because delay does not exceed 16 bits or 2 byte times. Some multiplexers cannot stay in sync with each other through a transparent DACS connection.

To preserve the superframe, the DACS can route "channel by channel." That is, switch "24 time slots" from link A to link B rather than "one T-1." However, a DACS then buffers channels. The total delay for the 24-slot connection could be as high as two frame times, or 48 byte times. Some bit interleaved multiplexers and all channel banks that use robbed bit signaling must connect this way.

An alternative is to lease fixed lines, and switch or route with customer owned equipment. The functions of a DACS are available, and more, from private customer premises equipment (CPE).

T-1/E-1 Conversion

When POTS was the major international service, the handoffs between countries were easiest to make at the analog level, for individual connections. E&M is an interface universally understood.

Rapid growth of international traffic, especially in data, revealed a need for DS-1 and faster interconnections. With approximately half the global traffic generated in the US (on the T-1 standard), that meant increasing connections between the "T-1" and the "E-1" parts of the world.

The main conversion point has been a DSX-1 DACS with both T-1 and E-1 interfaces. The T-1 DACS normally extracts robbed signaling bits from incoming T-1 lines, for insertion into outbound frames. A similar function in an E-1 DACS maps the signaling bits between subrate channels in time slot 16 (see Appendix D).

Fortunately for global communications, the signaling functions in T-1 and E-1 can be mapped to each other's channel associated signaling bits. The DACS performs this international gateway function at an office of an international carrier, or at one end of the international cable.

Switched T-1

A flurry of activity surrounded introductions of T-1 service in the early 1990s. Tariffs were set, but little resulted. The service never made much impact on the networking market as it was superseded by primary rate ISDN.

Interface specifications for switched T-1 were based on a subset of the DSS-1 signaling standard, the one defined for ISDN. However, the signaling channel was allowed to take any path, in or outside of the T-1 access line. This allowed a switched connection at 1.536 Mbit/s.

The later service based on standard ISDN that does almost the same job is Multirate Calling. That is, one call can set up a circuit-switched channel at a rate of Nx64 kbit/s, where N from

1 up to the maximum number of DS-0s (B channels) available on the access link. For Primary Rate Interface (T-1) N can be as large as 23, for a speed of 1.472 Mbit/s after saving one DS-0 for the signaling channel (D channel). On an E-1, there are at least 30 B channels, making 1.920 Mbit/s available in a single call. Some E-1 interfaces could support a connection of 31 DS-0s.

Inverse Multiplexing

When the bandwidth need is greater than a T-1, but less than a DS-3, you can combine multiple lines between two points to create a faster serial channel.

Imagine how a channel bank aggregates 24 (digital) inputs to make a T-1. Now turn the channel bank around, so the 24 interfaces point to the wide area lines and the T-1 goes to the terminal equipment. If the WAN links are all 64 kbit/s, the terminals at either end will think they are connected by a T-1.

Scaling up the multiplexers lets them use multiple T-1 lines to create a DS-3 or fractional DS-3 speed interface for the terminal equipment. Such inverse multiplexers are offered for up to about 8 T-1 lines. Higher speeds tend to justify an actual DS-3 installation, based on the relative costs of T-1 and T-3 service.

Software Defined and Virtual Private Networks

AT&T's Software Defined Network (SDN) is a dynamic reconfiguration within a digital CO switch like the Number 5 Electronic Switching System (5ESS). This service is not the same as DACS or ISDN. Sprint's VPN and MCI's Prism services operate similarly to provide a virtual network (VN).

VNs permit large corporations to establish the equivalent of leased virtual circuits, but set up on the PSTN. Circuits among scattered locations are set up only for the duration of individual calls. Therefore the charge is based on usage.

Typically a caller's phone is connected to a local PBX. The PBX accesses the long haul carrier over a leased local loop. At the caller's serving office, the dialed number — usually part of a uniform 7-digit numbering plan — is converted as necessary to 10 or 11 digits by a routing table. The call is then placed on the PSTN. At the called end, the access process is reversed. The ESS in the other serving office converts the incoming call to a specific extension and rings it through on a DID trunk to the customer's PBX at that site.

The customer gains control of many aspects of the company's internal phone system:

- Uniform numbering plan;
- Reduced charges based on usage;
- Restricted calling rights adjustable on an extension by extension basis
- Priority on the PSTN
- Functions like conference calls, forwarding, etc. similar to Centrex.

The flexibility is available because the carriers, for the first time, will allow users direct access to the common channel interoffice signaling (CCIS) system. It is anticipated that Prism, VPM, SDN and DACS/CCR will eventually evolve into the ISDN by changing the form of signaling to DSS-1, the ISDN standard.

Originally, virtual private networks were essentially voice services, a replacement for WATS or tie lines. Interfaces were individual analog lines or a separately proposed tariffed service. Later, virtual networks extended support to data services like 56 kbit/s digital, both leased (fixed) and dial-up. One version was named Software Defined Data Network (SDDN).

The T-1 access to SDN would be Megacom. This service consists of 24 PCM voice channels on a T-1 circuit direct to a 4ESS. The number 4 ESS is a toll office, not an end office. It looks for robbed bit signaling like that performed by the 5ESS on interoffice trunks. This signaling protocol is different from what passes between a PBX and a 5ESS. The customer's terminal equipment (channel bank, PBX, or bandwidth manager) must be compatible.

A second significant factor of virtual networks is installation cost. The minimum installation charge for SDN originally was over $100,000, but has dropped to about half that. Competition pushed initial costs still lower, but even now only large users of voice benefit from SDN.

Outlook For T-1 and ISDN

Carriers intended to demonstrate an Integrated Services Digital Network by the end of 1986. Only a few events actually came to pass by 1987. All were delayed and most reduced in scope from what had been planned. Further ISDN demonstrations continued for several years.

Implementation of Central office equipment followed over the next 10 years and more. (My own local CO got an ISDN-capable switch, a 5ESS, in 1997.) It may be as late as the end of the century before ISDN is available as a standard service on almost every local loop. But ISDN will come.

For twenty years, switch vendors prevented ISDN standards from being implemented in one way (CCITT published most of the recommendations). They tried to preserve commercial advantage with proprietary variations that would not interoperate with equipment from other vendors. Eventually they relented and agreed to migrate to one US standard: National ISDN.

ISDN gives a jump in bandwidth and multiple channels that dramatically change how we view the telephone. The goal is to provide every subscriber with a wideband digital connection to a worldwide integrated network for voice, data, and any other kind of information transmission desired.

The same local loop now used for analog voice and T-1 digital circuits will be adapted to carry two channels at 64,000 bit/s (the "B" or bearer channels) and one at 16,000 bit/s (the "D" or data channel). Signaling will be handled on the D channel. Specific message formats will request connections for the B channels. This separate signaling channel uses the same principles and procedures as common channel interoffice signaling (CCIS). But Signaling System #7 goes beyond CCIS in speed, capacity, and in adding features and services for end users.

As with ordinary phone calls, the caller dials to specify the destination. For data service, clear channels of a full 64,000 bit/s are offered outside the U.S. Initially in North America data on B channels ran at 56 kbit/s. Most areas now support 64K clear channels.

Packet data may be accepted on the D channel, in some areas. Throughput is a fraction of the 16 kbit/s channel bandwidth. The information may be sent anywhere, and reaches other packet nets through X.25 gateways.

The interface with "2B+D" channels is known as "basic rate interface." For large volume users, the "primary" interface offers 23B plus D. Here the D channel is 64,000 bit/s, so the composite circuit runs at the T-1 rate. The European format is 30 B channels and 1 D channel. A 64,000 bit/s time slot provides synchronization.

The primary rate interface ISDN line format uses the extended superframe for synchronization. This makes ISDN compatible with present transmission equipment. The biggest difference is in the local loop terminations. In the CO, the office channel unit (OCU) that interfaces with the subscriber loop takes on some functions of a multiplexer, as does the NCTE on the customer premises.

Existing local loops carry at least the basic ISDN service, and in most cases the primary rate as well. The huge installed base of 2-wire loops makes their continued use an economic necessity.

Why ISDN Is Inevitable

On it's hundredth anniversary, AT&T looked back at the original telephone operator, and came up with an interesting statistic. If we still had human operators, rather than automatic central office switches, there were not enough people in the entire country to handle the calls placed every day.

Until T-1 became popular, most new data circuits were installed and changed manually. Now look at the growth in data traffic: about 40% per year in 1990 and most years following. Data continues to grow much faster than voice, notably in the form of Internet traffic. Data exceeded voice in traffic volume about 1996.

For that growth in data to continue, the telephone companies must find a way to install and change digital connections with much less manual labor, even automatically. That is, the user must be able to "dial up" what he wants. That technology to provision data circuits, whatever it is in reality, will be called ISDN.

Equipment You Will Need At Your Office

In building a private communications network, the applications on the T-1 circuits will determine what types of customer premise equipment you will need. This equipment will be a critical component of the organization's operations — and can affect its very existence.

This chapter looks at features and functions of T-1 multiplexers. The focus is on generic information which should help in evaluating any brand of equipment. The goal is to prepare a manager to ask the right questions, to understand the answers, and to recognize what is important and what is not.

Here, then, is what to look for in T-1 customer premises equipment (CPE).

Interfaces

Most T-1 circuits are connected to multiplexers at the user's offices. The most basic function of a multiplexer or bandwidth manager is to combine "many" user circuits onto "few" transmission facilities. One measure of the equipment's value is just how many ports it has for connections to local low-speed devices (Fig. 5-1).

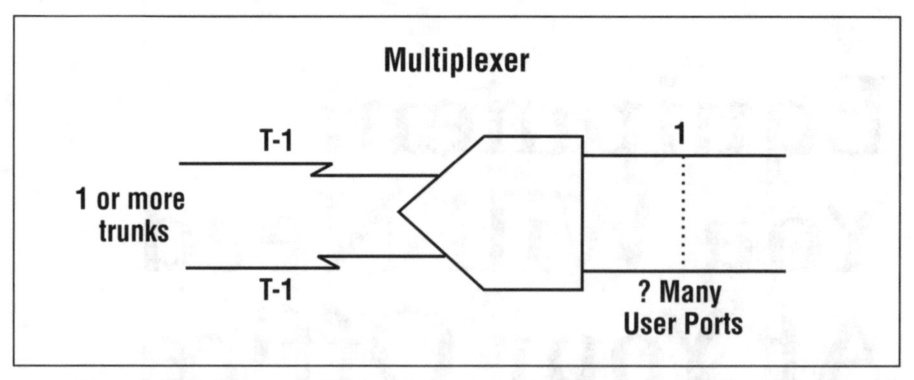

Figure. 5-1. The essential function of a multiplexer is to combine on one trunk many input ports — but how many? The actual usable number is more important than the maximum specification.

How Many User Ports?

Ever since T-1 was created for voice, with channel banks, the service has been thought of as having 24 channels (see Chapter 3). Some multiplexers support only 24 connections, data or voice. Newer multiplexers designed from the start for large networks offer hundreds of low speed data ports and/or many compressed voice ports on a single node.

Some multiplexers are still based on channel banks. They offer a small number of data or voice connections because they work within the traditional 24 voice channel format. In the worst case (Fig. 5-2), the count of data channels is 24.

Two to four data connections per time slot is common. This number may be fine for small sites or feeder nodes. However, if the application requires a large number of channels, especially low speed channels at a large central site, these designs require additional T-1 links and multiplexers simply to raise the port or channel count.

Some companies need to be able to expand the number of user ports on a node. This figure is important to an evaluation of CPE.

On some T-1 nodes, voice ports are available only by connecting external channel banks via T-1 links. That is, the processor or bandwidth manager has no analog voice module comparable to a data I/O port card. Even one voice channel occupies a full T-1 line. This method is practical for

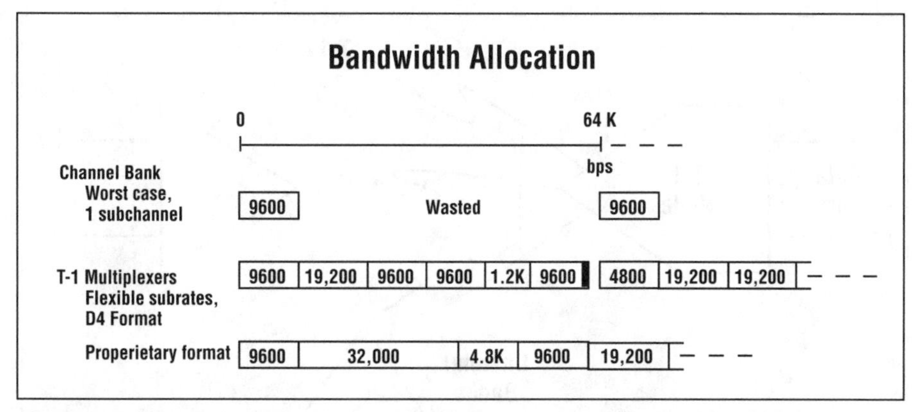

Figure. 5-2. *Voice channel banks and multiplexers based on them accommodate as few as one data channel in each DS-0 time slot (worst case). More recent T-1 multiplexers carry multiple channels per time slot, wasting some bandwidth, but remain fully compatible with the D4 format. Proprietary framing maximizes link efficiency, but at the cost of being incompatible with D4 format. Declining cost of bandwidth has reduced the importance of link efficiency.*

large numbers of voice connections, especially in multiples of 24. (Channel banks use PCM encoding for 24 voice connections per T-1 link).

Without voice compression, channel banks are not economical as remote concentrators when the T-1 line is expensive. (Fig. 5-3). Where line costs are moderate, an intelligent channel bank (under network management) makes a good feeder node. The traditional dumb channel bank will always be less attractive because it is not under network management and offers few if any remote diagnostics.

Link Efficiency

Like anything good in life, the number of T-1 links on a node is limited. An important measure of T-1 equipment is link efficiency: how well a multiplexer or bandwidth manager uses the links. Efficiency is defined as the aggregate throughput in usable bits per second, expressed as a percentage of the T-1 rate. The higher the efficiency, the greater use can be made of an expensive circuit.

Efficiency depends on how many time division multiplexer (TDM) channels can be created on a T-1 circuit, and the speeds

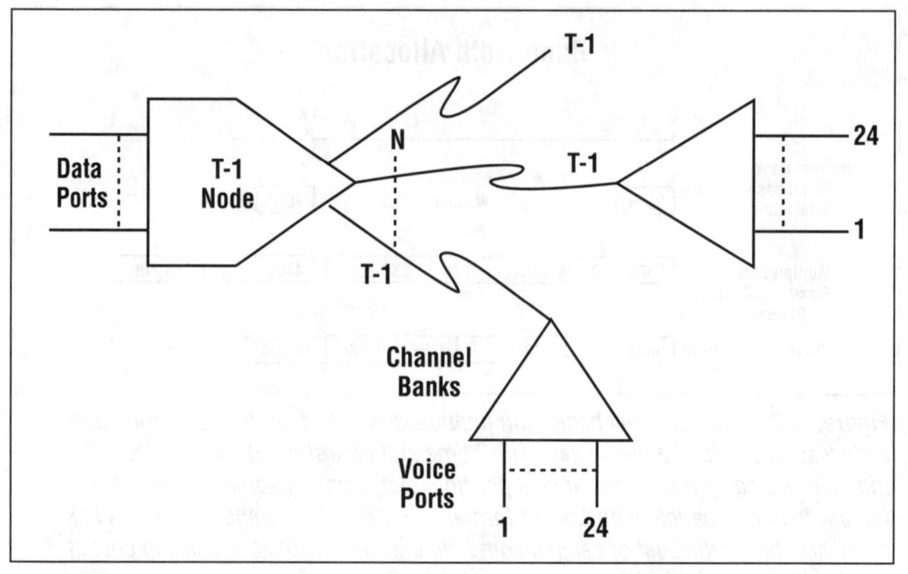

Figure. 5-3. *Connecting voice ports through a channel bank on a T-1 trunk line may be inefficient. For a low number of voice connections voice modules in the multiplexer chassis can be less expensive. The limit of 24 connections makes a channel bank uneconomical as a remote concentrator on the end of an expensive T-1 line.*

of those channels. Limits on the number of data connections, especially on low speed data ports, can have a devastating effect on efficiency. The most unfortunate aspect of a multiplexer that supports a limited number of data channels is the waste of bandwidth on the T-1 line. Beware of low link efficiency.

The 64 kbit/s divisions imposed by the DS-1 voice format may limit severely not only the port count, but also the flexibility and efficiency of a multiplexer in handling data. Even if a channel bank derivative accommodates multiple data channels within the DS-0 bandwidth segment, the 64 kbit/s divisions usually interfere with the free assignment of overall bandwidth. Only the latest T-1 technology seems able to offer both D4 format and high efficiency (up to 97%). Two examples will illustrate.

1. Among older channel bank-derived T-1 multiplexers, one model carries no more than 19,200 bit/s of data in each DS-0 channel. That is 460,800 bit/s total for 24 time slots, or about 30% of the full T-1 capacity. (Another product ran a maximum of 48 channels, two

per 64 kbit/s division, also at a top speed of 19,200 bit/s each. That's less than 60% of the T-1 bandwidth.) The user can't get access to the rest of the T-1 bandwidth he's paying for.

2. A new design is based on flexible framing within DS-0s or super-rate channels. This type mux can put up to six 9600 bit/s channels in each DS-0 by building flexible subframes within individual time slots. However there is still some waste. For maximum efficiency, narrow bandwidth segments are assigned within a super-rate channel (128 kbit/s, 256 kbit/s, 756 kbit/s, or any N x 64 kbit/s). This can be done without regard to the DS-0 boundaries within this channel. The result is capacity for additional logical circuits and higher link efficiency.

The routing of connections in the more sophisticated nodes — those that apply flexible framing in their multiplexer functions — requires only the bandwidth of the channel. That is, a 9600 bit/s connection occupies 9600 bit/s on the T-1 link, plus a small overhead. It does not take a DS-0 channel (64 kbit/s). The flexible frame allows modern multiplexers to carry hundreds of TDM channels, over multiple links, packing channels into each T-1.

As little as 30% efficiency (thus 70% waste) may be acceptable: if the T-1 circuit is company-owned twisted pair, short haul microwave, or rents from the telco for a few hundred dollars per month. The low costs of the equipment and lines justify the inefficiency.

A cross-country terrestrial circuit leased in 1985 for about $70,000 per month. Then, inefficiency could cost $45,000 per month — half a million dollars per year wasted on one circuit. In 1988 a cross country optical fiber link can cost as little as $15,000 per month, which declined to $7,000 by 1997. Continual change in line costs requires a new payback calculation at least once a year. Several multiplexer vendors offer software packages or services to perform network optimization and pricing against a tariff data base.

The TDM multiplexer or bandwidth manager designed from the data viewpoint looks at a channel as any end-to-end connection regardless of data rate. There is no limit to the number of channels beyond the total bandwidth and the physical capacity to hold I/O cards. For a data-oriented multiplexer the amount of user data packed on a T-1 line by an efficient nodal processor can exceed 99% of theoretical.

Emphasizing data, these multiplexers do give up full compatibility with the D4 format. Newer T-1 multiplexers derived from voice technology give up some efficiency (97% maximum) to retain the full D4 format.

The bandwidth needed for overhead to handle framing, control, and diagnostics also impacts efficiency. It can be under 1%. But be prepared for some muxes to take as much as a full DS-0 channel for overhead, the same as ISDN. Refer to the chapter on network management for additional discussion.

It is important when evaluating T-1 nodes to consider port count, T-1 line count, and aggregate input. But link efficiency may have the greatest impact on the economics of TDM networks.

Many Kinds of Inputs

Large numbers of ports and link efficiency help reach the ultimate goal of most private transport networks: to combine all existing transmission facilities into a single network. Unification and simplification help the user gain control. Flexibility accommodates future demands (Fig. 5-4). To get those benefits, many kinds of applications must interface easily with the network.

Ideally, the interface would be a port on an adaptable, modular component. Selecting from a range of I/O modules, the user could populate a node with any mix of port types. It should be possible to service a wide variety of data, voice, video, bulk file transfer, and other applications.

Among the types of interfaces available, a given user initially might need only one. Over the anticipated life of new equipment, we can expect at least several types will be required. Ideally, then, you should be able to install new types in the field, while the network is running.

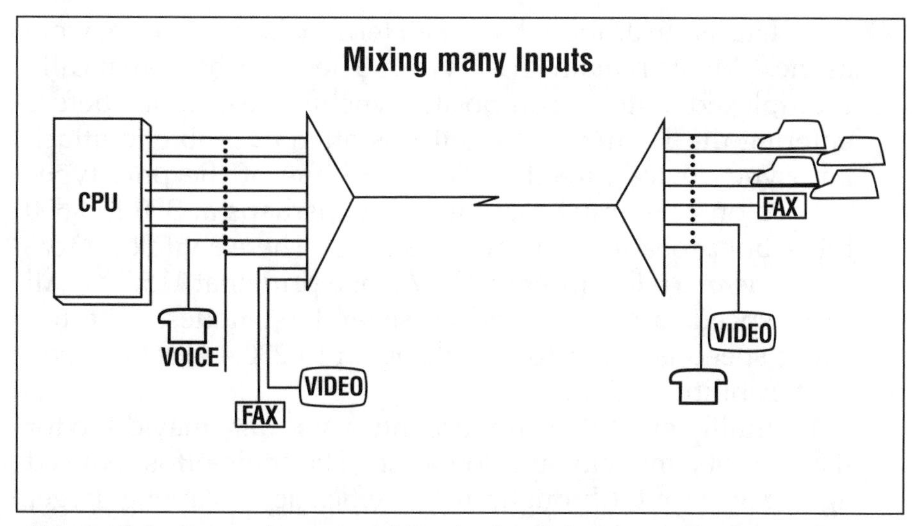

Figure. 5-4. For maximum usefulness in simplifying communications, a T-1 node should accommodate any form of business information.

Data Interfaces: What Kind?

Asynchronous Data is marked at the digital interface by two data paths, one for each direction, between two devices which operate independently at the same nominal speed (Fig. 5-5). The two speeds of the separate internal clocks can differ by a few percent, so the start-stop protocol uses extra bits before and after each character (the start and stop bits) to absorb the difference and thus keep the devices synchronized.

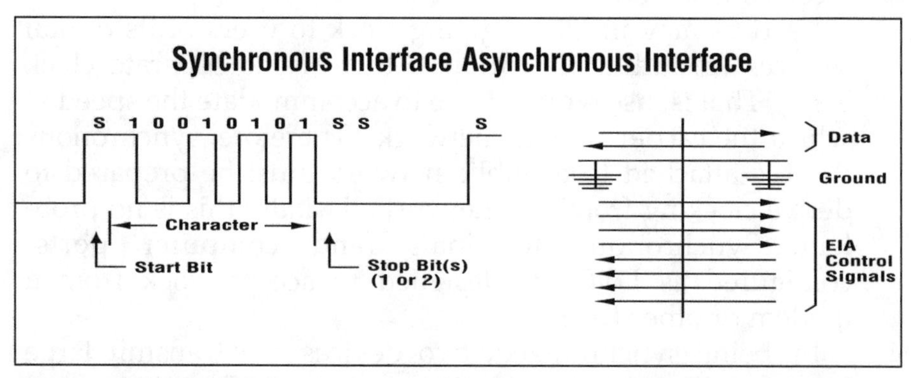

Figure. 5-5. Devices on either side of an asynchronous interface operate at approximately the same data rates, but may differ by several percent. The start and stop bits that frame each character compensate to keep both sides in step.

As late as 1990, about half of all terminals in existence were async. Many remain in service. They can be statistically multiplexed onto a composite synchronous signal before entering the T-1 network. But it is often desirable to attach an async device directly. Therefore one of the port types needed on a T-1 mux may be async, perhaps at 300 to 9600 bit/s, but probably more at 19.2 kbit/s. There could be a reason to tie in a telex line, at 50 bit/s, or a printer at 115,200. All these speeds are supported by several T-1 nodes. The top async speed is about to double again to 230.4 kbit/s. Look for flexibility in CPE.

T-1 multiplexers that support only async data may do so for the sake of simplicity and low cost. The trade-off is reduced aggregate input (throughput) or efficiency. Among larger bandwidth managers, some support async and some do not.

Synchronous Data is well suited to the time division multiplexing used in almost every T-1 multiplexer and bandwidth manager. The interface is characterized by a full duplex connection (separate data paths for each direction, operating simultaneously) and a clock signal passed from one device to the other (Fig. 5-6).

The clock signal is usually a square wave pulse train. It is what synchronizes the data terminal equipment (DTE) and the data communication equipment (DCE) to make sure the two are working together. Often the clock runs at twice the nominal bit rate to interpret more accurately the send and receive bit streams.

Carriers may insist on giving clock to users of its digital services (including T-1, though there is no separate clock lead). That is, users often have to accommodate the speed at which the carrier runs its network. Therefore, synchronous devices attached to a public network must be prepared to derive clocking from that network. Usually this is no problem. Synchronous terminals (and computer ports) configured as DTE are designed to accept clock from a modem or other DCE.

By being synchronized, two devices can transmit large blocks of characters in a continuous stream. There is no need for extra start/stop bits between characters to compensate

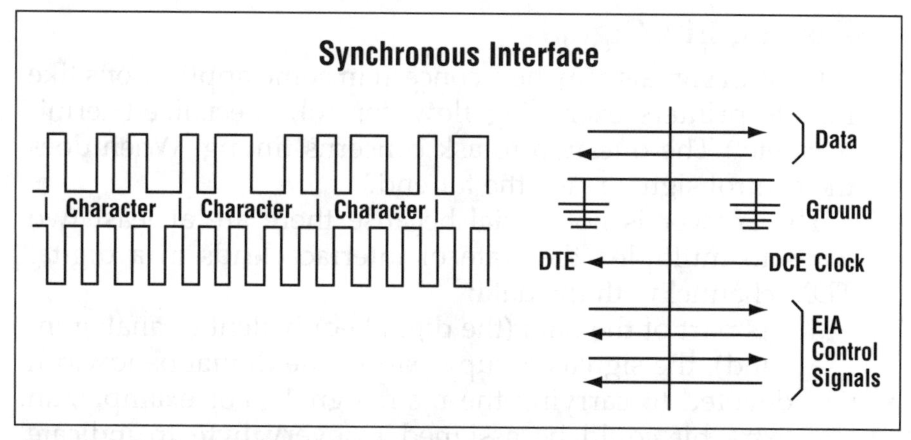

***Figure. 5-6.** A clock signal characterizes the synchronous interface. The clock keeps the two devices exactly in step. Characters can be sent continuously, in long blocks, without start and stop bits.*

for slight differences in speed, as in asynchronous. The clock signal locks the two together.

Synchronous is almost universally the mode for speeds above 19,200 bit/s. It is commonly used at 2400 bit/s and even 1200 bit/s. The absence of start and stop bits increases character throughput for a given bit rate compared to async. New applications that depend on a digital device tend to use synchronous. For example, some compressed video encoders transmit at 512,000 bit/s. Specialized computer to computer transfers may be able to use more than 10 million bit/s. Synchronous interfaces for these speeds are available.

Isochronous Data has been described as synchronous data without a clock lead. Characters are sent continuously, without start/stop bits. The timing is recovered from the transitions in the data stream. (This is the same as with T-1 signals at the DSX-1 interface.)

Isochronous data also preserves bit integrity end-to-end: what goes in is what comes out. It is bit oriented, transparent, and does not recognize control characters. As a result it appears often in military applications which require encryption. Isoc is seldom seen in commercial networks, but has some applications for "odd" async formats with long words (up to 256 bits) between start and stop bits.

Passing EIA Signals

Control signals may be a concern in some applications like remote printers exercising flow control, specialized terminals, etc. The question to ask concerns timing: When does the control signal reach the far end?

The answer is not trivial because there are at least two ways to multiplex the state of interface leads in a digital TDM channel with the data:

1. As part of the data (the digital equivalent of analog in-band), the signals occupy part of the digital bandwidth devoted to carrying the main signal. For example, an extra bit could be assigned to every byte to indicate whether that byte is regular data or signaling information. Clearly the EIA signal arrives with the data. But to ensure this precise timing relationship between data and signals consumes bandwidth: as little as 800 bit/s or as much as 1/8 the synchronous bandwidth.
2. On a separate channel (out of band), the signal lead status is statistically multiplexed with many others. This technique uses much less bandwidth. The signal channel is packetized so it introduces variable delay, an inherent feature of the method. Often signal lead changes will not arrive to coincide exactly with the data. This variation can lead to clipped data received at some kinds of terminals.

Analog Voice Features

Analog voice represents the largest portion of the networking requirement in some corporations. The most common need is to tie together PBX's in offices at different locations. The private network can carry this traffic over the equivalent of two- or four-wire analog tie trunks, with E&M signaling leads. This is the most common line between analog PBX's (Fig. 5-7). E&M is the only interface a T-1 multiplexer absolutely must support. There are five types — I and II are most common in the US (IV and V, in the UK). The more options the better.

The private backbone network need not switch phone calls, as there is a PBX at each location for that function. Nor need

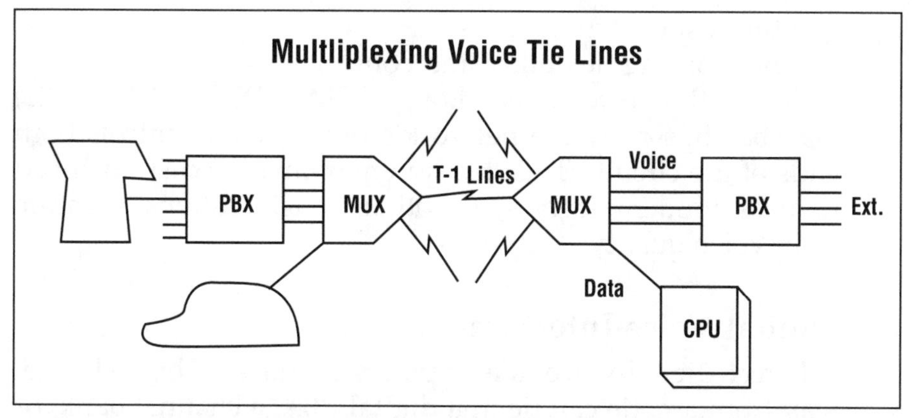

Figure. 5-7. *In place of individual analog tie lines, the T-1 network carries many voice connections. With voice circuits in digital form, they pass over backbone T-1 lines with data.*

the network, in many cases though not all, support "telephonic" functions like ringing or giving battery to a phone instrument. Again, the PBX usually does those things, but not always. To support foreign exchange lines (see Chapter 3) a multiplexer or channel bank needs other interfaces like:

- loop start, FXS and FXO,
- automatic ringdown,
- manual ringdown.

For specialized needs, there are adapters to convert an E&M trunk to these other telephone interfaces. Adapters can cost as much as the E&M ports on the T-1 multiplexer. As add-ons, not part of the T-1 node, adapters are not under centralized network management. Therefore it is always better if a private network accepts analog voice connections individually, and digitizes them at integrated I/O ports.

From there on, the voice signal may be treated like data until it is converted back to analog form at the far end. (See the discussion of "drop and insert" vs. "bypass," below.) As far as the network knows,

Data = Digital = Voice.

If private, the network need not meet any standard other than the link framing format (if any) imposed by the carrier. Far greater concerns will be the

- voice quality that satisfies the end users,

• the cost of terminations,
• networking flexibility and control.

The method of voice encoding (PCM, ADPCM, CELP, etc.) may be chosen by the network's owner who controls both ends of the circuit. This decision may involve cost, the need to pass modem signals, or other considerations separate from voice quality.

Digital Voice Interfaces

Direct digital voice is a popular interface. This T-1 gateway from a node can tie in a digital PBX, a channel bank, or the public switched telephone network (Fig. 5-8). Such interfaces have been available for PCM voice channels in several flavors over the years:

D4 — The U.S. standard, 24 channels with robbed bit signaling,
CPI — Computer-PBX Interface, also 24 channels,
DMI — Digital Multiplex Interface (DMI), 23 channels with a separate signaling channel, AT&T's specification,
PRI — Primary Rate Interface, the 23B+D of ISDN.

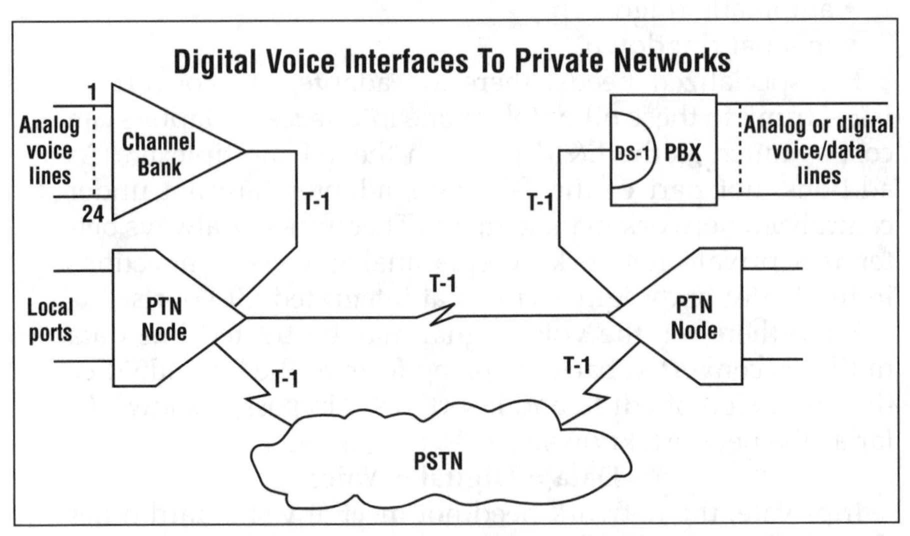

Figure. 5-8. *T-1 links between private networks and channel banks, PBX's, or the public switched telephone network (PSTN) are examples of digital voice interfaces. Access to the PSTN via T-1 might be for WATS, 800, or message toll service.*

DTI — Digital Trunk Interface (DTI) (Northern Telecom), and

DPNSS — British Telecom's version at 2.048 M bit/s.

These interfaces have evolved toward ISDN due to pressure from users to comply with standards and be compatible with public networks. Users want assurances that they can hook up to PSTN services — the one with common channel signaling is PRI.

For connecting a PBX to a private node, a digital T-1 interface is hard to beat on incremental cost — from the network's viewpoint. If the PBX has a DS-1 interface then the network need not worry about converting 24 analog voice signals to digital form and multiplexing them. This saves the T-1 node the cost and space for 23 interfaces to perform those functions. Naturally the saving is reflected in a lower cost to the network or for the node.

At the other end of the cable, the DS-1 interface may add expense to the PBX. If the PBX has no DS-1 interface, adding one can be expensive. The costs of hardware and software upgrades — in money and inconvenience — can offset or even override the savings gained by reducing the number of analog ports on the network node. Reducing the number of existing analog ports on the PBX produces little savings if they are already paid for. Some older analog PBX's, while still serviceable, will never have an integral T-1 interface.

The new intelligent channel banks offer a possible compromise. They join existing analog PBX ports to a DS-1 port on the node, and are part of the managed network.

Any cost advantage of the digital voice interface for a full 24 channels becomes far less certain at a node that serves only a few phones. Here, the analog voice interface can be more cost effective as it involves less equipment. A channel card in the mux tied to a standard analog port on the PBX avoids the more costly T-1 interface on the PBX. Channel banks can connect to E&M ports on the PBX, but ground or loop start ports usually take much less room in the PBX.

Note that PBX's with T-1 interfaces may be networked directly, point-to-point. The circuit would carry only the standard 24 PCM channels (with robbed bit signaling). A T-1 multiplexer could make a contribution by conserving voice bandwidth via ADPCM or other compression technique. To earn a place in purely voice networking — with no other considerations — the T-1 network with a PCM/DS-1 interface would have to compress PCM voice to a more compact format. That is, it would have to perform much the same function as a bit compression multiplexer (Fig. 5-9).

If the PBX does not have a T-1 interface, check carefully. The initial and lifetime costs of adding a digital voice interface may not compare well to the analog network interface, but they should. There are other considerations, like network control and digital to analog port connectivity.

Voice interface ports that are part of the node or exchange will be within the scope of network management. Dumb channel banks outside the node cannot be monitored or manipulated by the network management system at the central site (intelligent channel banks can be so managed). PBX's may have remote diagnostics, but they usually are maintained and reconfigured on-site.

For maximum networking flexibility, it should be possible to connect analog and digital PBX's. That is, a call entering the network via a DS-1 port should be able to leave the network via an analog E&M port. For that matter, it should also be able to connect to a telephone instrument (loop start FXS) or a central office trunk (ground start or — rarely — loop start FXO). This kind of connectivity will help preserve an investment in PBX's and other CPE.

In the future, the format of the T-1 signal may be other than D4 format. AT&T, for example, has a tariff for WATS and 800 via T-1 access. MEGACOM opens possibilities for two major variations:

- While it originally used standard PCM, Megacom also offers ADPCM channels on the T-1 line in the M44 format (the result of bit compression multiplexing).
- Since Megacom terminates on a toll switch, the signaling format is different from that customary on a PBX. Check for compatibility of CPE intended for this service.

Finally defined in 1988, another digital interface for voice is the ISDN basic rate interface. This has two 64 kbit/s "bearer" or B channels for voice or data and one 16 kbit/s "data" channel to carry signaling and low speed packet switched data.

Whichever voice interface is chosen, the node must be able to busy out any voice channels that are unavailable to the PBX. For example, say a T-1 circuit is lost, and with it the voice tie lines between Chicago and New York. A user in Chicago dials NYC from the Chicago PBX. The least cost routing feature prefers the tie lines, which are down. Unless the T-1 node explicitly tells the PBX that the channel is not available, the PBX will keep ringing on a dead tie line (high and dry) until the caller hangs up. If a dead line is busied out (E lead grounded on the individual channel interface, or the appropriate state of the signaling bits are set in the DS-1 interface) the PBX will select an alternate route to complete the call.

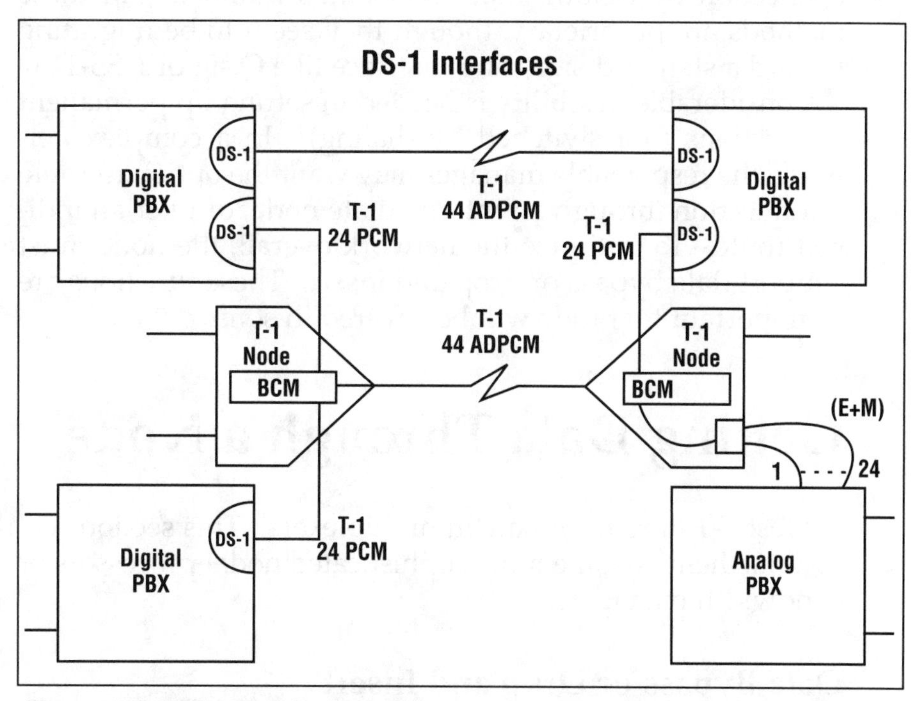

Figure. 5-9. *PBX's with DS-1 interfaces may network directly with each other. The node or multiplexer earns a place in the scheme by compressing voice channels, to save bandwidth, and by providing connectivity between digital and analog PBX's.*

Channel Routing Flexibility

Ideally, any port on the network should be able to connect to any other port. A normal restriction is that both be of the same type (voice or data, speed, etc.). Most often much less than universal connectivity is required. Many virtual circuits or connections are fixed and dedicated to specific applications. Voice tie lines between offices, for example, might be assigned permanently to the PBX at each end. Even in the case of dynamic bandwidth allocation (where only the active tie line connections are given bandwidth on the T-1 circuit) the connection is either there or not — its end points are not switched by the network.

In the 1990s some TDM multiplexers began to recognize dialed digits for the purpose of setting up connections. In these cases the caller, typically on a voice service, determined the path through the network and the far-end point on the connection by dialing that end point's number. All these methods are proprietary, though they seem to be migrating toward a standard signaling interface like Q.sig or DSS-1.

Considerable flexibility is needed in setting up permanent connections (not switched by dialing). In a complex network, the responsible manager may want the option to route a connection through an intermediate node, or over an indirect route. To optimize the network overall, the node must support data bypass or drop and insert. These functions are so important that they will be covered in some detail.

Getting Data Through a Node

Most T-1 lines terminate in multiplexers. This section will assume them or some more sophisticated nodal processors or bandwidth managers.

Data Bypass or Drop and Insert

Even the simplest single-link multiplexers (starting with channel banks in the 1960s) can pass data through a site. Individual channels are demultiplexed at a local port on one

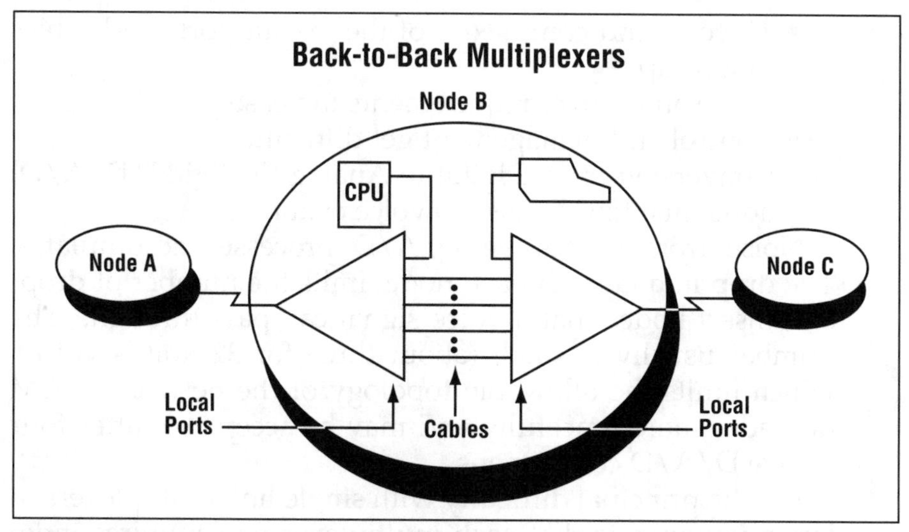

Figure. 5-10. Back-to-back multiplexers require two I/O ports and a cable to move data between T-1 lines.

mux then patched with a cable to a port on a second mux (Fig. 5-10). This arrangement is known as "drop and insert." It has drawbacks which get more serious with increasing network size:

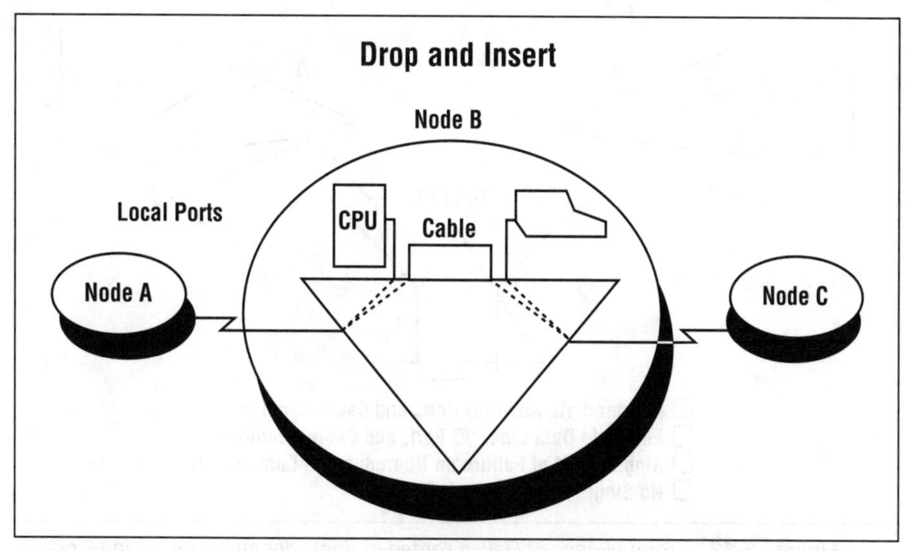

Figure. 5-11. Drop and insert from a data viewpoint means that all channels from two separate T-1 lines must terminate in local ports. Channels between A and C must be patched at B with a cable that requires two ports on the single multiplexer.

- The cost and complexity of the in/out ports and cables mount up.
- Space and power requirements increase.
- Control and management get difficult.
- Conversion from Digital to Analog to Digital (D/A/D) adds quantizing noise to voice channels.

Noise from successive D/A/D processes accumulates. The drop in quality at each node limits the number of drop-and-insert nodes that a voice signal can pass through. The number usually is small (about three for 32 kbit/s voice), which limits the allowable topology of the network. PCM degrades more gracefully and may be acceptable after four or five D/A/D conversions.

But the principal difficulty with single link multiplexers in large networks is that each multiplexer pair requires independent management. Any diagnostics built into multiplexers work only on a "per pair" basis. For unified control, another layer of equipment must be added.

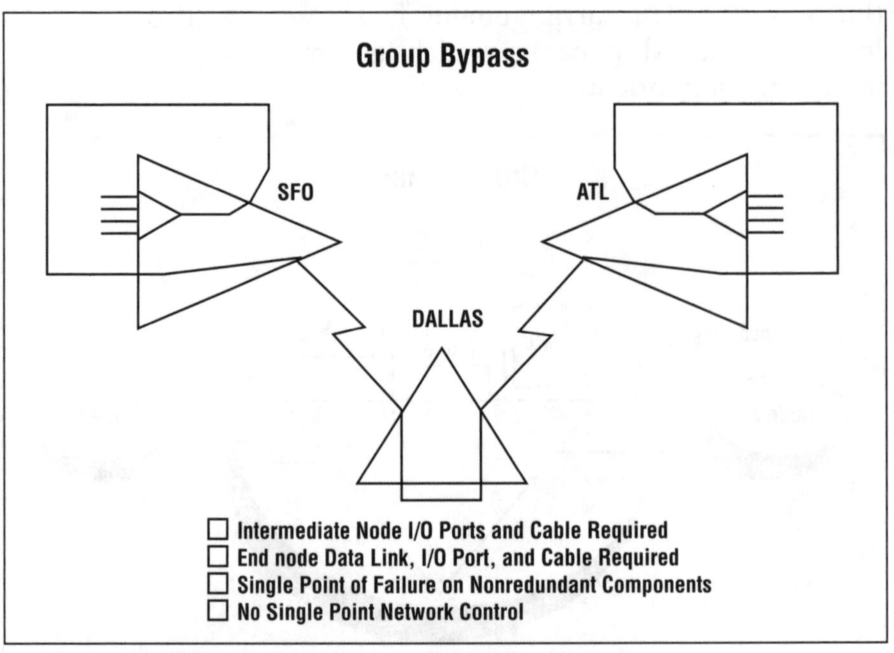

Figure. 5-12. *Group bypass at Dallas routes multiple channels, voice and/or data, out of one I/O port, over a single cable, and back into a second port. While saving ports and cables, it limits the ability to reconfigure connections remotely as cable patching may be necessary.*

Multiple Data Links

A (1981) type of T-1 multiplexer supports two data links. In most respects, however, this design acts as if it were two single-link muxes in one cabinet. While both links enter a single multiplexer, all channels from both links must terminate in local ports (Fig. 5-11). Data passing through the node still requires two ports and a cable, to drop and insert.

Some data multiplexers performed drop and insert on groups of logical channels or connections. By looping a data link back to a port (Fig. 5-12) a group of connections totaling 64,000 to 512,000 bit/s could be dropped and inserted. One version is done internally. Group bypass reduces the amount of equipment needed, but restricts flexibility in routing those channels. Usually they must be routed as a group, over a cable; individual switching is not available for a channel in such a group.

Properly applied, group bypass can be a powerful feature. It works well where a large number of slower data channels are permanently connected between fixed points. Treated as a unit, all connections in the group will be made together by an automatic alternate routing operation — faster than setting up each connection individually.

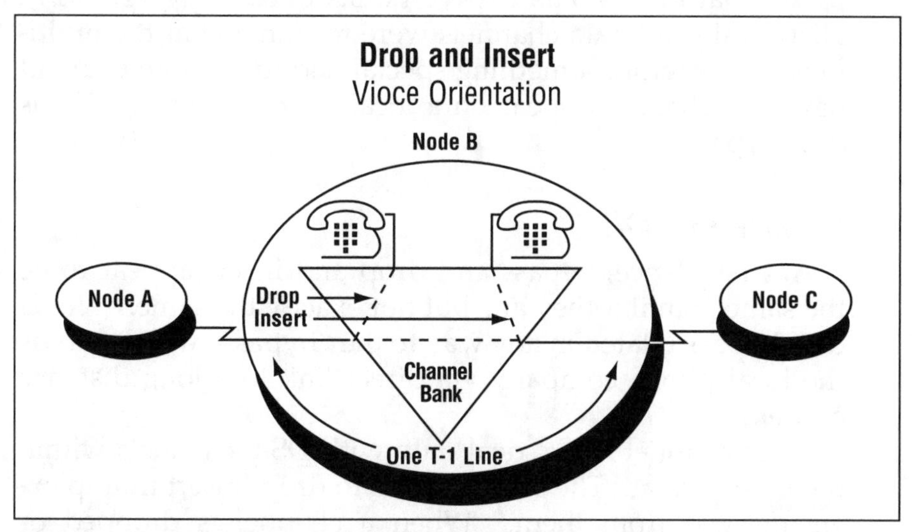

Figure. 5-13. *Drop and insert in the voice world means something different from the same term in the data world. A channel bank thinks of itself as sitting in the middle of a single T-1 circuit. A connection from the left circuit that drops locally leaves room on the right to insert a different connection. This is also called "add and drop."*

There is another class of multiplexer that offers a different view of "drop and insert," also called "add and drop." This is a smarter multiplexer based on a channel bank design for the voice world. It can pass channels between data links internally (Fig. 5-13). To voice people, the T-1 line at Chicago comes in from SFO and goes on to DAL. Even though the two branches, to SFO and DAL, appear as separate 4-wire circuits at CHI, and attach to the multiplexer by separate connectors (and are now tariffed separately), they were considered historically to be only one line.

The normal voice approach, then, is to consider the SFO-CHI-DAL (San Francisco-Chicago-Dallas) path as a single channel. The channel bank accepts each time slot from one data link and immediately passes it through to the same time slot on the other data link. An exception must be made when a channel is to be "dropped" to a local port. A channel dropped from one side of the T-1 line leaves a time slot open on the other, where a local channel may be inserted. In reality, some D/I multiplexers are two channel banks, with a connection between their buses for passing channels between T-1 links.

From the data viewpoint, there is a comparable multiplexer that moves data between links internally (Fig. 5-14). Historically, all data channels were terminated at the multiplexer. Therefore something special had to be done to avoid having a channel appear on a local port: that something is data bypass.

Bypass vs. D/I

At first glance, bypass and drop and insert appear to be the same. Similar they are, but not exactly the same. (Nor is data bypass related in any way to the "bypass" which avoids the local phone company via direct link to a long distance carrier.)

Voice channel banks deal only with DS-0 channels within the DS-1 format. The same applies to drop/insert multiplexers derived from them. When a channel is dropped or inserted, it is 64,000 bit/s. And since the channel bank sees the data links as one continuous line, there are connectors for at most two T-1 circuits.

In contrast, a data oriented device views each T-1 circuit and each connection as a separate entity. Therefore there is no reason not to have more than two data links. CPE vendors have offered 6 to 256 T-1's from a single node intended for private networks; larger devices are available designed for the public network.

A data multiplexer doesn't limit to 24 the number of TDM channels that can be bypassed. Bypass is performed by a control module or internal switch module, without involving I/O ports. One device can bypass hundreds of data or voice channels within each node. Each TDM channel can originate and terminate on any link — even the same one.

"Data bypass," then, is more flexible than voice "drop and insert" in dealing with large numbers of channels running at various data speeds, to and from many T-1 data links. Part of that flexibility is software control, from a remote command site.

The bypass feature of a network node or data multiplexer should deal with any standard speed from 50 bit/s (for telex) up to (and beyond) 64,000. Channel banks deal only with the DS-0 rate of 64,000 bit/s. In the worst case, this means that to bypass a 9,600 terminal connection or a 32,000 bit/s voice channel requires a full DS-0 time slot as shown in Fig. 5-2.

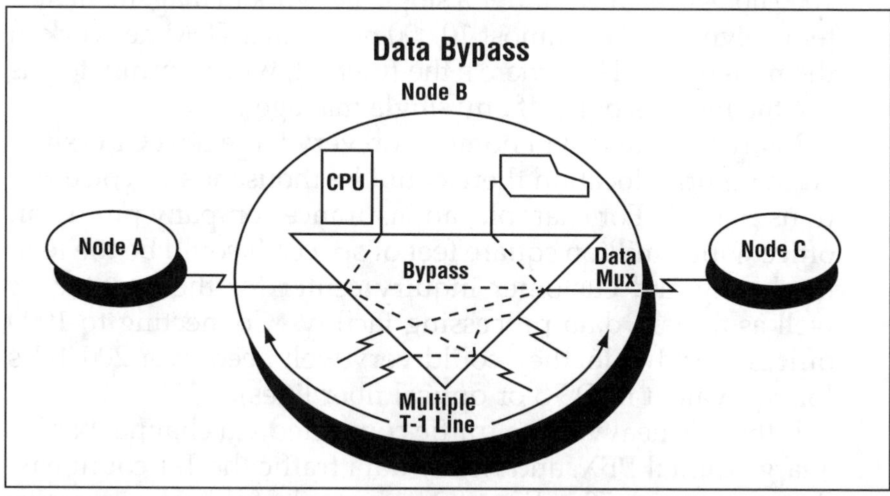

Figure. 5-14. *Data bypass resembles the drop and insert function of the voice channel bank. What distinguishes the two is that "data bypass" offers variable bandwidth allocation and additional flexibility in routing connections among many more T-1 links.*

Even in a node which can carry multiple data channels per voice time slot, it is possible that only increments of 64, 8, or 2 kbit/s can be switched.

The caution here is to determine exactly how much bandwidth is used for each kind of connection, what that means to link efficiency, and how the link efficiency affects the number (and cost) of T-1 circuits. This is particularly important when dealing with "32 K" voice, as pointed out in the chapter on voice, and subrate data circuits. The actual bandwidth may be significantly higher than the nominal speed.

Migrating to a BIG Node

How could anyone possibly use 144 or 256 T-1 lines into one site in a private network?

Not easily. But the trend is solidly toward larger and larger networks. Just as IBM underestimated (twice) the possible size of SNA networks, most vendors have guessed far too low when setting the maximum size of a digital backbone network size. By the early 1990s there were many users with more than 1000 nodes installed under a single network management system. Nynex linked almost 10,000 nodes in a TDM network in the mid-1990s. Then there's the Internet, which fortunately is not the responsibility of any single manager.

Many large nets will home in on very large host CPU sites. At the central location there could be thousands of voice stations as well. For example, an insurance company put up an office with 1 million square feet of space. It could be the telemarketing and customer inquiry center for the country, as well as a main data processing facility. Connecting to 1000 offices worldwide, they could very well need over 200 T-1's (or equivalent in DS-3 or optical fiber lines).

If there is heavy voice traffic connected via channel banks, a large digital PBX, and heavy data traffic the T-1 count easily can exceed 50. Every 24 channels (or just one voice connection that required another channel bank) adds a T-1 link from the node. CAD/CAM, graphics, and videoconferencing consume huge chunks of bandwidth.

Vendors who compete seriously for a leadership position in T-1 networking have announced very large nodes. They range up to 256 T-1 ports, nominally. In reality their practical capacity may be much less, for two reasons:

1. Backplane throughput; there is little point in having 88 T-1 ports if you can cross connect only 44 of them.
2. Slot shortage; if the node is fitted with the full complement of T-1 ports, there are no physical slots left for anything else (like voice or data inputs, and perhaps even the processors to run all those T-1's).

Ignoring the first limitation can be very misleading. There is no way to use bandwidth-limited equipment. If you can't connect anything to the T-1, what good in having it?

Slot limitations, though important, leave equipment useable at maximum T-1 capacity. A node with all T-1 cards is a DACS if there is no blocking for DS-0 or subrate cross connections.

But how do you get to very large nodes? Is it necessary to change out all your equipment? Or can you keep what you have and add to it? Adding on will generally prove less expensive for hardware. But an additional, very significant saving is in wiring. What would it cost (time, money, inconvenience) if you had to change all your low speed data and analog voice wiring in the communications room? Wouldn't it be better to leave that wiring in place and add another rack or shelf of new equipment? Look for an easy migration path from what's needed today to what might be needed tomorrow (Fig. 5-15).

Almost certainly a future need will be 45 Mbit/s DS-3 connections; or 155 Mbit/s on fiber. Ideally your new T-1 equipment will migrate, gracefully, toward a larger size, perhaps with a DS-3 or an OC-3 port card. For full networking, each node should support at least two (preferably 4 or more) trunks at those speeds. Then you will be able to build net topologies similar to those of traditional T-1 networks.

How Is Switching Done?

We must be very careful of the term "switching." There are two very different meanings. As with other terms having dual meanings, one comes from voice and one from data:

Voice switching refers to the function of a PBX or CO switch. This kind of switch accepts dial digits as instructions on where to connect the calling phone. The average phone can't be connected unless you dial it every time.

Data switching, in TDM nodes, means setting up a connection to a pre-defined destination. The caller has no choice about the called location. This form of switching is used to assign T-1 bandwidth only to active terminals, for example. Voice people call this function automatic ringdown (ARD), not switching. You see it on "courtesy" reservation phones at car rentals and hotels.

All T-1 nodes perform "data switching:" automatic ringdown. Only a few provide "voice switching" with the ability to select the called port based on interpretation of dialed digits.

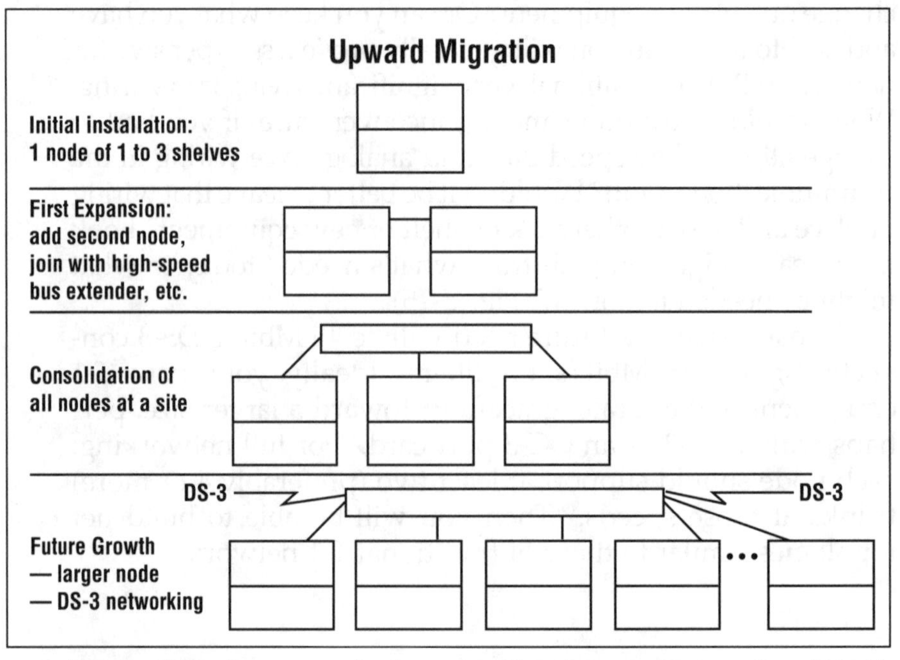

Figure. 5-15. *Future growth of a node will seldom require more than 100 T-1 ports. But a mix of T-1, data, voice, and DS-3 connections could grow to need a bandwidth of hundreds of megabits per second. Ideally, the early equipment would be incorporated into a larger node — without changing the voice or data wiring and without discarding installed equipment.*

The "ARD" form of switching does not mean that the path through the network is always the same, just the end points. In fact, the automatic alternate routing (AAR) feature changes the path to avoid failed lines and nodes — but the end points remain the same.

In some cases vendors have adopted the term "switching" to describe AAR without defining it explicitly. By letting the market think what it may, these vendors have confused ARD and voice PBX type switching in the minds of many people.

Previous chapters assumed that each node can make connections independently. That is, each node can act as a distributed switch element. Some T-1 multiplexers operate this way; others centralize the connection function and distribute only the multiplexing. At least one does some of each.

The T-1 Node as Voice/Data Switch

In an emergency, automatic alternate routing switches circuits on a T-1 to different connections, which are standing by. This is important. Most users hope they never need it. Similar switching to alternate connections is needed routinely on a daily basis in some companies. For example, some organizations want to reconfigure their circuit connections and redistribute bandwidth among applications — switching from all-voice during the day to all-data during the night, etc. Ideally, these changes, after they are set up, are done automatically, and need not concern the operator under normal conditions.

It is the individual channels needed on demand ("Gimme a videoconference circuit to Dallas in 23 minutes," screams the president) that test the switching flexibility of the T-1 bandwidth manager. Obviously, only a soft-configured device that lends itself to remote programming will give the central site staff the control over the entire network that it wants and needs (especially if it is to keep the president happy).

Part of network control is flexibility and convenience in assigning bandwidth to channels. The minimum required to set up a connection is to configure one end port and designate the other end port. The network management system then finds a path with available bandwidth, assigns the needed bandwidth, sets up tandem connections in any intermediate nodes, and configures the second port to match the first. Less advanced control systems require additional input from the operator.

What may the central staff be called on to do? They might install new channels within the network to satisfy short term needs. For example, end of month financial data entry requires a temporary clerk on an extra terminal for two days. An efficient switching function in the network could find sufficient spare bandwidth to create a 9600 bit/s channel to the computer.

The speed with which a private network can define an additional channel between sites already on the network stands out when compared to the delay involved in getting it from a telephone company — called also, telco provisioning. The cost stands out too (see Table 5-1).

Installation on a private T-1 network involves the incremental cost for additional ports, one at each end. DSU/CSUs exist on the T-1 lines; therefore there is no additional cost. The delay in installing the on-net portion of the line is in the area of minutes. Physically stringing a cable at either end may take longer.

On the other hand, the new telco line must work its way through the order backlog, including the lengthy coordination procedure between local and long distance carriers. The new line will also need CSUs (Channel Service Units) at both ends.

On-demand connections between analog voice ports are being supplied more in T-1 nets via nodes with ARD features. The destination is pre-assigned, at set-up, so the connection (call) is made automatically. No dialing is necessary or, for that matter, possible. Routing information is entered, off-line, via the supervisory port or network man-

PRIVATE 56 kbit/s LINE INSTALLATION

	Order delay	Installation cost	DSU/CSU cost
On Private T-1 Network	Minutes	$800 or less	none (cables only)
Telco Supplied	Weeks to months	about $1,500	Up to $1,500 (plus cables)

Table 5 -1

agement system. Changes take effect the next time the connection is requested, or at a specified time in the future.

The way to obtain the effect of circuit switching within T-1 nodes is to employ a "connect on demand" feature. A data port is inactive until the attached terminal raises a signal on the interface connector. This might be RTS (Request To Send) or DTR (Data Terminal Ready) on an EIA data port, the M lead (Mouth lead) on a voice interface from a PBX. Only then does the control module in the bandwidth manager give that I/O port any bandwidth on the T-1 line. The control module creates a new frame on the data link to accommodate the channel. An ideal application would be a Group 4 facsimile machine used intermittently. It could benefit from large bandwidth for fast document transfer, but need not occupy that bandwidth all the time.

If user terminals are shut off when not in use (some terminals have a time-out feature to drop RTS or whatever signal is being used) then that terminal will relinquish its bandwidth. This means the total bandwidth of devices attached to the ports on a time division multiplexer can exceed the aggregate available on the T-1 lines.

Users receive service on demand as long as bandwidth is available. Some T-1 nodes will recognize different priorities assigned to ports. Those in queue with higher ratings will be connected, as bandwidth becomes available, ahead of ports with lower rank.

Features like those of a data PBX, particularly terminal-specified connections, appeared briefly in the T-1 arena. Switching statistical multiplexers and data PBXs on the T-1 backbone provided connectivity among async terminals and hosts for some years. Those functions have largely been replaced by LANs and routers.

Distributed vs. Centralized Switching

Central connections appear in some statistical multiplexers as well at T-1 units. It implies that every logical circuit must pass through a central or regional node. If a connection is requested between two ports on the same remote node, that traffic must travel over a data link to reach the switching center, then return — using double bandwidth on the link. This double trip on the data link is not required by nodes that distribute the connection function.

If the T-1 link to the central site goes down, so does the logical circuit based on centralized connections.

Distributed switch architectures can maintain local connections between ports on a node even if that node is isolated from the rest of the network. If a large distributed network is broken by a major failure, the resulting segments continue to operate independently.

The switching architecture, distributed or centralized, does not have to match the control architecture, which can also be distributed or centralized. For example, see the chapter on network management for a discussion of centralized control of distributed switches. The importance of distributed vs. central switching depends, to a great extent, on the cost of the T-1 circuits and the importance of reliability.

Reliability

In survey after survey, network users and operators list reliability as the prime concern in communications. It is mentioned, typically, more than twice as often as the next most important item, service and support. In numerical ratings of importance, reliability always scores 9.99 out of 10.

The topic of reliability has been held until this late in the chapter so that we would have a common basis for the discussion — after some of the primary features in multiplexer nodes had been described. As we go over this most important topic, related ideas will again be drawn in.

Redundancy

In case a light bulb burns out, keep a spare handy. It works well — unless the only bulb in the garage burns out at midnight — during a snow storm. Then you wish you had another in place, ready to go. When an entire company depends on continuing communications, "keeping a spare handy" simply isn't good enough. The spare must be on line and ready to take over immediately if the primary unit fails.

Hot standby redundancy is offered in most major T-1 multiplexers and bandwidth managers for all the common modules. That is, those whose failure could bring down the node or an entire T-1 circuit. These include control modules, power supplies, line drivers, and CSUs.

But having the spare on line is still not enough. There is a law of communications (derived from Murphy's) that says: "Any backup module not continuously tested will be dead when it's needed."

Therefore, all redundant modules must be powered up and monitored continuously. This testing could be a background operation of the self-diagnostics that monitor the on-line modules.

Though technically possible, redundancy on the low speed port modules has not been shipped. One vendor talked about it with no promised delivery date. Users have accepted non-redundant I/O because the loss of one unit here disrupts very few ports, typically one to four. The additional cost and bulk have not been justified by the small risk, as these modules can be changed individually. As the number of ports per module increases, the value of redundancy here will increase too.

Changing the Bulb

If you have a second bulb lit in the garage, you don't want to pull the main fuse during a house party simply to replace the burned out one. In a digital network, most managers cannot afford to interfere with active traffic. This is a potential problem when changing a failed module. The circuits on a module that send data from that module into the main unit (the bus drivers) could affect data flow on the backplane as a connection is broken or made.

Therefore installing a replacement port module (or any redundant module) is not as simple as "pulling one out and sliding one in." Not only must the nodal software allow changes "on the fly," but also the system must avoid introducing errors during the exchange.

One way to avoid glitching the data bus is to put the module's bus drivers in the same electrical state as the bus for every lead; on or off. In practice, this is impossible due to the high data rate. An alternative is to build the drivers from "tri-state devices" — semiconductors that can be put in a high impedance or "disconnected" state as well as on or off. This third state electrically isolates the module's bus connector so it cannot interfere with the bus.

But to be in the third state the module must be powered — while disconnected. To resolve this contradiction, an auxiliary cable in some devices may power the boards and put the bus drivers into the high-impedance state for plugging in or pulling out. Bus connector pins of staggered lengths do the same thing. Staggered pins have replaced external cables in new designs.

This feature allows a move, addition, or removal of any board or module without resetting the node or affecting traffic. It permits modifying the node or reconfiguring the network without bringing either down. These designs make it relatively easy to add T-1 data links and nodes to the network. This is also how a failed module is replaced while the backup unit is operating. The goal is to make board failures and replacements invisible to network users.

Power Supplies: A Special Case of Redundancy

Power supplies work so hard they are among the components that fail most often. However, the only sure way to check them continuously is under load. For example, a spare power supply often is kept powered on but not active. Though it holds a steady output voltage it may not be able to deliver its rated current when suddenly switched into use. It could fail completely. Also, switching large currents usually produces transients that can "glitch" the bus and introduce errors in any or all data channels.

Rather than have a standby ready to take over, the answer for reliable power is "load sharing power supplies." That is, multiple power supplies in parallel deliver equal portions of the total current requirement. Each supply runs with some spare capacity. Any one supply may drop off (fail completely) without causing the remaining supplies to exceed their ratings. For small nodes, the minimum configuration is two power supplies, either one of which can handle the entire load.

The phone companies would call this "one for N sparing with automatic protection." An additional benefit is that load sharing operation reduces the stress on components and extends the life expectancy of each power supply. So supplies that share a load, other things equal, last longer than one with an unloaded backup.

Clocking: A Critical Redundancy Factor

As networks get larger, clocking becomes more complex and more important. It requires a few definitions to clarify some terms and concepts often bandied about but seldom pinned down in connection with T-1:

Take clock from a line: a node adjusts its internal oscillator with a phase locked loop or similar circuit that adjusts the local oscillator to match exactly the bit rate from the line. When so adjusted, the node can send bits at exactly the rate it receives them. The source of clocking must be smooth, that is, operating at a constant rate. A PLL cannot lock to a high-speed clock that is interrupted periodically to achieve a lower nominal speed.

Give clock to a line: the node places average pulses on the line at the rate of its internal oscillator, disregarding the rate of incoming bits (which should be the same, as the other node should be taking clock from the line, or be loop timed).

Loop timing: the node sends bits to each line at the same rate it receives them from that line. In well behaved networks, the rates of all lines are the same as the node's internal clock. With multiple carriers and satellites, some lines may be loop timed but not synchronized to a node's clock. This mode is pleisiochronous.

A T-1 multiplexer or node must accept the clock rate of any digital services that pass through a switch or DACS (see Chapter 4). Many networks, especially those without T-1 lines, go out of their way to have at least one such line, just to get this highly stable clock signal. The digital line then becomes the master timing source for the whole network.

At any one time a node can derive its clock signal from only one source. What happens when a node with multiple lines loses the line supplying clocking? The requirement to sync to the carrier's switches hasn't changed, so the node must have a fallback arrangement that derives clocking from a carrier line that's still working, if one exists, or from another source.

Large nodes with up to 10 or 16 or 30 or a 100 digital lines must have a plan for every node to find an alternate clock source if the prime clock signal goes away. What could that alternate be?

1. Another T-1 service trunk line. The simplest fallback would be to another T-1 line that terminates on a switch or DACS. The two links could be from the same or different carriers. Being closely related to the trunk that was lost, the transition probably would be easiest. But lines from the same carrier might also be taken down by the fault or disaster that took down the first T-1.
2. DDS tail circuit. The 56,000 bit/s and subrate DDS services work from the same master clock as the AT&T T-1 service. A smart node can look at the slower speed and still maintain T-1 synchronization. Working from the lower speed requires more sophisticated hardware. DDS is a solid choice in some areas, as it keeps the private net synchronized to the public net. But some 56 kbit/s lines suffer excessive jitter, and many 56K lines are not DDS but rely on the relatively poor clock in a channel bank.
3. A satellite carrier. The satellite might derive its speed indirectly from a master clock. This fact is seldom volunteered, and is not guaranteed. If the satellite channel takes its clock from any stable master clock (some satellites carry atomic clocks in space) it is usually workable as a secondary source for short periods. However, a satellite signal tends to shift phase over time and may appear to shift frequency slightly.
4. A link to another node. If that node has access to a CO switch or DACS, it can lock onto the BSRF from the CO. A T-1 line out of this node could then be clocked accurately. This implies that the link joining the two private nodes is cable, microwave, etc., or is supported on standard M13 facilities. (If a line goes through a DACS or

terminates on a CO switch it may be used as a clock source directly.) This line can then transfer clock from the first node to the second.

5. Station Clock. Some T-1 nodes accept a special clock signal, for example 8 kHz. It is possible to derive this rate from many stable sources, including primary standards. However, the most attractive option is a time base derived from Loran or satellite broadcasts.

Loran, a radio navigation aid, allows a receiver to identify its location within a few dozen feet. The calculation depends on signals from multiple land-based transmitters. To get the accuracy, the transmitters must be ultra-stable: Stratum 1. A special radio receiver at the site of the T-1 multiplexer picks up the Loran broadcasts and generates a clock signal from them. This timing is quite good.

The Loran stations get their timing from satellites. The Global Positioning System broadcasts time information to special receivers, telling them how to adjust their internal

Network Clocking Strata

(based on Compatibility Bulletin 147)

Clock Stratum	Short Term Accuracy[1]	Long Term Stability[2]	Synchronization Ability
1	$\pm 1 \times 10^{-11}$	Not applicable	Primary standard, like atomic clock; needs no calibration; always runs free.
2	$\pm 1 \times 10^{-10}$	$\pm 1.6 \times 10^{-8}$	Must track or lock onto any stratum 2 clock.
3	Less than 256 frame slips/day on link to other Stratum 3 device.[3]	$\pm 4.6 \times 10^{-6}$	Must track or lock onto any stratum 3 clock.
4	Not defined	$\pm 32 \times 10^{-6}$	Must track or lock onto any stratum 4 clock.

[1] Change in rate over one day.
[2] Total drift over 20 years while running free.
[3] About $\pm 5 \times 10^{-7}$

Table 5-2

oscillators. The bit rate of the transmission is only 50 bit/s so the receiving antenna is quite small, about the size of a standard desk telephone. The GPS satellites receive their information from the National Bureau of Standards. Therefore GPS receivers can offer true Stratum 1 clock signals. The cost of GPS receivers dropped sharply in less than a decade and are affordable for most private networks.

6. An internal free-running oscillator, usually crystal controlled. In the event an entire section of the network becomes isolated from all DDS and T-1 sources, a node in that section generates its own clock by falling back to an internal oscillator. It must be accurate to within +/- 50 bit/s in 1.5 Mbit/s. This is the limit of +/- 32 parts per million (ppm) allowed for the transmission rate of the network, but it should be at least as good as a Stratum 4 clock.

Clock precision is defined at four strata (Table 5-2). Stratum 4 is good enough to fall within the range accepted by the standard M13 multiplexer. However, in larger networks it is important to have a more stable clock source as the master. Variations in clock rate propagate through the network and can cause disturbances, including errors. If every node (or most of them) had stratum 1 internal clocks, most synchronization problems would disappear.

Note that clocking option 4 places a considerable burden on the nodes. The normal arrangement (Fig. 5-16) requires node A to derive clock from one T-1 line and pass it to another link. (Node B takes clock from this link.) After A loses its clock source, it must tell B it needs clock then prepare to accept timing from that link. Node B, if it had been taking clock from the link, must now accept it from its local line and use it to clock itself and the T-1 link.

A similar situation can occur in a ring network (Fig. 5-17) with one DACS-switched line and the rest free-running or asynchronous on M13 multiplexers. Normally, node A gives clock to B, which passes it to C. If link AB fails, node C must recognize that good clock is still available from node D, and arrange to pass timing to link BC and thus node B.

As the examples indicate, alternate clocking schemes must be planned closely with automatic alternate routing patterns. Not only must AAR maintain connectivity among critical nodes, it must also provide paths necessary for the clock source to reach all working nodes. Then the difficult part is ensuring that the nodes know which line carries reliable clocking.

In the past the nodes have held fall-back lists for clock sources. If the line used as the primary clock source went away, the node looked to the next line on the list. At the bottom of the list, usually, was the internal oscillator. As the topology of networks gets more complex, vendors are offering schemes to allow nodes to pick the best available clock source dynamically. Expect precise station clocks to change the rules in the future.

A few node designs operate isochronously on the T-1 data links; that is, they will sync up with several lines running at slightly different speeds. There being no magic left in the world, something must give if the inbound TDM channel runs faster than the outbound. Buffers can control the difference for a time. But inevitably the buffer will overflow, the frame will slip. At that point there is an error in the channel. A slip does give the buffer a chance to empty so it again can accommodate the difference in speeds.

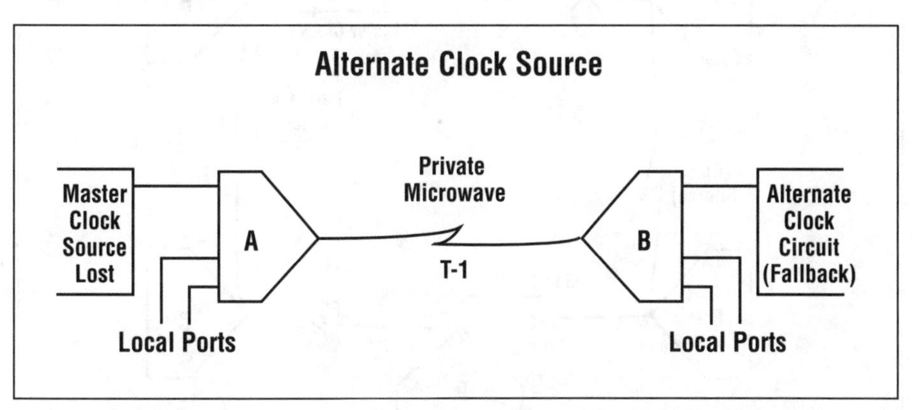

Figure. 5-16. Loss of the working clock source causes a shift to an alternate source at another node. In this case that means reversing the direction of travel for the clock signal over a link. This maneuver requires considerable intelligence in the nodes or network control system.

For small speed differences, frame slippages (and the resulting errors) at buffer overflow can be infrequent. The error rate may be quite acceptable to certain voice applications, and unnoticeable in channels that are error protected by a higher level data protocol. If there is unused bandwidth on the faster, incoming line, the node may be able to slip frames while the buffer contains only null characters. One packetized algorithm slips on empty packets when possible.

The Effect of Nodal Architecture

Early T-1 multiplexer designs stressed the inherent redundancy and reliability of fully distributed processing for network control. Every node, even every signal handling module, had its own microprocessor. Every function to run the network was handled in the nodes.

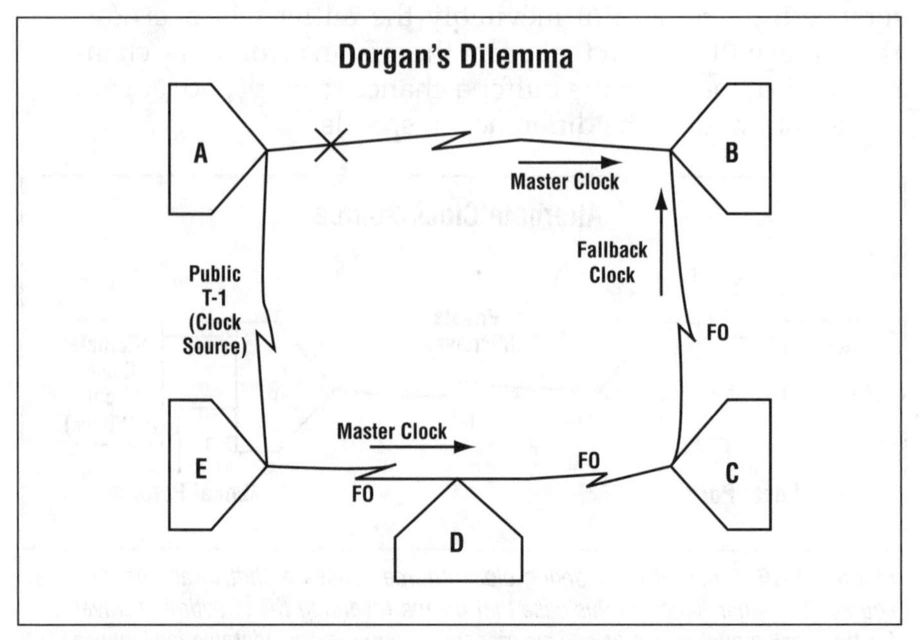

Figure. 5-17. *An occasion that can reverse clocking flow is the invocation of automatic alternate routing in a ring that loses a line (X) near the clock source.*

This meant that every node incurred the expense of sufficient processing power, memory, etc. to run the largest anticipated network, regardless of the size of the network installed. For that price, users got a highly resilient network that, like some worms, could live on in all its parts if it got cut into segments.

Where Is Control?

Another design philosophy limits nodes to local functions. That is, nodes deal only with internal operations and with what is connected directly. Network-wide functions (like automatic alternate routing) are handled by one or a few specialized processors. This allows network designers to put more capacity for these networking functions into a few places than is practical to put into every node of a distributed network. It also removes significant cost elements from the majority of nodes in a net.

These so-called Center-Weighted Control designs were introduced after distributed designs, about 1988. It appears that they will support more nodes in a network (beyond 10,000) and allow easier expansion of a net. See the chapter on Network Management for further discussion.

Internal Bus Architecture

Within the nodes themselves at least two vendors have moved away from the traditional parallel-bus backplane. In all older designs, the modules in a node plugged into a common bus, just like the add-in boards in a personal computer. The alternative approach is to link the common control card to every other module with a serial bus, or perhaps two or three serial buses. The control card contains the switching fabric for the node.

Multiple serial buses per physical slot offer a higher aggregate throughput for the entire node. One brand node in its minimum size supports 32 serial buses at 2.048 Mbit/s each, or 64 Mbit/s. These are for data only; control information uses separate buses. Therefore the 64 Mbit/s is fully available. It becomes the equivalent of 80 or more Mbit/s on a parallel bus where control and data must contend for bandwidth. Larger nodes run the serial buses at 10 Mbit/s or faster, to support ports at DS-3 and OC-3 speeds.

Multiple serial buses offer a form of redundancy as well. They can operate independently, to provide multiple paths between the control card and each trunk or port module.

Linking each slot with serial buses requires that there be a special slot for the control card. Logically, the control card sits at the center of a star configuration, with the other cards at the points. Any type of card can go at any point of the star. The entire structure is symmetrical.

As a result of the symmetry, there is no high speed or low speed side to this type of multiplexer. The control card can set up connections between any modules (even locally between low speed data ports in one node). Any data port can be an aggregate.

For a comparably symmetrical device refer to the discussion of the DACS elsewhere in this book.

Access Multiplexers

As T-1 is used more as an access path to specific services, the nature of the T-1 CPE has expanded into a simpler form. The traditional large T-1 networking node is intended to build backbones, support automatic alternate routing, and provide a management platform. It is far more complex than needed for simpler tasks, like drawing local dial tone, reaching the Internet, or providing integrated access to multiple services.

Depending on the application, the proper CPE to terminate a single-ended T-1 access line may be much simpler because it is not a "networking" device.

Internet Access

If your company has electronic mail, a computer that hosts a site on the World Wide Web, or you gather or distribute much information by file transfers, you probably want a T-1 to your Internet Service Provider. Many small to medium companies find a 56 kbit/s line is sufficient, if occasional (or more than occasional) slow response is tolerable.

For heavy users, or if you want quick transfers, a T-1 at your site ensures that slow response will not be for lack of bandwidth at your end. As the popularity of T-1 access grows (most business access added lately is at T-1) the expectation of your correspondents is that you will not allow a bottleneck at your end by connecting with less than a T-1.

The T-1 line to the ISP is usually provided by the LEC. It is a point-to-point connection, unchannelized.

The equipment you need depends on what the ISP is using, but most often is a router or a FRAD with routing functions. Routers don't have integral CSUs at this writing, so an external CSU is needed to terminate the T-1 local loop. "FRADs" are available with the T-1 CSU in the same box. Integration saves the extra box and a $100 V.35 cable.

Most ISPs offer "SLIP accounts" for businesses. The Serial Line Internet Protocol is a way to send IP over a V.35 interface and a serial line like a T-1. The Point to Point Protocol (PPP) is a later protocol that is more functional, carries other LAN traffic types (IPX, AppleTalk), and is preferred in most cases. PPP is needed if your ISP accepts IPX or AppleTalk from you.

PPP (or SLIP) runs between the FRAD or router you have and a serial port on a router at the ISP's office. The connection is permanent, full time, and requires no adjustment.

If the ISP accepts calls on an ISDN interface, your router or FRAD needs an ISDN interface too. The T-1 transmission line is used for the Primary Rate Interface (PRI), which is 23 bearer (user) channels and 1 Data (signaling) channel, or 23B+D. Most ISPs use equipment that accepts a call on one DS-0. Some ISPs can bond two calls on two DS-0s into a single logical channel for access at 128 kbit/s. By extension to multirate calling, a PRI on your equipment will some day be able to place a call at Nx64 kbit/s, to get any bandwidth you need up to almost 1.5 Mbit/s.

Fractional T-1 Access

The local loop is always a full T-1. So far no carrier has offered another transmission rate in the local loop, though the technology is available (see Digital Subscriber Line). You need a special CSU to isolate the data from the bundle of DS-0s in use, and present that data stream on a serial interface (usually a V.35).

The minimal CSU for FT-1 has a single line interface (RJ-48 jack connector) and one serial port. It is programmable in software or from switch settings to map the data between the two sides.

FT-1 CSUs have as many as six serial ports. Each port is assigned to a configurable bundle of DS-0 channels. These multiport CSUs allow one T-1 local loop to carry many channels that the long distance carrier transports to different locations.

As CSUs, these devices default to 'loop timing' which synchronizes to the clock from the central office. That clock is divided to create the transmit and receive clock signals for the V.35 ports.

One form of CSU adds a DSX-1 port on the terminal side. As many channels as wanted can be mapped directly from the T-1 local loop to the DSX. This allows, for example, voice channels to be passed to a digital PBX.

Local Dial Tone

A digital PBX can accept a T-1 directly, terminating all 24 channels or as many as desired. As described directly above, a multiport CSU performs this function and also can drop some bandwidth into serial ports.

Channel banks are the original way to terminate a T-1 line. They still work as designed, converting digital transmission service into analog voice grade interfaces for phones, facsimile machines, and PBXs.

Most channel banks need a separate CSU.

How To Build A T-1 Network

▶ Previous chapters covered technical aspects of T-1. This chapter looks at practical network design. It is non-technical. The most critical thing to understand about network design is that it is always a question of trade-offs. You can get the best quality, best-sounding, never-busy network in the world. You may not, however, be able to afford it. Network design is thus always a tradeoff between what you want and how much money you have. This makes network design as much an art as it is a science.

"Network design" is also broader than just buying equipment and circuits. It includes everything that supports your network: space, test equipment, people, training, administration, documentation, billing, etc. To look at just the size of the bill from the carrier is clearly misleading. You could easily increase the carrier bill with these other "overheads" by 50%.

Making Tradeoffs

The choices come down to a "best available" balance between desires and resources. Service level versus budget. You want to get the best result for what you spend. Vendors want not only the best price possible, they want as much of the job as possible. Therefore, they usually offer opinions that favor their products and services.

Telephone companies tend to design networks based on central office equipment and usage-sensitive circuit facilities (as against flat-rate tie or private lines). This is The Public Network Option. By contrast, a hardware vendor tends to favor putting the processing equipment at the customer's sites and suggests the least-expensive leased lines—The Private Network.

You can go too far in either direction. Here's an extreme example. You could either

(a) Save all the money. Rip out all your phones. Put two pay phones in the lobby. Or,

(b) Give everyone three CO lines in a hunt group, two handsets, an answering machine, a personal computer with telecom software, and a smart auto dialer. No caller will ever get a busy and all your employees will always have a line free to place calls.

The first alternative will save you lots of money and put you out of business. The expense of the second alternative may also put you of business—unless your employees are all currency traders, each earning $4 million a year in commissions and making you $20 million a year. In which case, give them the best telecom that money can buy. They deserve it.

In network design, the best tradeoff is typically "somewhere in between" pure public or pure private. In short, a hybrid network. The point here is that tradeoffs are a management decision. Your top management should decide which end of the expenditure/service tradeoff spectrum they want. And they should do that before you go further in network design.

If your management wants to get specific, here are some of the major tradeoff questions, which you need to answer to reach the "optimum" solution. Each has its own tradeoff:

- Justification: Is there reason to install T-1?
- Ownership: Public (carrier-based) vs. private (corporate)?
- Network topology: star vs. multidrop vs. mesh, etc.?
- Traffic mix to be placed on network: synchronous data, asynchronous, voice, etc.
- Network reliability and availability requirements

We will now go through these points in detail.

T-1 or Not T-1?

Simple Break even: Let's say you have some voice grade lines between point A and point B. Should you swap over to T-1? Table 6-1 shows the simple tradeoff between the distance and the number of voice grade analog lines that cost the same as a T-1 channel. You will notice the answer depends on the length of the circuit (based on 1989 prices). Be aware prices change almost daily due to competition in the telecom business. And be aware that these "break evens" are based on tariffed rates. Many carriers will cut their tariffed rates if you give them more than one circuit or guarantee to have your circuit/s for at least one year. The rules are changing to favor T-1 even more.

Twenty four analog voice circuits are standard in a T-1 link. Pricing in current long distance tariffs makes the T-1 line cheaper than 24 analog lines for distances under about 900 miles. Fractional T-1 offerings have been priced to compete with small numbers of voice grade lines at most distances. FT-1 should be considered for any location with more than six circuits of any kind.

For more than 24 voice circuits, consider voice compression to alter the break-even ratio. Then the tradeoff is between voice quality and line costs.

Analog lines require a PBX port per voice channel; a T-1 interface requires only one port for 24 channels. Usually the one T-1 port is cheaper than the 24 analog ports and may justify a higher line cost for a T-1. But not always. Some PBX vendors charge dearly for software upgrades to handle T-1 directly. Look at the total cost to upgrade the PBX to T-1, not just the T-1 interface card.

Rule of Thumb
Break-Even Points Between T-1 and Voice-Grade Lines

T-1 Line	VG Lines
Within a wire center (both ends served by one CO)	2-5
Within a LATA (entire circuit from local carrier)	5-10
100 miles, interLATA (on a long distance carrier)	8-12

Table 6-1

If your analog lines carry data, the economics shift in favor of T-1. The most sophisticated bandwidth managers can eliminate modems, handle data in digital form, and cram far more data channels into a T-1 than they can into 24 equivalent voice channels. For example, a 9600 bit/s connection will occupy just 9600 bit/s of bandwidth (plus a small overhead) in a T-1 stream. Carried on a voice-grade line it will be equivalent to 64,000 bit/s—a standard voice channel.

If you ran 9600 bit/s data on 24 leased voice lines, you'd get only 24 data channels. If you run them on a T-1 in digital form, you can get up to 150 synchronous data connections at 9600 bit/s. The T-1 backbone of a T-1 network often carries both voice and data. In many cases either data or voice cost-justifies the entire network. The other "rides for free."

Dollars & Necessity

If you spend $1 million a year on a voice network, you're looking at a T-1 network. Spending $1 million a year means:

1. You have sufficient traffic. You can save money with your own T-1 network.
2. You need frequent changes in your network. You can make changes more easily if your corporation has a private T-1 network.
3. You probably have applications on your existing network which are absolutely critical to your business.

These applications in such industries as airlines, banking, hotel reservations, require your network to be available 100% of the time. If it goes down, you lose business—perhaps lots of business. Phone companies will reimburse you for line outages, but they will not reimburse you for loss of business.

Should Your Network Be Public or Private?

Two Rules of Thumb:

1. For a few large sites, build a private network.
2. For many small sites scattered geographically, best get service from a public network.

Most large corporations are a mix of both public and private networks, commonly called a hybrid network.

Virtually all the services you can buy from the carriers (switching, contention, E-mail, protocol conversion, etc.), you can do for yourself in-house. You can also choose which tasks you want to do and which tasks you want to let others do for you.

What your corporation chooses has to do with top management's "corporate culture" of owning versus renting, as much as it has to do with the strict logic of the particular situation. For example, companies headed by entrepreneurs prefer to buy. Companies headed by bureaucrats tend to lease.

The Privately-Owned Network

Plain-vanilla leased lines and customer-owned nodes that perform switching, multiplexing, etc. can do the same job as public providers. This was not always true. But the economics of telecom have become much more favorable with solid-state technology, VLSI, etc. Consequently, many companies are becoming their own long-haul carriers.

The relative costs of "bought" versus "made" networks depend on the size of the organization. Large size will usually favor an in-house network. For the largest organizations, the economics will practically force the decision to build some form of private digital backbone network.

The Public Option

A phone company will take full or partial responsibility for a corporate network. Some users say "We're in the insurance/car/fertilizer business. We don't want to also be in the telecom industry. Therefore let us do what we do best. Let the phone company do what it does best."

The public network solution supplies all the equipment and personnel as well as the lines. You and your company invest very little up front. The overall cost might or might not be higher. Most large companies don't figure the total costs, but simply the "out of company" costs. This often favors internal (privately-owned) networks. It often depends on who's trying to build which empire.

Of course, once hired and trained, internal expertise becomes a corporate asset. These people continue to pay dividends by periodically optimizing the mix of services and equipment. In most cases this will mean that a knowledgeable customer will design his own least expensive network to meet his own specific and on-going needs. You cannot expect the same diligence towards cost-cutting if the network is managed by a public carrier. That would cut their revenues.

Hybrid Networks

In a hybrid network, the backbone trunks may be privately-managed. The spur trunks to smaller sites are then managed by the telephone company. But like all generalities, this one is not great. In some hybrid networks, the backbone is a shared public facility and the spurs are privately- owned short-haul microwave carrying T-1 trunks from outlying branch offices.

The point of a hybrid network is that it is meant to "have the best of both worlds." Ownership is good where it makes sense. Taking monthly service from the carriers is good where that makes sense.

Obscuring the economics (and thus making it difficult for the author to make generalizations) are "new" events, like the willingness of some public carriers to guarantee low T-1 line prices and high T-1 line quality for five years. Also, some public carriers are even allowing customer-owned equipment in telephone company-owned central offices.

Whose Job?

Who will take the lead responsibility? Who will design the network? Who will maintain it, with fault isolation and repair dispatch? Who will take charge of the inevitable reconfigurations? Will it be the local carrier? The long distance carrier? The switch or multiplexer vendor? An independent consultant? Or you, the user?

Major corporations typically opt for their professional telecom manager. This person reports to the head of MIS, one step from the president.

To rely on outsiders is to place an important, perhaps vital, competitive company resource in the hands of people whose own business can come before the goals of the network users. Therefore, top management often wants the management and diagnostic skill needed to run a network to be developed, nurtured, maintained, and guarded in-house.

Should data be mixed with voice?

Rule of Thumb: Voice is easily transported on a digital T-1 line. It is harder and more expensive to integrate data into a voice PBX. Therefore, you generally find voice and data integrated for transmission, separated for switching.

Key Questions:

- Which internal group is responsible for voice? Which for data? Will these departments allow them to be mixed? Or will each insist on "control?"
- Which type of traffic will dominate on the T-1 backbone? Will one department allow itself to be dominated by another? Will they both tolerate the creation of a third department to run the backbone network?

Practically no company fully integrates voice and data. The most common situation is where a digital backbone network of T-1 circuits offers the equivalent of tie lines to the voice and data services. Telephone calls are still switched by PBX's. Data circuits still terminates at computers.

Fortunately, network management software with partitioning offers a way to please everyone:

- the MIS manager who wants control of the data lines;
- the telecom manager who wants to control voice lines; and
- the president, who wants to save the most money by combining all of the company's communications on a single, efficient network.

(See Chapter 8 for a discussion of partitioning.)

Which Network Topology;?

Rule of Thumb: Tariffs generally favor a star topology within LATA's. For interLATA networking, one node per

LATA is favored in a mesh layout. A mesh offers many alternate routes for reliability. While mainframe computers often require multidrop lines, these logical connections need not determine the physical topology.

What might the network look like: star vs. mesh; multidrop vs. point to point; etc. The distribution of user sites may determine the topology, but in most national corporations there are several possibilities:

Point to Point

The simplest possible network is a single line between two nodes. This has been a very popular arrangement in voice grade lines—a modem and one other device at each end. But few terminals or computers can use a full T-1 rate. Almost all T-1 lines end in hardware that splits the bandwidth into slower, standard rates. This device commonly is a T-1 multiplexer (Fig. 6-1).

Any T-1 multiplexer can perform in this topology. Only one data link is needed. A second, standby link is available on some multiplexers and channel banks for higher reliability. T-1 multiplexers are starting to look like data switches or a voice PBX. Also data switches and voice PBX's are now available with direct (i.e., no channel banks needed) T-1 interfaces.

The cost of even one T-1 line can be so significant it must be planned carefully. Examine routing options to determine the best trade-off between local and long-haul charges. A normal practice of interexchange carriers is to route a line through the

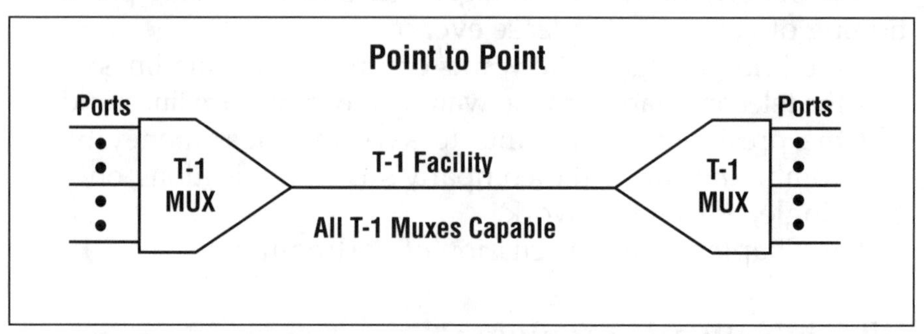

Figure 6-1. *Point-to-point network is the simplest possible. Most T-1 links in this form terminate in multiplexers to split the large T-1 bandwidth into the slower rates usable by standard equipment.*

closest serving office (SO) on each end. This route minimizes the intra-LATA mileage charge for the local loop. However, it may increase the length and cost of the long-haul IOC (Fig. 6-2) when the closest serving office lies in the opposite direction from the next customer site. It may be less expensive to request service through a serving office located "as the crow flies."

In other words, there are "tricks" in getting a line from your office A to your office B across the country. These "tricks" are akin to airline flying. For example, going New York to Las Vegas is often cheaper flying New York—Los Angeles—Las Vegas. Even though you "backtrack" nearly 20%, you can save 40% over a "direct" fare routing New York-Vegas. "Tricks" like this in networking are becoming more common—especially with the breakup of the Bell System, and the willingness of Bell operating companies to "bargain."

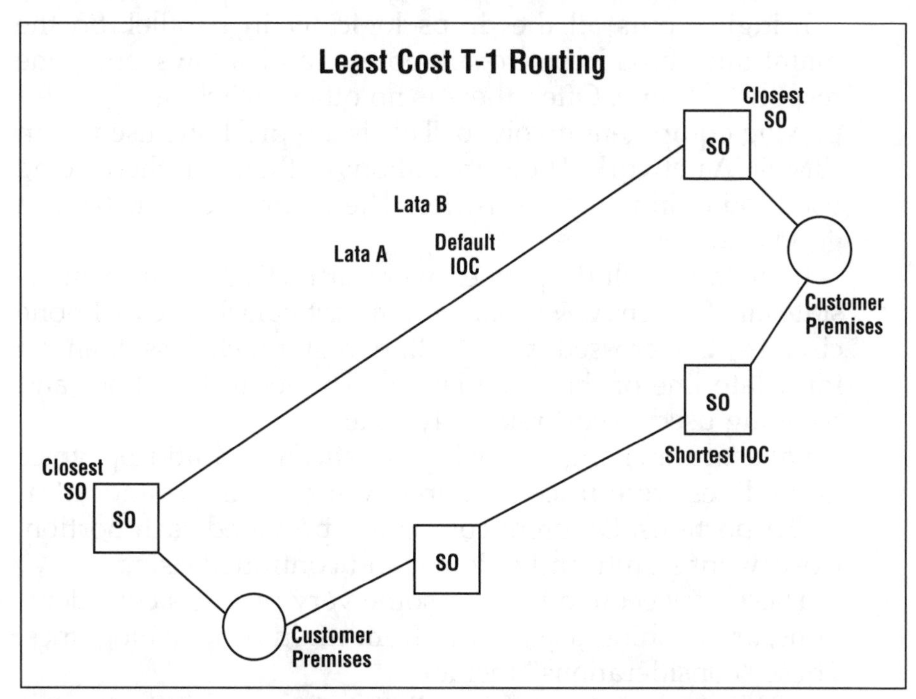

Figure 6-2. *The least cost routing for a T-1 circuit may not pass through the closest serving office of the inter-exchange carrier. That serving office (SO) may increase the interopffice channel length (IOC) and result in a higher total line cost. Least cost routing in telecommunicaitons is not always intuitive, or the way the crow flies. This is because of strange variations in the prices for interstate, intrastate, and intra-LATA calling.*

Multidrop

For many decades the dominant network design was "multidrop." A multidrop line is often used with a front end processor (FEP) on a mainframe computer. A multidrop line starts from the FEP as a leased analog line. It extends through multiple central offices. A "bridge" junction in a CO connects this main line with a local loop which runs into a customer site (Fig. 6-3). Everything was analog originally, but DDS offers digital versions.

Each drop circuit typically connects through a modem (or CSU/DSU) to a cluster controller (star configuration controller) which connects to several terminals. A cluster controller responds only to messages addressed to its terminals. Thus there is no need to separate traffic on each drop line. All drops hear every message from that port on the FEP.

Bridging puts all the drops logically in parallel. So the mainframe broadcasts to all points, and allows only one response at time. Often there is no other switching or multiplexing equipment involved. This is a typical line used in an IBM SNA network. There are zillions of them out there being used today in many industries. The insurance industry is a good example.

Interstate multidrop lines were attractive because interstate tariffs from AT&T made them particularly cheap. Phone circuits that crossed a state line cost much less than an intrastate line of the same length. The local drop lines and bridging used to cost relatively little.

Divestiture changed pricing drastically. Multidrop voice grade lines were priced separately for the inter- and intra-LATA portions. Different companies provided each portion. Costs went significantly higher, and continued rising.

These new economics, and some very practical considerations, are pushing users away from multi-drop analog lines. These "considerations" include:

- The coordination of installation is more difficult, now involving as many as 10 vendors of lines and equipment.
- Trouble-shooting leads to finger pointing more often, and the large number of vendors increases the number of "others" to point to. (Exponentially.)

- Growing demand translates into additional drops on existing lines; more drops slow the response for all users.

What is replacing analog multi-drops? For many large users, a T-1 backbone network is the answer. Early users found no multidrop arrangement available in T-1 nodes, so they converted to logical point-to-point connections from the FEP to various cluster controllers. The change required addi-

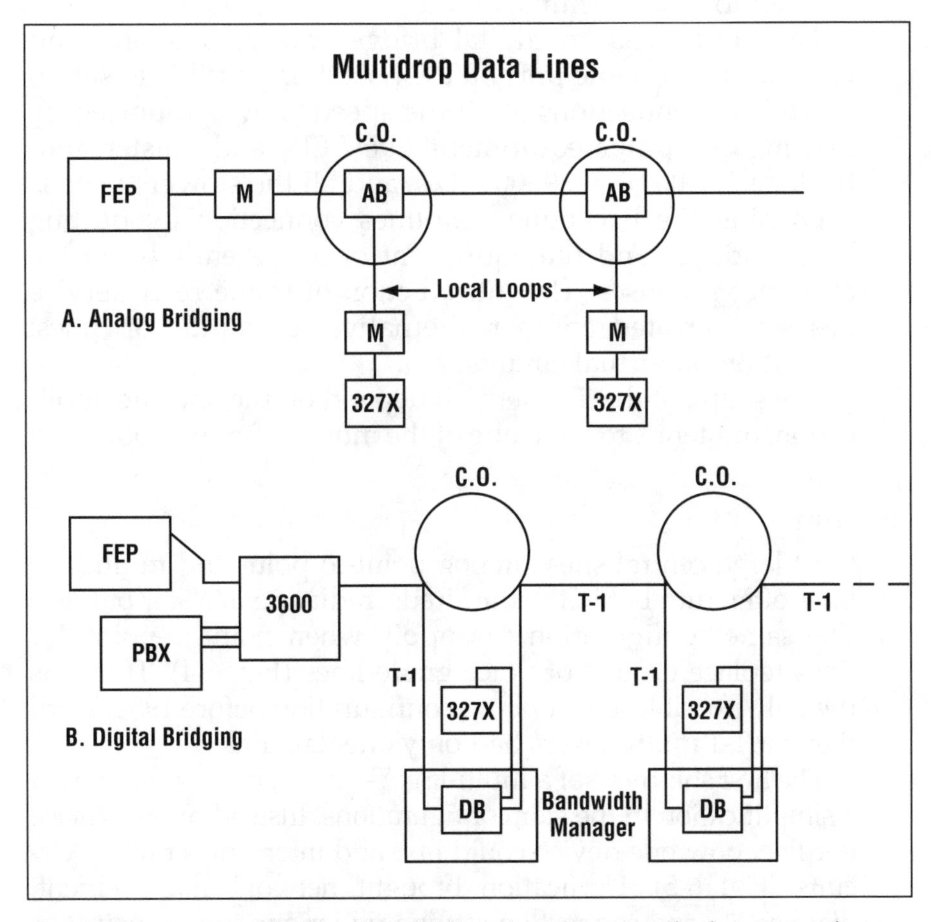

Figure 6-3. *Analog multidrop lines (a) have long been popular between mainframes and remote terminals. The front end processor broadcasts data to all drops, but each cluster controller responds only to messages for its terminals. Cluster controllers transmit back to the FEP one at a time, not to each other. Therefore no additional multiplexing or switching is needed. Digital emulation of analog multidrop lines (b) builds on the T-1 backbone. Bridging is in the node. This circuit behaves in all respects like its earlier analog counterpart.*

tional FEP ports and (horrors!) generating a new build of system software for the CPU and FEP.

Now you can get digital bridges in the T-1 nodes. They emulate the drops in analog lines:

- all the terminal controllers receive all broadcasts from the host mainframe computer.
- digital bridges pass data from terminals to the FEP, but not to other terminals.

Thus terminals on digital bridges can all respond and transmit to the same port on the host. It is possible to set up multidrop connections at all the speeds now supported by existing computer equipment like FEPs and cluster controllers: 56,000 bit/s, 9600 bit/s, and all the slower speeds.

Owning the backbone simplifies connection by putting local bridging and the equivalent of drops entirely on the customer premises. The introduction of frame relay service has since created another alternative for multidrop lines: emulation on virtual circuits.

The shape of the T-1 net will depend on the specific application, but tends to have one of the more complex topologies:

Star

At large central sites, analog point-to-point and multidrop lines often number in the hundreds, radiating in a star pattern. The same configuration can apply when point-to-point T-1 lines replace groups of voice grade lines (Fig. 6-4). This was the only possible T-1 hubbing configuration before 1984. Until then all T-1 multiplexers had only one data link.

The first shipment of a multi-link T-1 node, in 1984, permitted a simplification in the star configurations. Instead of one device per line, now one device could hub and interconnect many circuits (Fig. 6-5). Unification brought network management, diagnostics, and connection routing under one control point.

Digital Access and Cross-connect Systems (DACS) and other major hubbing nodes tend toward star formations. For very large sites these multiplexers may be daisy chained at one location to expand the number of T-1 connections, or the larger models of the equipment installed.

Economics favor stars in some situations. In the past, inter-

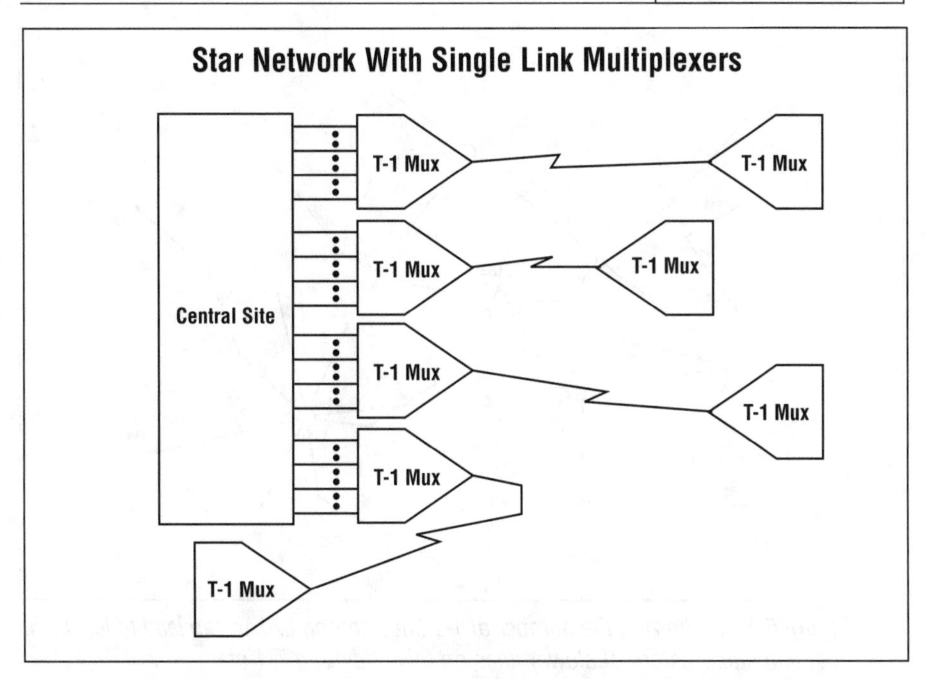

Figure 6-4. *Star network with single link multiplexers resembles the arrangement of multi-drop lines common with mainframe computers.*

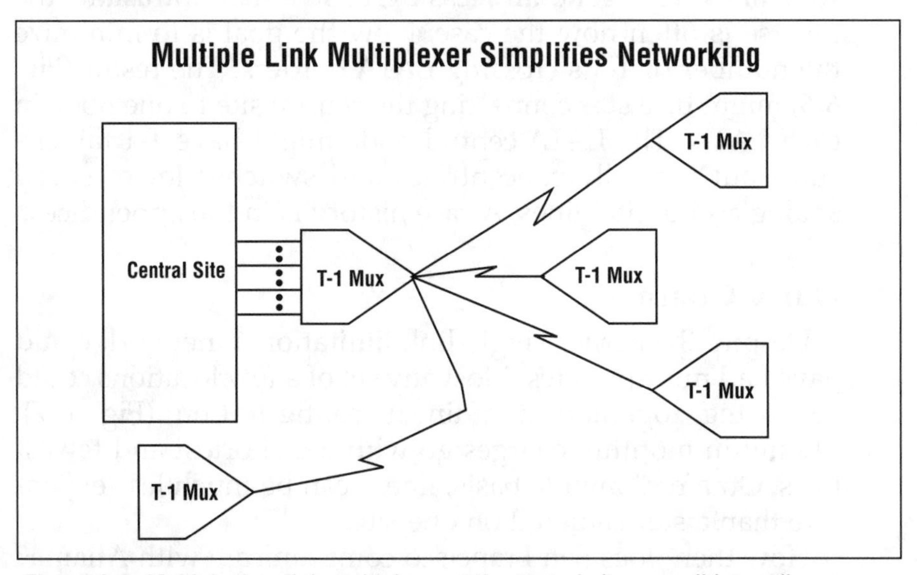

Figure 6-5. *Multiple data link multiplexer at the central site consolidates all communications. Integration makes network management easier. Bypass at the central node also permits remote locations to communicate with each other without involving the host computer.*

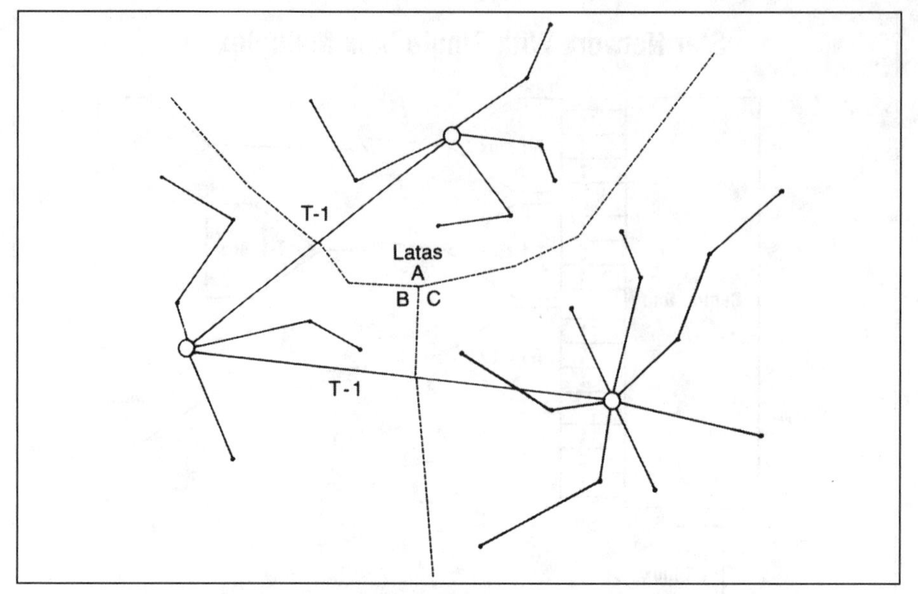

Figure 6-6. Minimizing the number of T-1 lines among LATAs can lead to least line cost and star clusters of slower lines on thle ends of a T-1 star.

state lines were generally less expensive than intrastate, the reverse is often now the case. Now the goal is to minimize the number of T-1's crossing LATA borders. The result (Fig. 6-6) might be a star connecting the central site to one node in each LATA. The LATA central node might have T-1 tail circuits, but typically concentrates and switches lower speed analog and digital lines. A case history is in the appendices.

Daisy Chain

Despite the earlier single-link limitation, a network could have had multiple sites. Most any set of a few locations could be strung together in a linear configuration (Fig. 6-7). Minimum monthly charges go with the shortest and fewest lines. On a nationwide basis, linear can be much less expensive than a star centered on one site.

How then does San Francisco communicate with Atlanta? There is no need to run a separate line between SFO and ATL if the nodes at CHI and NYC can take data coming in on one T-1 line and send it out on the other. As mentioned in the earlier discussion of CPE, the taking in and sending out is done with "data bypass" or "Drop and Insert" (D/I). This discussion will assume them. As covered elsewhere, the intermediate node could also rely on back-to-back multiplexers, but this involves unnecessary demux/mux operations which require additional equipment, space, etc. Multiple muxes also fragments the network. One-piece nodes can be gathered under a single control system.

Dual link multiplexers with bypass or drop/insert are ideally suited for linear networks along "thin routes" like pipelines, railroad tracks, borders, etc. Each station can connect to the adjacent stations on either side.

Ring topology

The linear shape can be economical, but a failure in any line segment splits the network at the break. The ring network (Fig. 6-8) overcomes this drawback. A ring provides a path between nodes even after the loss of a circuit. If the ring

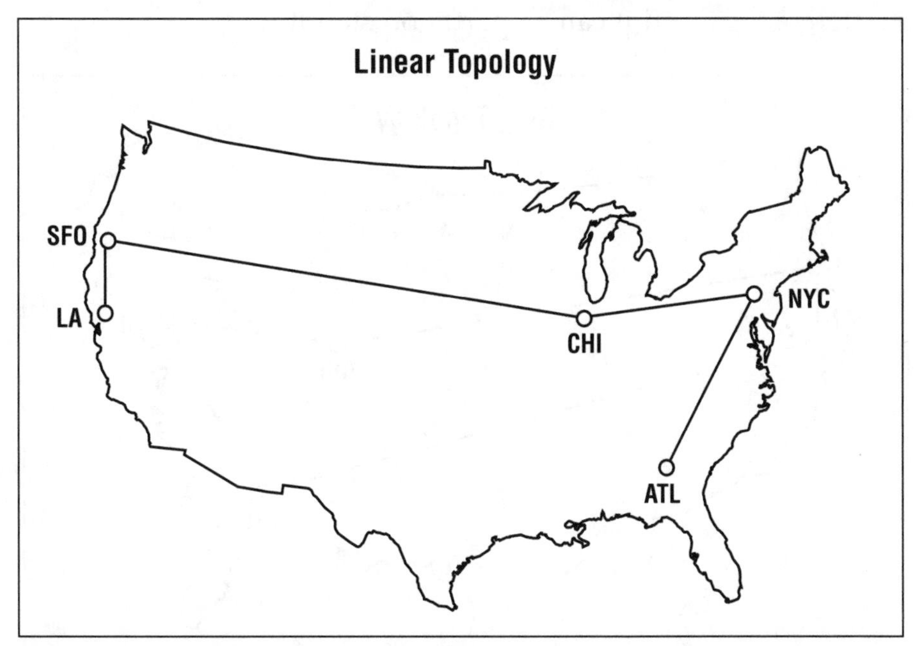

Figure 6-7. *A linear network connects multiple sites with a minimum number of circuits, an advantage when circuits are expensive. Traffic between distant nodes must be passed through intermediate nodes.*

is broken by a line failure, there will be an alternate path by following the ring the other way round.

Connections (the virtual circuits carried by a link) will be broken if that link fails. To take full advantage of the alternate path, the nodes must offer Automatic Alternate Routing (Fig. 6-9). AAR is a switching function made possible by data bypass. Recognizing a failed circuit, the network automatically initiates a search for replacement paths. Most networks will start making the new connections within seconds. Quick reconnections can prevent losing calls in progress or "timing out" of computer sessions. See also the chapter on network management.

If bandwidth on the remaining links permits, broken TDM connections are re-established over them. The simplest way to ensure enough flexibility for alternate routes is to build multiple rings. This requires some forethought. Is the need for reliability so great there must be at least two T-1 circuits at each node? Would a 56 kbit/s line be adequate for emergencies? Would dial-up lines serve? The ring may be up and running at all times, but it can run at different speeds on different legs, and it can be partly on standby.

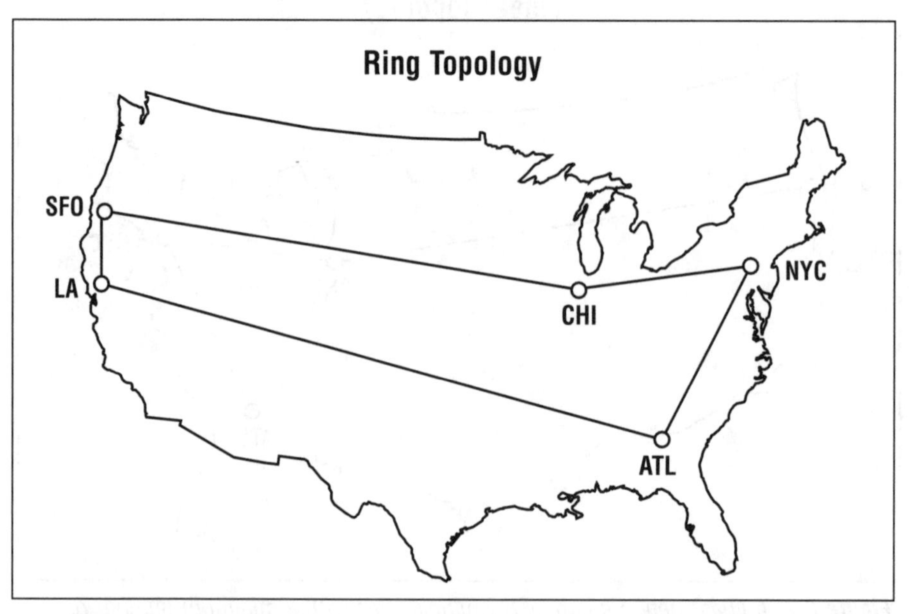

Figure 6-8. *A ring topology ensures a path between any two nodes even after the loss of a T-1 circuit.*

Mesh

Another way to connect ATL and SFO would be to set up a direct (point-to-point) circuit. Carried to an illogical conclusion, every pair of sites would then be joined by a direct line. The result is a fully connected mesh. This is expensive. Very expensive.

Practical topologies for national organizations often fall between a ring and a mesh (Fig. 6-10). Where the backbone has T-1 links, a sparse mesh or interlocking rings offers the most economical arrangement that keeps redundancy of routes among important nodes.

For additional protection against loss of a line, normal traffic between two nodes should be dispersed over as many alternate routes as practical. This way, the loss of a line interrupts only a fraction of the connections between any pair of nodes. For example, the traffic between Atlanta and Chicago would be split. Part would go through New York, part through Dallas and Denver. If the NYC-CHI line were lost,

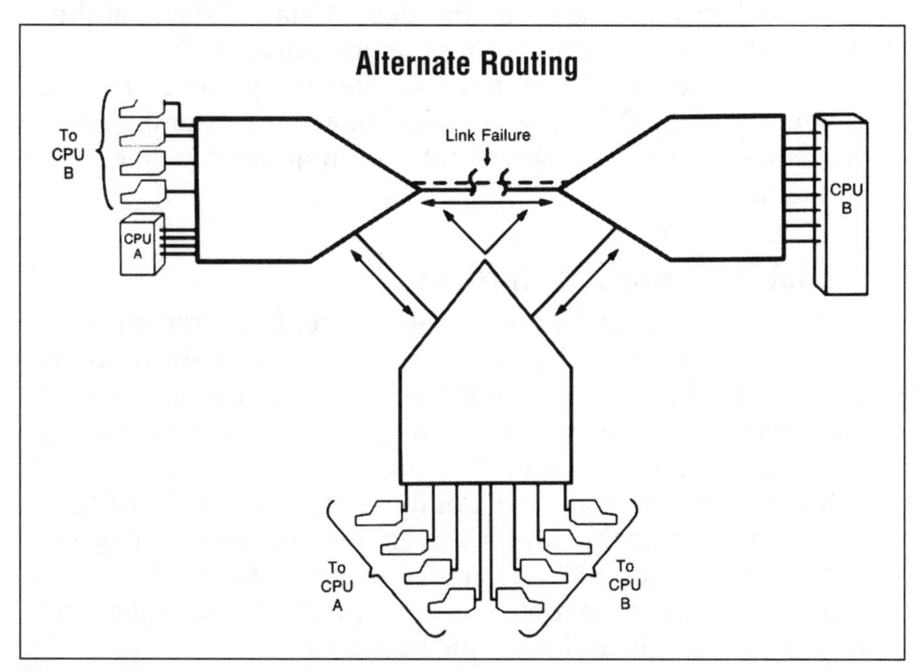

Figure 6-9. *Automatic alternate routing feature finds a path for connections, broken by a link failure, over remaining links. Channels are bypassed at the intermediate node. Users may be restored within seconds.*

half the traffic would continue uninterrupted.

Route diversity is one reason larger users may not always want to let their T-1 nodal processors pick the route. These nodal processors could put all of the traffic between two points on a single path. If that path were to fail, all the connections supporting an application or an entire site could be broken at one time.

Design Effects On Network Topology

The management objectives of each network determine which design factors carry how much weight. First comes the necessity to do the communications job. The need may be for connectivity through switching, mixing voice and data, or reconfiguring the network to match shifting demands.

Second in importance at the design stage is availability. This is the first concern during the operations stage.

Economy, especially initial cost, generally comes further down the list. Deciding to consolidate an organization's functions on a T-1 backbone network requires a major commitment.

Reliability and Availability

One part of reliability is low error rate. Each transmission medium has an inherent error rate: terrestrial wire, microwave, fiber optics, satellite, etc. Most important is what the carrier will guarantee, with financial compensation if errors rise above a set level.

The other part of line reliability is the percentage of time the circuit is available for use. Even the most reliable carriers will not guarantee 100% uptime. For applications where outage time is very expensive, the prudent manager will arrange for single and multiple backups.

Redundant T-1 lines, a working circuit and a hot spare end-to-end (Fig. 6-11), offer the ultimate in full-capacity reli-

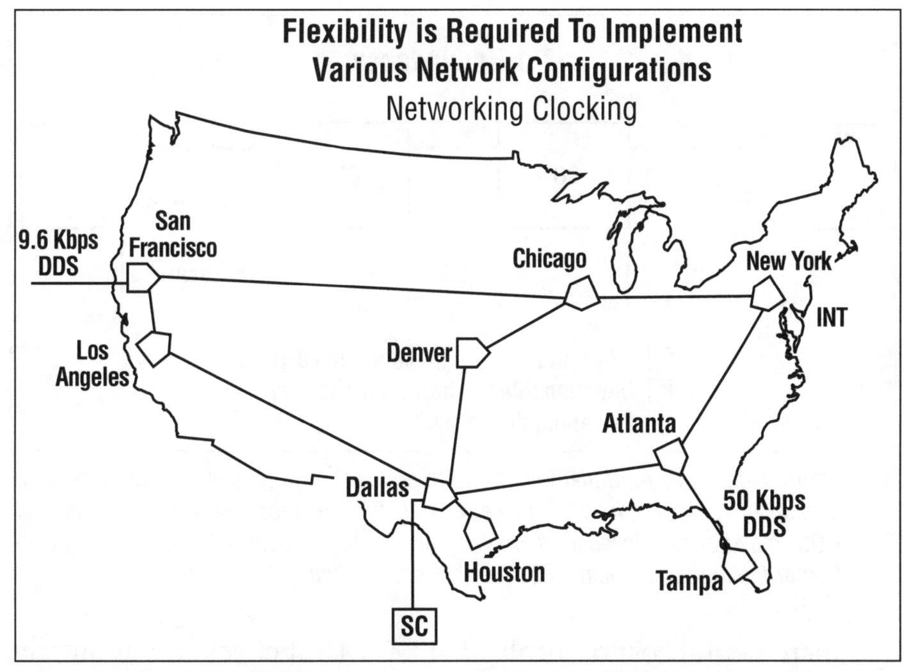

Figure 6-10. *Practical networks for large, dispersed corporations often combine multiple rings to approximate a mesh. Rings offer alternate routes to reestablish connections lost by a circuit failure.*

ability — and the greatest expense. Fortunately, the carrier normally protects the inter-office channel (IOC) or long distance portion of all circuits with spare capacity and automatic backup. (But you'd better check.) So usually the network operator need worry about only the trip from the inter-LATA carrier's serving office to the customer premises. This may involve the local loop and a LATA distribution channel (LDC) between central offices.

Since the local loop is the line most vulnerable to disruption, it deserves an evaluation for potential outage. A second T-1 local loop can cost as little as a few hundred dollars per month. For less than the utmost reliability, redundancy on only the local loop part of the route could be a very reasonable cost.

The cost of the local distribution channel (LDC — the one between central offices) is based on distance and can be significant within larger LATA's. If the LDC legs are part of a carrier's backbone, they may be protected sufficiently for the application. The gain in reliability may not be worth the

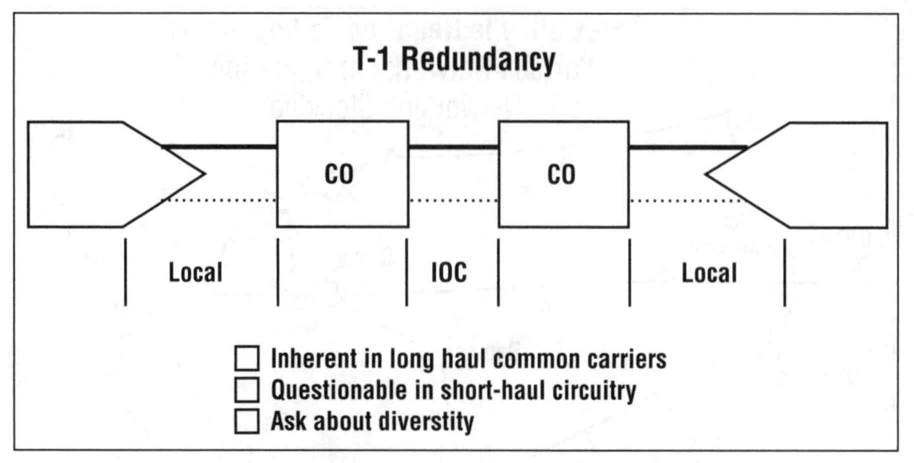

Figure 6-11. *Fully redundant lines end to end is the surest way to have a working backup in case of a failed T-1 line. Paying for the duplicate interoffice channel (IOC) may not be necessary as the carrier normally provides backup for that portion of the network. Local loops are the responsibility of the user.*

incremental cost of duplicate LDCs. (But check. Can you pay a little extra for carrier redundancy?)

To apply either type of local line redundancy requires a central office switch to transfer the IOC (or LDC) between local loops (Fig. 6-12). This capability is tariffed separately, and may be either manual or automatic. Automatic backup protection recognizes loss of signal. It then goes to the spare without intervention. At some telcos this service will also switch over when it measures a high bit error rate.

This service works well with some models of T-1 nodes and CSUs that fall back to a hot spare by themselves on loss of signal. Otherwise, a second switch is needed at the customer site. Having the node recognize a fault and switch to the spare T-1 link will save the charges for the CO transfer switch and an extra analog line to activate it.

The manual version, as its name implies, requires an action by the user — to close a contact. This contact is a simple SPST (single pole, single throw) normally open switch (supplied by the customer). It bridges a separate leased local loop (priced separately) made up of continuous twisted pair from SO (serving office) to CPE site. The network manager can then choose which T-1 loop to use at any time.

Automatic fallback by the multiplexer offers a way to reduce the cost of line redundancy in some situations. If both T-1 local loops are bridged to the long-haul carrier at the CO (there are digital bridges that perform the same function as analog bridges), and the two local T-1 lines take different paths, then communications should survive the loss of the working circuit.

These "if's" are fairly serious, however. Digital T-1 bridges exist, but which carrier, local or long distance, will provide it? At what cost? Diverse routing may not be readily available for the practical reason that alternate cables do not exist. The point in buying a spare circuit is obscured somewhat if that "spare circuit" is part of the same cable or travels in the same conduit as the primary circuit. The vast majority of local loop failures are caused by the destruction of cable. Putting multiple circuits in one cable provides practically no redundancy. Some users back up the copper local loop with microwave or infrared transceivers.

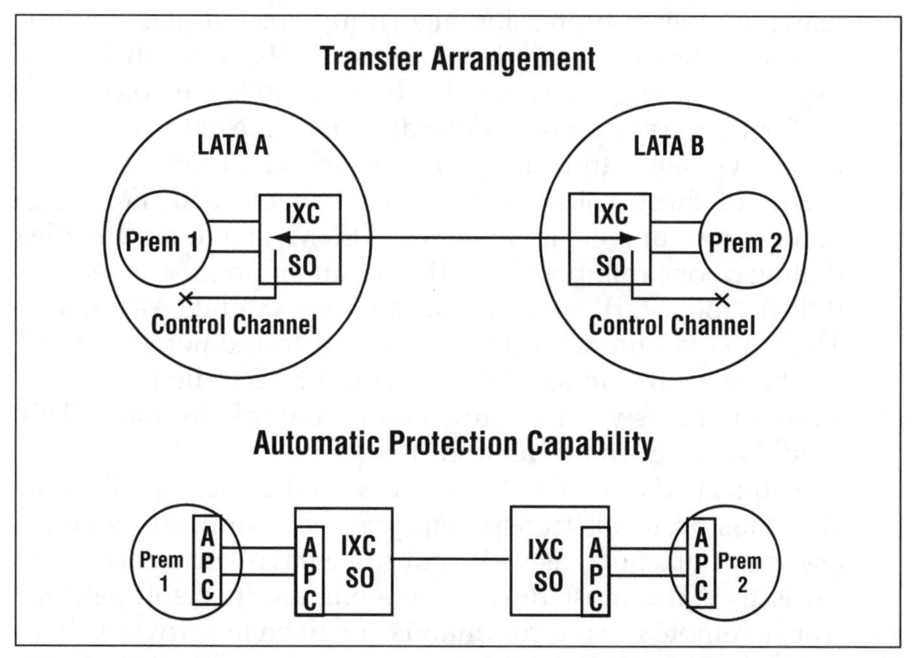

Figure 6-12. *A transfer switch at the central office brings the backup local loop into the connection between the IOC and the customer site. The switch may be manual (A) or automatic upon loss of signal on the operating circuit (B).*

CPE Multiplexing, Switching, Recovery

Most of the nodal processors discussed in this book are called T-1 multiplexers by the trade press. The name is not quite accurate as they usually do far more than "multiplex." They also "switch." The need for connectivity on demand has been met by varieties of virtual circuit switching that resemble a digital matrix switch. And because the network carries the very life of a business, the better quality T-1 nodes are able to recover from faults. They heal themselves, to an extent, to maintain network availability and the flow of information.

The essence of any switching or multiplexing node is flexibility. The network operator needs as many options as possible in types of inputs; their number and speeds; and the allowable mix.

In terms of networking, voice inputs offer a prime example. Depending on the application, the optimum treatment may be 64 kbit/s PCM, 32 kbit/s ADPCM, or a lower speed using another compression algorithm. For internal communications, or over long distances where bandwidth is very expensive, it might be desirable to trade off some voice quality for a large saving in bandwidth. Here one of the newer 16 or 8 kbit/s algorithms might be the best choice.

Any of those voice interfaces is available with E&M signaling, the universal standard. (E&M signaling handles dialing phone numbers, etc.) But for an off premise extension (OPX), "loop start" for a standard phone should be available. Thus, a small branch office connected to the network need not have a PBX on site. Phones could be handled as extensions on the switch at any other network location. This would resemble a private Centrex.

Being TDMs in almost every case, T-1 nodes handle synchronous protocols transparently. Some manufacturers offer asynchronous ports as well, and some have isochronous I/O cards too. The most flexible versions all offer soft settings from a remote site (all commands are given in software; there are no hardware switches).

These topics are covered in detail in the previous chapter.

Similar Central Office Functions

Some years ago, a local loop to a central office could be connected to an analog port on a switch for entry into the PSTN (Public Switched Telephone Network) or another analog line as part of a leased circuit. Then came digital services, like DDS, that correspond to the fixed analog leased line. Then came the transfer arrangement to switch between a main and a standby line. Then came more services and functions, growing to a flood in reaction to competition.

Telephone companies are now offering central office multiplexers and switches. They perform many of the functions available from CPE (i.e. your own equipment). They give the network operator additional options in designing and running a complex network:

Central Office Multiplexing

M24 multiplexing (24 voice channels in a T-1 circuit) is the standard voice channel bank. There is also a digital version to pass data channels. As an information funnel, the M24 channel bank can be used facing either direction, that is, with its "neck" or aggregate side toward the network or toward the customer loop.

In the case where it faces the customer (Fig. 6-13), the M24 mux could simply fan out 24 voice or data channels from a T-1 local loop, letting those channels access various local and long distance services. AT&T has tariffed several options like this under the name "Integrated Access." There is no need to use a full T-1 from the interexchange carrier.

As all digital channels are at the DS-0 level, or in analog form, they can be routed one at a time to almost any service desired, including dedicated 56/64 kbit/s circuits, switched 56 kbit/s digital, SDN/VPN/Prism, voice grade private lines, WATS, 800 service, etc.

Dealing with the local telco is different. Until Bell South tariffed integrated access in early 1988, all telcos had insisted

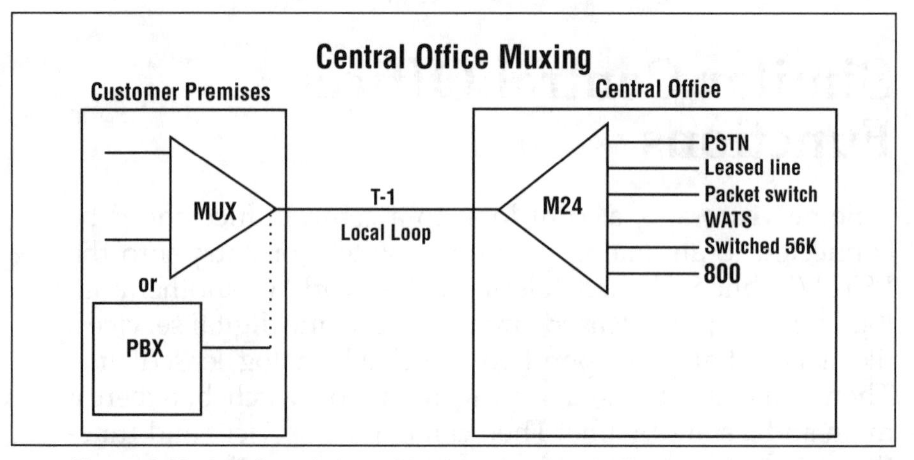

Figure 6-13. *A dedicated channel bank in the central office is one of the newer services inter-exchange carriers offer. The local T-1 is fanned out into as many as 24 voice and data channels which may be routed separately, in other services. These can include any mix of WATS, PSTN ports on the switch, leased analog lines, switched 56 kbit/s service, or any other service available through that serving office.*

that each class of service (leased lines, long distance access, local dial service, etc.) be on a separate facility. Special construction tariffs were available to integrate access, but that route is too expensive for most users who need it. Market pressure should have lead most local carriers to offer this service by the early 1990s, but some didn't understand the concept at this writing (1997).

Operating Companies responded quickly to demands for Fractional T-1 in the local loop. That flexibility and response to market demand should see integrated access a reality eventually.

AT&T has tariffed Megacom, for 800 and WATS, over T-1. MCI, Sprint, and the other long distance companies have similar services. AT&T's Megacom gives a user T-1 access directly to an AT&T 4ESS toll switch for WATS and 800 services. The responsibility for getting his T-1 line to the AT&T toll switch is the user's. It can be arranged through the local phone company. It can be arranged through a microwave link. Or it can be arranged by the long distance carrier through a "bypass" carrier. Most large cities have or will soon see a metropolitan bypass carrier based on optical fiber. In some jurisdictions the local cable TV franchise can become a carrier by allowing users to attach RF modems for a T-1 channel.

Turned the other way at the central office, the M24 channel banks combine many voice grade local loops onto a T-1 for the inter-LATA portion of the connection. For example, take a cluster of your offices within a LATA. None warrants a T-1 individually, but if taken together, by hubbing on the central office (Fig. 6-14), they can support (i.e. economically justify) a T-1 circuit. Even two or three voice grade lines can justify fractional T-1 from the interexchange carrier.

This arrangement can reduce the inter-LATA line charges. The local loops connecting to that T-1 connection could be tie lines to PBX's or leased analog lines carrying data as modem signals. In some cases you can get DDS-like data ports on the channel bank at the central office to provide digital service to a CSU at your premises.

You would normally place your larger offices on the main T-1 network. These sites therefore could usually justify an on-site T-1 node which would take advantage of voice compression and also handle data channels more efficiently.

Voice Compression and ADPCM

M44 multiplexing is mass conversion of PCM to ADPCM (Adaptive Differential Pulse Code Modulation). This function is performed by a Bit Compression Multiplexer (BCM), an ADPCM transcoder, or by dedicated hardware chips built into a T-1 multiplexer node. The BCM converts two PCM encoded T-1 links to a single ADPCM T-1 circuit. AT&T's encoding scheme carries up to 44 or 48 voice channels, full duplex on each T-1.

AT&T's original 44 voice channel scheme for ADPCM reserves four voice channels for signaling. Each of these delta channels (32 kbit/s each) carries signaling for eleven 32K voice channels. There are four such bundles of 12 channels in a T-1 frame. In CEPT countries, the 2.048 Mbit/s service carries five bundles created by an M55 multiplexer.

Therefore, individual ADPCM channels cannot be switched — they would become separated from essential signaling information. The four ADPCM channels devoted

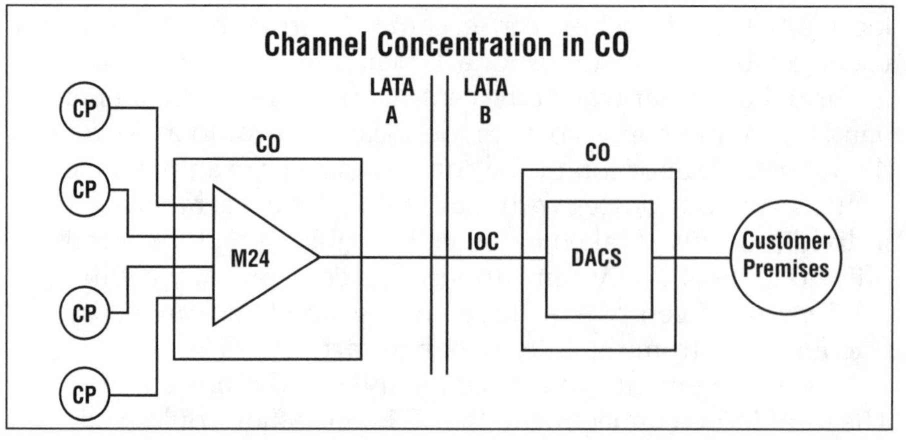

Figure 6-14. *Many DS-0 and analog lines may be collected at a serving office for transport over a single T-1 circuit to a distant customer premises. The channel bank is provided under tariff by the long haul carrier.*

to signaling and synchronization mean the M44 cannot quite double the 24 channels of an M24. AT&T's later BCMs supported 48 channels in a mode that depended on "bit robbing" rather than separate signaling channels.

The Bell System has had a published "practice" on ADPCM for years. The M44 is based on it. M44 multiplexing has not been widely used outside of telephone companies, who use the technique on voice circuits to maximize the efficiency of installed cable — and to avoid installing more wire — particularly on international circuits.

Other vendors' implementations of ADPCM may put slightly more or slightly fewer channels in a T-1. What, in effect, is voice compression (44 channels on one T-1) is now offered to customers by carriers as well as equipment vendors.

M44s (and there are now "M96" boxes) may be located in two central offices (A in Fig. 6-15), or the CO function may be mated to CPE (B in that Figure). If there is an M44 on customer sites at both ends of a T-1 (C) the digital signal is passed by the central office transparently. There is no need for an M44 in the CO.

To switch one of these channels (from M44 compatible CPE) onto the PSTN or to route it separately through a DACS, a second M44 is required in the CO to put the signal back into DS-0 format. M48 format, which includes signaling "in band," allows a DACS to switch or route pairs of voice channels among "M48" multiplexers.

ADPCM, being designed for voice compression, does not handle data as well as the M24 or a T-1 data multiplexer. Some M44s, or transcoders, can pass 56 kbit/s data but require the equivalent of two ADPCM voice channels to do it; that is, 64K bit/s. Modem signals above 4800 bit/s generally will not pass through an ADPCM channel, so the M44 is not appropriate to concentrate leased voice grade data lines, and will slow facsimile transmissions. The M44 was designed for pure voice and has its best applications there.

Later algorithms like CELP can compress voice to narrower channels. Because of the hardware involved, a CELP transcoder will often "demodulate" facsimile signals to a data stream and thus offer a normal fax transfer speed (9600 bit/s).

DACS and Switching

Another internal telco function available as a separately tariffed service is switching through a DACS (Digital Access and Cross-connect System). The basics of its operation are covered in Chapter 4.

DACS with CCR combines with M24 and M44 multiplexing to create network hubs in central offices (Fig. 6-16).

Fractions of a T-1, in multiples of 64 kbit/s, are routed among any of the customer premises. Within the limits of the bandwidth under lease, the number of 64 kbit/s bundles on each path can be set arbitrarily. By drawing on switched 56 kbit/s and T-1 services, CCR can accommodate occasional peaks in traffic demand beyond the permanently leased bandwidth.

For the utmost flexibility, all of the long haul portion of the network could be on-demand or switched digital services. Leased circuits would then be in place only between customer premises and long-haul serving offices. The economics favor leased long-haul lines except for less than about two hours of connection time per day over long distances. Shorter distances favor leased circuits even more strongly.

The drawbacks of CCR lie in its shared nature. The central control computer acts on instructions in the order received. The time delay in the queue may be 20 or 30 minutes. While a pend-

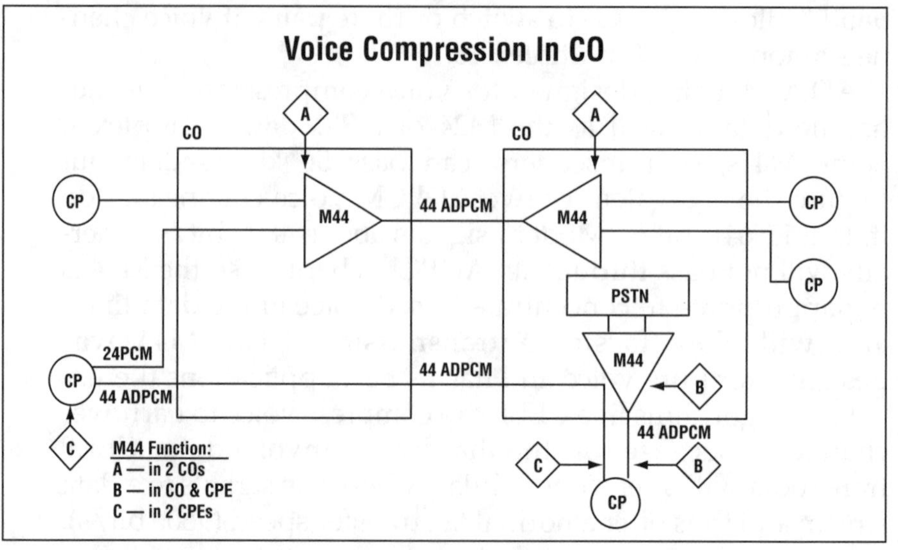

Figure 6-15 *Central office M44 combines two local T-1 loops for efficient interoffice transport (A). When CPE compresses voice to ADPCM, the CO M44 must split the bit stream into two lines with PCM format (based on DS-0 signals) if individual channels are routed by switches or DACS (B). If CPE at both ends compress voice in any way, and the entire T-1 is dedicated to these two sites, no M44 function is needed from the carrier (C). However, if individual channels are to be switched in any way, they must be brought to DS-0 form.*

ing change can be canceled before it takes effect, there is no way to know exactly when it will take effect. Of course the largest user will be the carrier itself, principally using it for provisioning and testing. These functions may have to claim priority.

Single-Ended T-1 Access

Early T-1 networking was based on point-to-point T-1 lines between multiplexers on customer premises. That is, the user asked the carrier to provide a transparent pipe; what went into that pipe was determined entirely by the user.

There was always a small proportion of T-1 "lines" that ended in the local central office, most often on the class 5 switch providing dial tone for POTS voice.

Carriers have added more and more services, some based on entirely new networks. Frame relay and the Internet are two examples that proved very popular.

As a result, hundreds of thousands of user organizations find they need the capacity of T-1 bandwidth to get the services they want. By 1997, the number of T-1 local loops exceeded 1 million.

Where are they being used?

Internet Access

The large graphics files and sheer quantity of information available on the World Wide Web and USEnet make 56 kbit/s barely adequate for even a medium-size business location. An office that hosts a popular WWWeb site (server) almost needs a T-1.

The nature of the Internet, based on Transmission Control Protocol/Internet Protocol (TCP/IP), usually leads to an Ethernet LAN at a user's site. Most providers of Internet service (ISPs) bundle a router, the local loop, a leased line to their router, and the Internet service itself into one monthly charge.

The cost of the router is often higher than necessary. Only the simplest routing functions are needed at sites that have a single LAN and no connections to other sites. For example, since there is only one possible path for most users to reach the Internet, the router can have a single entry its address table: the ISP's router. One major ISP turns off the Routing Information Protocol (RIP) that lets routers learn about available paths and the addresses of nearby devices.

In most cases, the smallest, simplest router will work.

Frame Relay Access

An attraction of frame relay service is its ability to carry all data protocols, LAN and legacy, to most cities in the world. As a packet-based technology, and the likely successor to X.25, frame relay offers all the benefits of statistical multiplexing, shared resources, and carrier maintenance of the backbone network. In addition, it has proven quite reliable, stable, and offers a relatively low latency and transit delay compared to X.25 services.

Add in the postalization of rates (no mileage charges), and the picture is hard to beat.

So the popularity of frame relay is understandable. When a popular service takes on all the communications needs at a site, the bandwidth requirement quickly moves up to T-1.

Frame relay is offered at port speeds of 56 kbit/s, fractional T-1, and full T-1 or E-1 speeds. This is the port speed, for which there is a scheduled charge.

Fractional and full DS-1 'ports' are provisioned on T-1 (or E-1) local loops. The loops are exactly the same as those carrying point-to-point leased T-1 lines. The customer needs to provide a standard CSU.

It is the "terminal equipment" that speaks frame relay. FR is synchronous; the interface most often is V.35 electrical.

FR access devices (FRADs) are available in sizes from a single port, serial or LAN, up to something quite large comparable to a T-1 multiplexer. They often offer internal CSUs, for T-1 or 56/64 kbit/s.

FRADs are designed for a wide range of protocols, old and new. Even though no frame relay network provides a multicast function, FRADs can emulate multidrop lines like those used by SNA, Burroughs, and other polled protocols in mainframe computer networks.

When voice capability is added to the data protocols, the FRAD takes on many of the aspects of the T-1 multiplexer. It is natural that the access speed to FR be T-1.

If the networking requirement is limited to LAN protocols, the CPE can be routers. Most routers offer a frame relay compatible serial interface, though not an integral CSU.

Dial Tone

From voice T-1 came, and to voice it often returns.

The number of trunks needed by a business can exceed the number of people in the office: not only does everyone talk at once, they have facsimile machines and data modems going at the same time.

To deliver the number of trunks wanted is often hard for local telcos. They don't have the wires in the ground (or on the poles). The wires they have don't go to the right places.

The answer can be to use T-1 technology to deliver 24 trunks on only 2 pairs.

Carriers deploy the equivalent of a channel bank, the remote terminal, in a cabinet by the side of the road. The customer then pays only the normal charges for analog lines, because that is the form of the connection that reaches the customer's office.

However, if you want to get the T-1 line extended to your PBX, Telco Logic takes over:

- The telco saves the cost of the remote terminal, and
- There is less (no additional) cable to lay, and
- the "digital handoff" at T-1 may even eliminate the cabinet; BUT
- You pay more for the two pair than you pay for 24 pair, about $200 per month in Virginia on top of all the charges for the 24 trunks. Ask Dilbert to explain.

If you persist in taking voice trunks on a T-1, for reasons of reduced PBX cost or other reasons, be prepared to struggle. The regular business office still doesn't know about T-1. The specialists are overworked, or something, but they don't call back quickly. Try to work with a regular telco salesperson if you have one; that person has a better chance of finding the right T-1 expert and getting a response.

You will have to tell the order taker what kind of signaling your equipment uses: E&M, and what type it can handle (wink start?); standard loop start. The switch is configured to provide a specific service type; you match it with a PBX interface or a channel bank.

A separate channel bank may have more flexibility. If it has a DSX-1 port, it can still hand off a T-1 to the PBX. At the same time, it could offer a data port to get you into integrated access.

Integrated Access

Think of IA as "all of the above." One T-1 local loop lets you reach the Internet, access a frame relay service, and draw dial tone on multiple trunks. CPE and a multiplexer function in the CO split the T-1 into some number of channels. Each channel can be any speed from 56K to

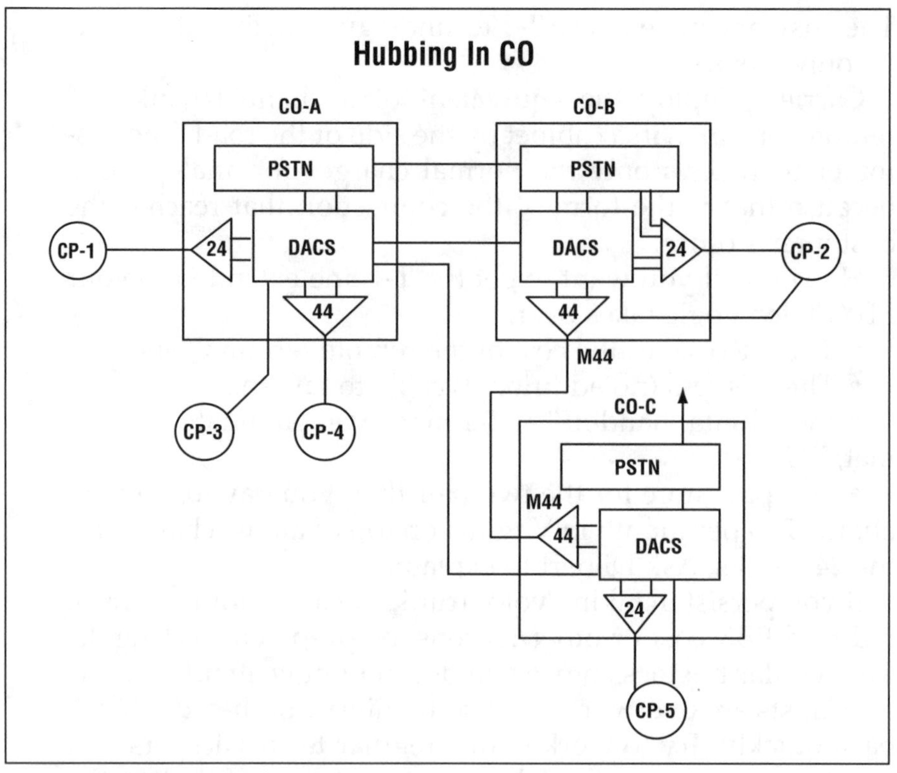

Figure 6-16 *DACS hubbing collects and routes DS-0 connections through the central office. ADPCM channels must be "decompressed" to PCM format for switching.*

Nx64K. That is, if the local telco will cooperate and provide the CO mux.

AT&T lead the movement to integrated access, offering it to its long distance customers as an extra service. To provision IA, the local telco supplies the T-1 loop from the customer to AT&T's point of presence (POP, their central office location). AT&T provides the multiplexing to split the T-1 bandwidth, and also the cross connections to get each channel to its target service.

Southern Bell has tariffed IA, but not all RBOCs. Bell Atlantic itself won't provide integrated access, even to its own services. But you can get trunks that draw dial tone, frame relay, and Internet if you first pass through a long distance company. Odd.

Project Management For T-1 Installations

A high speed digital network will take longer than planned (and cost more). Therefore it is important to manage the process tightly. Loose control can lead to extended delays and cost overruns.

Site Preparation

"Unprepared equipment room" is the leading cause of delay in putting a network into operation. You must coordinate with your carriers and CPE hardware vendors so you will know exactly what they need, including:

1. Physical access between the outside world and the room:
   - Loading docks, elevators, ramps for hardware. Are the doorways high enough for equipment racks on pallets and/or dollies?
   - AC power, in the locations needed, with the right number of individually fused circuits, with the proper amperage rating, and with the proper connectors. 30 amp plugs will not fit into 20 amp sockets.
   - Air conditioning, usually lots of it.
2. Enough space for equipment:
   - Allow clearance to get at front and rear of free-standing cabinets, including space to swing doors.
   - Plan for the bulk of the cables needed to make local connections within the room. Raised floors are one way; overhead wire troughs, another.
   - Proper ventilation requires space for the air to flow through cabinets.
   - Room to change cable connections, replace modules, and in general troubleshoot an installation.
   - Room to grow. Every successful network expands.
3. Special requirements:
   - A plywood board on the wall for your carrier's terminal block, your own "distribution frame" of punch down

blocks, and perhaps some wall-mounted channel banks or similar equipment. Larger installations require free-standing relay racks of punch down blocks or connectorized patch panels (or both).

- A solid earth ground for the equipment. The electrical grounds inside buildings are often several volts above earth potential. They can be loaded with AC noise from elevator motors, production machinery, etc. What happens if someone is arc welding in your building? What happens to your expensive communications equipment? It happens, more often than you'd think.

Staff Preparation

Assume all vendors perform flawlessly. They deliver everything in perfect condition and on time. Then what do you do?

Even in this fantasy, you must be prepared to run the equipment, put up your applications, troubleshoot, and start planning for changes. This means training your staff in all aspects of the CPE and carrier services.

Most vendors offer training, either on your site or in formal classes at their locations. Take all you can get in the beginning — it's often "free" or included in the bid price. Sometimes you can negotiate continued training into your overall purchase agreement.

Train as many people as you can afford to spare from your offices. The inevitable turnover in your staff will make extra knowledge very helpful later on.

If your CPE vendor can set up your network at one of his sites before delivery, get some hands-on training on your own equipment. This "staging;" is the only time you will get to see all nodes operating in one place.

Staging is also a good time for checking documentation. You want to have "as built" drawings, cable lists, equipment serial numbers, etc. Perhaps you can get it automatically as a data base built by the network management system.

Due to modifications during staging, the documentation may be delayed. Be sure to ask. Get all your documents before you make the final payment.

The constant change in your requirements will almost certainly bring up possible changes in the network even while it is being built and installed. Resist changing the work order during this period. Your vendor will accommodate you. But he'll charge you more. And there will be additional delay — and plenty of excuses for it.

To meet performance clauses in the purchase contract, the vendor may want to get an acceptance of a staged network meeting the original specifications before modifying it. That means that both parties are assured the network is doing what it's supposed to do. Changing the requirements at the last minute could free the vendor from an obligation to meet promised delivery dates, or the networks specifications.

"Real Time" in Turning Up a Node

From the experience of many large corporations, real time runs more slowly than the calendar you normally plan by. While the T-1 line may be "turned up" on the date requested, don't plan to use it immediately. Set your schedule based on an "installation date for lines" but use that date as a fixed marker, not the day you want to run real live, important applications on the new network.

In the early days of T-1, it was common to spend a year designing and installing a network. At that time the process was not yet familiar to most parties, users or vendors. With practice, the we've gained the confidence and knowledge to work faster.

Today, the practical progression for relatively good service runs over about half the time of a decade ago The 12-step process goes something like this:

1. Installation date minus 6 months.

• Start to plan the network.

• Review CPE and carrier service options.

2. I minus 5 months.

• Finalize the network design.

• Put out the Request For Proposal, to carriers and CPE vendors.

3. I minus 4 months.

• Order the T-1 lines.

• Order the T-1 CPE.

Lines may be had from some carriers in some locations within a few weeks. The backlog is nowhere near the year-long delay found shortly after divestiture in 1983. Long distance circuits still take more engineering and coordination.

CPE used to need more time than the lines. But as installations have speeded up many hardware vendors delivered faster. Some now build to forecast, rather than order, so may have equipment on the shelf.

4. I minus 2 months.
- Complete planning and start implementation of site preparation: physical, electrical, etc.
- Confirm delivery of lines and CPE with vendors.
- Ensure at least 3 phone lines will be working at each site during install (to call carriers and other vendors).

5. I minus 2 weeks.
- Confirm how T-1 lines will be terminated at your site: your CSU, test equipment, carrier's temporary CSU, etc.
- Obtain necessary equipment, if any.

6. I DAY (Installation day for lines).
- Provide access to premises for carrier.
- Provide personnel to support installers.

7. I to Plus one week.
- Let lines run without live traffic.
- Monitor signal quality, perform bit error rate tests, etc.

8. I plus two days.
- Start install of CPE, T-1 nodes, etc.
- Establish connectivity among nodes.
- Put CPE in idle mode and let nodes talk to each other; start without applications running on them.

The rate of infant mortality among lines and CPE is higher than anyone would like. Every T-1 line installation, from every carrier that uses electrical transmission, seems plagued with problems for at least a week, often up to a month. Some reports indicate new T-1 repeaters need to burn in. Once operating, optical fiber appears to have fewer problems. For whatever reason, the average T-1 line used to settle down to much better error rates after about two weeks.

9.I plus one week.

- Start putting up connections for real applications, low priority stuff first.
- Monitor how these initial connections work.
- Gradually move additional connections to the new T-1.

10. I plus two weeks.

- Finish migrating traffic to the T-1 network.
- Evaluate quality of circuits and CPE.

Most users report the main problems in T-1 lines occur in the local loop. Within a month, they can tell if the line is going to settle down or not. If not, the only solution may be a different local loop. In changing local loops, it may pay to add a new one, then cancel the old. The cost for a second loop is a gamble. It may also be no better than the first.

Problems with equipment usually involve swapping boards or modules. Any problems at this time are under warranty.

11. Installation Day plus two months.

- Start keeping track of possible future expansion.

12. Installation Day plus six months.

- Start the cycle over again.

Beyond T-1: Desktop to DS-3

Not long ago the technology of customer premise equipment limited the concept of a T-1 network to just the T-1 level. Now, the digital network includes everything from the 1200 bit per second RS-232 port behind your desktop PC to a 155 million bit/s or faster optical fiber backbone.

As late as 1988 it was common to plan a large corporate network in several separate layers (Fig. 7-1):

- The T-1 nodes served the major sites only, with emphasis on synchronous data.
- Asynchronous data travelled on an overlay network of statistical multiplexers.
- Voice switched by PBXs shared the T-1 transmission path only among major sites.
- Analog tail circuits carried voice and data between the backbone and smaller sites. These lines were invisible to the network control and management system of the T-1 nodes.

It was easy then, even logical, for a person planning a private T-1 network to think in terms of ten, twenty, or fifty nodes. This was true even if the company had hundreds of locations. Most of the offices were served by leased analog lines for voice and data. Some locations justified 9600 bit/s or 56 kbit/s digital data service when a T-1 wasn't economic.

That kind of thinking made sense when:

- T-1 nodes were very expensive, at an average starting cost of $50,000 to $150,000 per site.

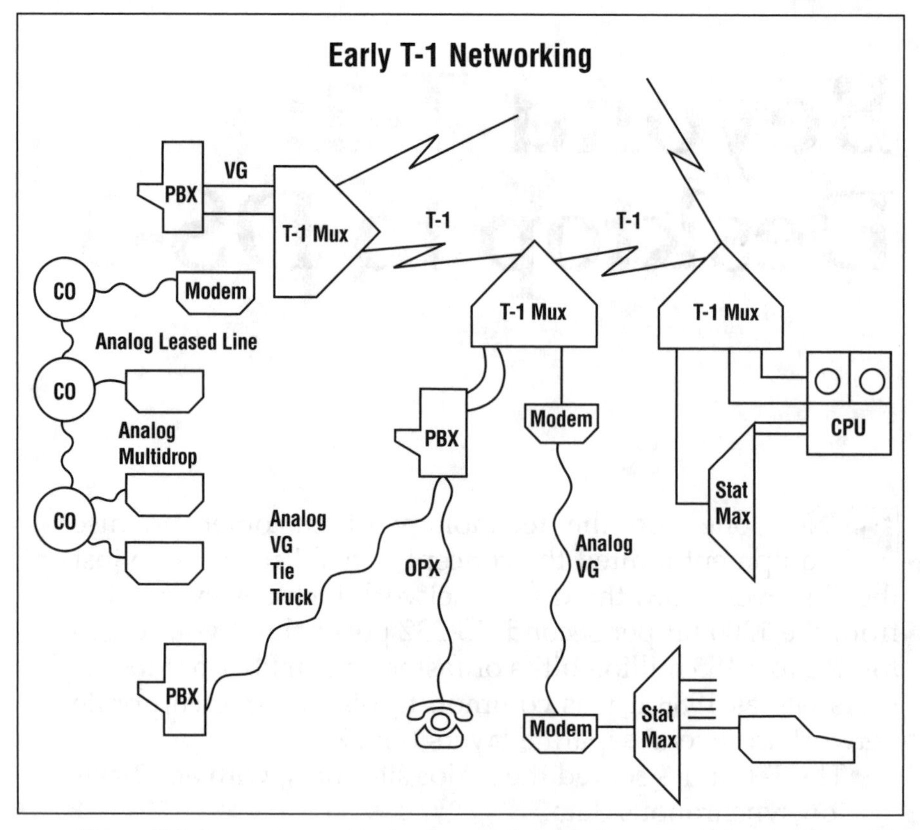

Figure 7-1 *Economics and technology both tended to make T-1 networks rather limited in scope during the first five years. Analog tail circuits served remote sites.*

- T-1 lines were much more expensive than today.
- 56K multiplexers were not fully compatible with any T-1 nodes.

The new networking

Things have changed—dramatically. The new form of a private network is all-digital, at every site (Fig. 7-2). Two factors contribute to the change:

1. less expensive ingredients make it economic,

- customer premises equipment (CPE);
- facilities, not just T-1 and 56K but also 155 and 622 Mbit/s digital lines.

2. Customer Premises Equipment that
- remains manageable with thousands of nodes in a network; and
- cost so little that it is affordable at all those sites.

Saving Money

Trends in tariffed costs are up for analog facilities and down for digital facilities. As time goes on, it takes fewer analog lines to justify a digital line. In many locations, as few as 3 or 4 leased voice grade lines cost as much as a 56 kbit/s digital service. As few as two or three 56K lines cost as much

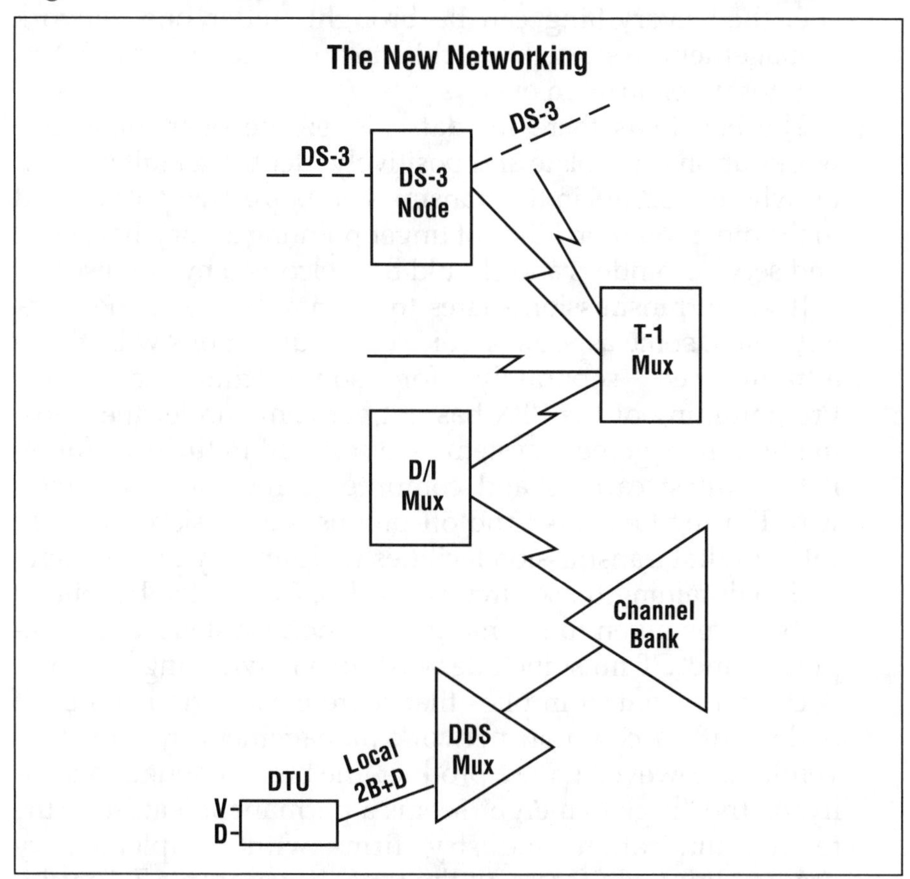

Figure 7-2 Lower costs for digital lines and for customer premises equipment have made all-digital networking not only possible, but attractive. Every location is on the digital "backbone," though smaller sites may be served by 56 kbit/s or other data link speed less than a full T-1.

as a T-1. The result is that you can cost-justify digital service (56K or T-1) in more and more locations.

That is not to say in every location. You can expect some sites to show a very long (or even non-existent) payback period for an all-digital solution. However, it still may make sense for other reasons to include those sites in a general conversion from analog to digital. The justification (savings) will come from reduced maintenance costs rather than from reduced line charges.

Unifying Control

If the network is all-digital, from the largest to the smallest site, then everything can be brought under one network management system. Does this mean voice and data? Yes, but not necessarily in every sense.

The benefit is there for data alone: one operator at one workstation can isolate and positively identify a fault no matter where it occurs in the transmission path. This puts an end to the most common kinds of finger pointing among hardware and service vendors and should be welcomed by any user.

If voice transmission shares the digital backbone network with data, some aspects of voice communications will still be administered separately for some time to come. Programming of the PBX has not yet come under the transmission management system — but it will in the near future if the largest carriers and computer companies have their way. Putting tie trunks and off-premises extensions on dedicated digital transmission facilities will simplify provisioning and maintaining those circuits—today. Computer Telephony is the name given to the merger of phone systems and computers, and CT must include wide area networking.

Unifying control implies that there exist various types of nodes with a common network management system. Few vendors, however, have a product line broad enough to qualify. As the "urge to merge" seems a permanent feature of the telecommunications industry, firms with complementary product lines join forces all the time. For example T-1 multiplexer vendors have merged with statistical multiplexer and LAN companies. Invariably, part of the merger announcement is the statement of intent to combine network management systems for all the products into a single master control system.

An earlier edition of this book predicted that there was such unanimity behind the concept, that we would see unification of management from everyone (eventually). It has happened, under the products of software companies (or software divisions of companies like Hewlett-Packard). In fact it has happened twice:

- H-Ps OpenView network management system has been adopted by a large number of CPE vendors and even many software companies who build add-in products and management applications to run on top of OV, which uses SNMP
- Simple Network Management Protocol (SNMP) came out of the Internet, a product of router vendors. But it proved universally applicable to practically any device and is incorporated into almost all types of networking equipment.

Integrating Equipment

A growing number of vendors offer what might qualify as a line of products broad enough for "The New Networking." At minimum, this implies, in addition to T-1 nodes,

- a comprehensive network management system, or at least manageability by a third-party NMS workstation;
- 56/64K multiplexers with multiple voice channels (preferably compatible with the T-1 node);
- compatible local distribution of data within a customer's building (LAN, modem, ISDN technology, etc.);
- statistical multiplexers or some form of sub-rate data multiplexing (including frame relay or ATM) to raise the number of logical channels on a T-1 to at least a thousand; and
- multiple direct DS-3, OC-3, or faster interfaces on a large node.

This is the key set of features that has emerged from market requirements. As expected in the last edition of this book (1990) more vendors offered this range of products, either by designing their own, reselling some one else's products, or acquiring suitable companies.

Up to DS-3 and Beyond

The telco central office long ago realized that T-1 was not fast enough for major routes. Engineers gradually increased the speed of the transmission span:

- doubling the T-1 capacity in a T-1C;
- doubling again into T-2 with a total of four T-1's;
- jumping by a factor of seven to DS-3, 28 T-1's.

This is the point where, in 1990, private networks generally stopped. Even then, Telcos continued up the scale of multiplexing ratios to aggregate speeds for a single optical fiber of greater than 1 gigabit per second (10 to the 9th power).

In less than a decade, all of those speeds became available to end users, either in the wide area or for local area connectivity.

DS-3 Options

There were few choices in 1990 when it came to DS-3 formats. The long established standard in the public network is still called M13. But for technical reasons there were new proposals for formats at about the same speed called Syntran and SONET. Syntran didn't succeed; SONET won in a big way, but mostly at 155 Mbit/s and above.

Classic M13

The name M13 comes from the fact that this multiplexer connects the DS-1 to the DS-3 level. One DS-3 contain 28 DS-1's.

They are combined in two stages. The M13 process first joins four DS-1 lines into a DS-2 bit stream at 6.312 Mbit/s. This step is more than a straight time division multiplexer function. "M12" bit-interleaves the four inputs, but it also adds 136 kbit/s of overhead and "justification" or bit stuffing.

The extra bits accommodate some variations in the clock rates of the incoming T-1 lines. At the far end the stuffed bits are removed to restore the original bit rate on each T-1.

Refer to the earlier discussions in this book on clocking and synchronization. The M13 is the key element that accommodates T-1 clock rates set by a user's equipment when that rate is not locked to the primary standard of the public network.

The second stage within M13 combines seven DS-2 streams into one DS-3. Like M12, the M23 process is bit-interleaved, with justification. That is, the DS-3 clock need not be synchronized with any DS-2 or DS-1 line speed. M13 overruns the DS-2 clock, and "bit stuffs" or adds extra bits to fill out the DS-3 rate of 44.736 Mbit/s. By comparison, 7 times 6.312 is only 44.184 Mbit/s, or 552 kbit/s less.

In "olden times" this approach was highly desirable. It was very difficult to run all levels of transmission equipment from the same clock source. In most locations, all the T-1 lines were point-to-point between channel banks, which ran on their own (not very accurate) internal clocks.

There were no station clock sources and no digital CO switches to set precise frequency standards. Nor was there any networking, as we now know it, which requires precise timing to cross connect bit streams. So it was unnecessary, then, to lock the many digital devices to the same clock, and easier to accept the differences than to lock them together.

Whatever the original reasons for the two-step asynchronous multiplexing, it creates two problems for us today:

1. Discarding the overhead bits, and especially the random stuff bits, during demuxing introduces irregularities which create jitter and wander (phase changes with frequencies above or below 10 Hz).
2. There is no reliable way to tell where a given T-1 or DS-0 will fall within the DS-3 bit stream. This makes it very difficult (previously not worth doing) to extract just one DS-0 because it will have no definite location within the DS-3 frame.

To get at an individual channel you will typically reverse the two-step process and demux the DS-3 down to individual T-1's. The need for a DS-3 "DACS" is clear. To get simple switching capability among DS-3 lines requires a better format.

With accurate clock sources available, a natural step is to synchronize the DS-1 and DS-3 levels. Then the stuffing bits are always the same, and the C-bits that controlled the stuffing become available for other functions.

AT&T uses C bits for an end to end parity check (C bit

Parity). Thus the second way to put 28 T-1's in a T-3, M28.

It was once possible to obtain framed, unchannelized DS-3. However AT&T migrated to C-bit parity service, discontinuing the older M13 and unformatted services for a while. Demand for fast access to certain services brought back the unchannelized DS-3 service.

Syntran

Bellcore (Bell Communication Research Corp., the laboratory formerly owned jointly by the seven Bell regional holding companies) proposed Syntran (for synchronous transmission). Syntran describes how to multiplex DS-0 and DS-1 signals into a DS-3, in a fully synchronous format.

The resulting bit stream has the same rate as the standard M13, but each DS-0 has a fixed position within the frame. The C-bits provide a data link and an error check, but not compatible with M28.

Syntran offered two important features:

1. Synchronous format allowed easy DACS switching of DS-0 and DS-1 channels among DS-3 ports.
2. The aggregate bit rate was compatible with the large number of existing DS-4 and higher transmission facilities that them made up the Pleisiochronous digital hierarchy in North America.

Unfortunately for Syntran, DS-3 never was installed much outside of North America. And while M13 and Syntran are defined as electrical interfaces, the new installations of facilities near that speed are mostly optical fiber.

SONET

The T1 committee of the Exchange Carriers Standards Association developed a format for an international optical interface. It is called SONET (Synchronous Optical Network). Syntran provided some of the ideas that went into SONET, and Bellcore eventually became a sponsor of SONET. (The ECSA was a member organization of ANSI and the forum for the T1X1 and T1D1 committees that worked on these topics.)

But SONET is much more than DS-3. The goal is a standard for interconnection of different national networks at speeds from about 150 Mbit/s up to several gigabits per second. The DS-3 level is only one portion of the standard, intended for network access from customer premises.

The first stages of international agreement took shape in 1988. The exact STS-1 and OC-1 speed was set at 51.840 Mbit/s. This satisfies the Europeans who wanted a format (OC-3 at 155.52 Mbit/s) to accommodate their 34 Mbit/s DS-3 as well as the North American 45 Mbit/s. A compromise within CCITT settled on a framing pattern, signaling, etc. Balloting during 1988 resulted in publication of a few final standards for the basic transport services. How the overhead bits would be used was left for later decisions, for example order wires, operations channels, and higher-level protocols for provisioning and testing.

Vendors have a way to encapsulate the M13 electrical format within the slightly faster optical fiber interface of a SONET OC-1. As expected, some networking nodes developed the ability to offer both M28 and SONET, with cross connections between them.

Networking At DS-3 and Faster

The higher levels of the digital hierarchy are following a progression in networking complexity similar to what we saw in T-1:

1. Point to point terminals; M13 multiplexers with 28 T-1's in and one DS-3 out.
2. Multiple M13 multiplexers in a single chassis; something like a 48-port D4 channel bank that followed the 24-port D3.
3. Networking of multiple DS-3 lines into one hub with the ability to cross-connect DS-1's and even DS-0's and subrates within that hub.
4. Multiplexing many DS-3 "inputs" into a higher speed aggregate, for example OC-3.

Level 3 generally used to be possible only with one of the synchronous formats. Progress in integrated circuits now makes levels 3 and 4 practical. Devices with this performance are installed at telephone operating companies and corporations with large communications needs.

Major Internet Service Providers (ISPs) are leading proponents of networking at this scale. They need a backbone with trunks running at over 600 Mbit/s in order to offer DS-3 access to the Internet.

Network topologies for DS-3 and faster backbones have evolved from point to point up to the partial mesh or interlocking rings familiar from T-1 nets. There are many fewer sites that can cost-justify multiple DS-3's than there will be T-1 hubs, but the numbers for both are increasing rapidly. The combination likely to be most common is a sparse backbone of fast trunks surrounded by a T-1 network (Fig. 7-3). The "new networking" concept includes subsidiary connectivity via 56/64 kbit/s lines or ISDN basic rate.

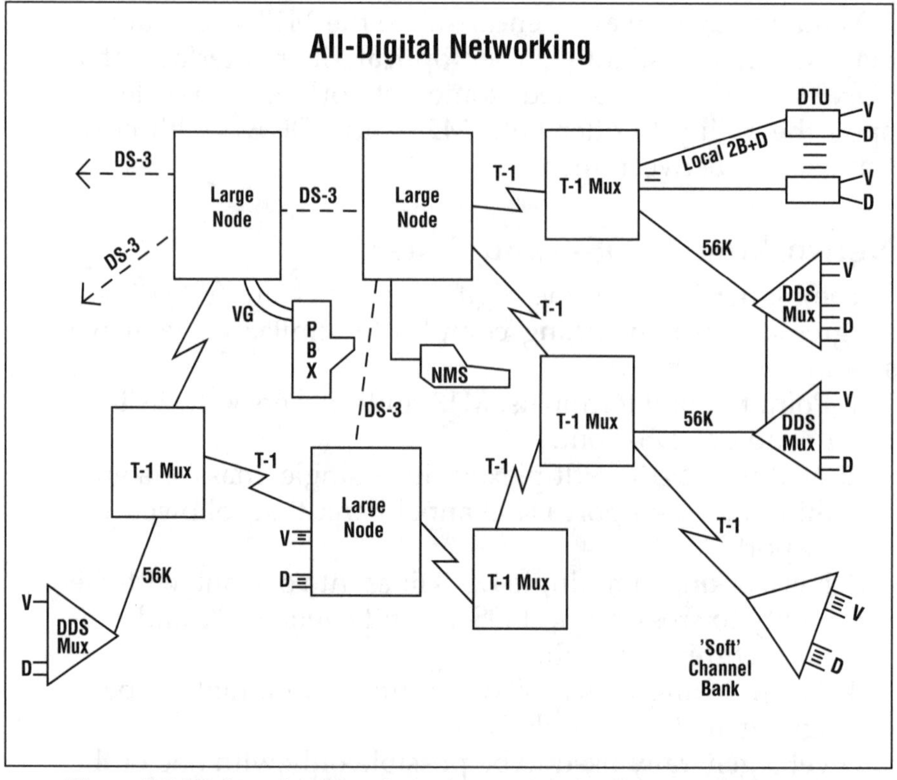

Figure 7-3 *DS-3 portion of typical private network will be a sparse backbone to surrounding T-1 nodes and lines. 56 kbit/s lines will extend to smaller sites for full digital connectivity.*

Down to the Desktop

You don't see too many T-1 nodes in private offices where the phones, terminals, and PCs are actually working. The T-1 equipment is locked away safely in closets, or maybe in the computer room (guarded by MIS).

But your PC has an RS-232C interface that's good for only 50 feet (officially) between devices. How do you connect a terminal or PC to the T-1 node at the far end of the building?

Various vendors offer many different extensions to their data ports. Other than "long reach" (low-capacitance) cable, most older techniques are built on modem technology. Newer approaches tend to rely on LAN or ISDN technologies.

Modem

For tail circuits outside of the building that houses the T-1 node, you will still see regular modems applied (Fig. 7-4). They most often use analog voice grade facilities provided by telcos. Usually these are four wire modems on leased (fixed) lines, which can be of any length. Dial modems are also used.

Many modems offer the advantages of a management system. In many cases, however, that modem management system assumes that one modem of each pair is at the "central site" or co-located with the management computer. That is,

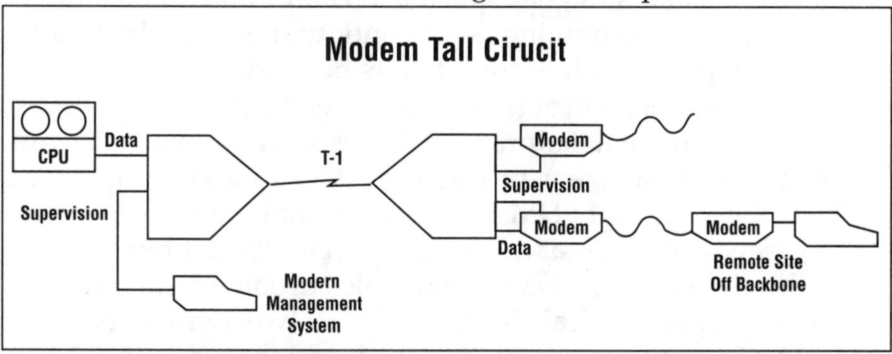

Figure 7-4 *Regular modems on voice grade lines appear in most T-1 networks as tail circuits. Management systems, available for many leased line modems, may have to be extended to the remote sites as dedicated supervisory channels to each remote master modem.*

the management system is designed to be cabled directly into one modem of each pair. If the modems are on tail circuits off a T-1 backbone, most of them will be located at remote sites. Perhaps none of the modems is located at the central site.

You can manage these modems by setting up a dedicated logical circuit from the central site to each "master" modem at a remote site. In this way the modem management system doesn't realize that the equipment is dispersed geographically.

To minimize the amount of bandwidth consumed for modem management, you can operate the supervisory channels at the slowest speed possible. Then multiplex many of them within a DS-0 on the T-1 line. Here is where efficient sub-rate data multiplexing (SRDM), statistical multiplexing, and/or sub-rate switching in the T-1 node will pay off. Proprietary formats are acceptable here because all the traffic will terminate on your own equipment, typically at your central or main control site.

Limited Distance Modem

An LDM is a line driver that increases the allowable distance between RS-232 devices (Fig. 7-5). Often they draw power from the interface pins, so no external power supply or connection is needed. They are relatively simple, reliable, and can be inexpensive (under $100 each end for async up to 9.6 kbit/s for a few thousand feet). Synchronous transmission, longer distances, higher speeds, V.35 and X.21 interfaces, or downline programming and configuration capability can raise the price to a hundred dollars per end.

Being simple, however, means low-end LDMs have no way to report internal failures, or to respond to external commands. The management system of the T-1 backbone network cannot be aware of LDMs. Therefore troubleshooting from the T-1 viewpoint stops at the RS-232 port on the T-1 multiplexer.

Configuring an LDM, if possible, usually means setting dip switches. Typical designs require two twisted pairs (4 wires) to carry one channel up to several thousand feet. Most often this approach applies within a building. More expensive models drive signals up to about 10 miles if the line leased from the telco is all copper, end to end.

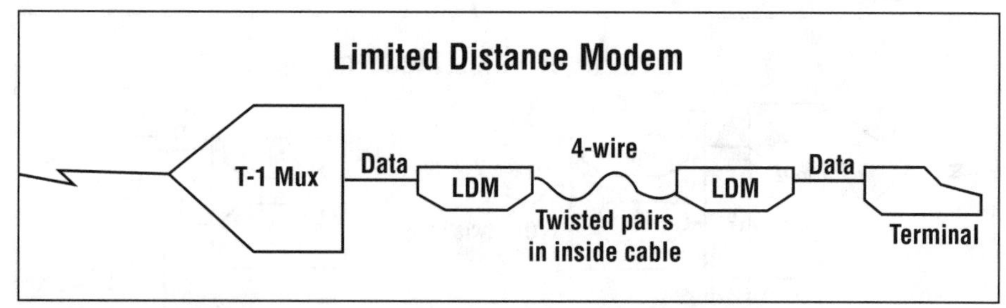

Figure 7-5 *Limited distance modems (LDMs), or line drivers, increase the cable length allowable for an RS-232 interface to as much as several miles. They almost always lack any remote diagnostics or control.*

Data Over Voice (DoV) Modem

If your phones offer only POTS (plain old telephone service) you are using only a small portion of the bandwidth on your twisted pair house wiring. This is basically the same wire used to carry T-1, so obviously there is more bandwidth available than 300-3300 Hz.

DoV's take advantage of the higher frequency portion of the bandwidth on 2-wire connections to establish a data path using analog modem technology (Fig. 7-6).

In a typical application, the end user replaces his phone with a DoV modem. The phone is then plugged into the DoV, as is the terminal or PC. That is, the DoV goes between the phone and the (RJ-11) connector to the phone line in the wall. The DoV has a data port, typically a DB-25 connector with the RS-232 electrical interface.

Internally, the DoV consists of a low-pass filter and a modem. Unlike the standard modem, which produces tones within the audible range (300-3300 Hz), a DoV modulates a carrier frequency of about 100 kilohertz. Some designs use two carriers, perhaps 80 kHz and 140 kHz, each to carry data in one direction. Thus DoV's are full duplex.

A high frequency carrier signal makes it relatively easy to separate the data functions from POTS. Wide separation in frequencies also simplifies filter design. These factors are of considerable importance to avoid disruption of voice service from any form of failure in the data portion of the circuit.

To ensure complete separation, the filters should be passive, that is, only coils and capacitors or other circuits

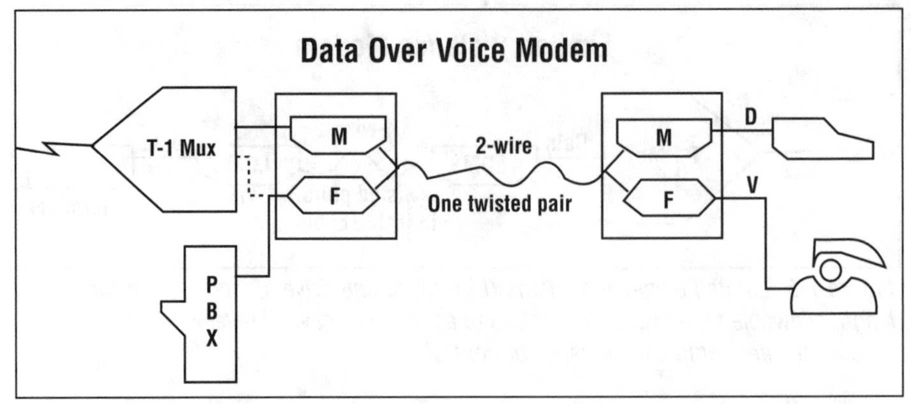

Figure 7-6 *Data Over Voice modems use frequencies in the range of 75 kHz to 150 kHz for data channels that do not interfere with normal POTS voice functions. Filters (F) at both ends keep the modem (M) energy out of the telephones and the voice switch (local or remote).*

powered by the voice energy alone. Then, if the data modem is disabled by a power loss or circuit failure, the phone will continue to operate on the PBX's battery back-up or on a central office line. When telcos offer data service using DoV's they insist that the voice path survive any failure in the data path. Curiously, this requirement is not imposed on ISDN equipment like terminal adapters.

At the central site, another DoV unit filters out the data carrier from the voice path and shunts it to the modem. After demodulation back to a standard interface, data is cabled to a host computer, a multiplexer, a data switch, etc. The standard voice-grade bandwidth is passed to the voice switch.

Central site DoV packaging favors rack-mount shelves for greatest density. Shared power supplies and cabinets also minimize the cost.

The voice path needs no set up—it looks like pure copper wire to the standard telephone and PBX line card. Configuration of the data port commonly requires setting of dip switches at each end. Simple DoV units have no network management beyond front panel alarm lights. More sophisticated types offer remote configuration of port parameters (speed, parity, etc.) and centralized alarm gathering.

The reach of a DoV system can be up to about 5 miles.

ISDN Terminal Adapter

After 20 years of trials (and tribulations) ISDN central office switches have arrived in most US locations. The technology for basic rate access (2B+D) not only works reliably, it is one of the better ways to obtain a digital circuit:

- Wide availability of ISDN, though certainly not everywhere.
- Reasonable cost, comparable to a pair of business voice grade lines, though some RBOCs apply surcharges for "data" calls.
- Inexpensive equipment (CPE), mostly routers and voice to start with, but now spreading to legacy protocols.

The idea of ISDN for most smaller applications is to carry three channels on the local loop: two for user information, the B or bearer channels, and one for signaling data, the D channel. This happens over the standard 2-wire local loop of one twisted pair. Each B channel is 64 kbit/s, the D channel is 16 kbit/s, and there is another 16 kbit/s for overhead, making a total of 160 kbit/s. This is called the "U" interface (Fig. 7-7).

At the customer's site the U interface plugs into a "Network Termination" and then into a "terminal adapter" that converts the format to RS-232 or to another data or voice format.

One reason that this system works well is that the multiplexing functions needed to format the U interface have been reduced to commercially available integrated circuit chips. More of these chips will be appearing continually.

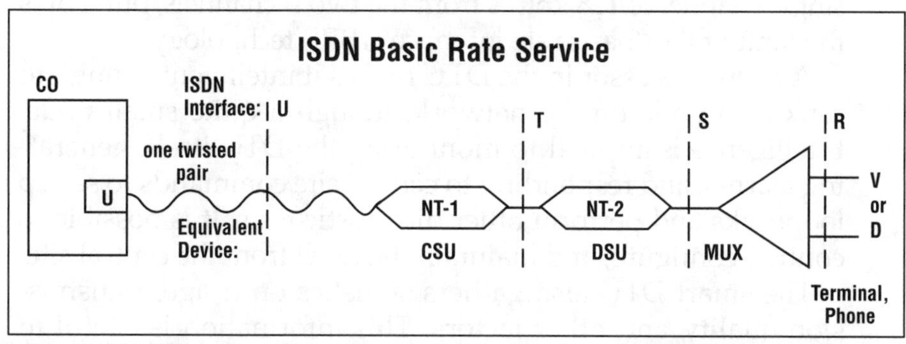

Figure 7-7 *Technology exists to transmit 2 bearer channels (2B) and a signaling data channel (+D) on one twisted pair, called the U interface in ISDN jargon. The device that converts U into the more familiar RS-232 or voice interface (at R) is the Terminal Adapter (TA).*

ISDN chips for 2B+D have been used to extend the reach of T-1 nodes within the customer premises. In the node, multiple chip sets are built into cards or modules that plug into in the T-1 device. The U interface circuits are physically smaller than standard data circuits. For example, one card carrying up to 24 DS-0 channels occupies a slot that might otherwise hold a card with only two to six data ports.

The same chip set is in the proprietary terminal adapter or data termination unit (DTU) that goes on the desk top. A single twisted pair joins the DTU to the T-1 node (Fig. 7-8).

ISDN chips implement the essentials, leaving the equipment manufacturer to worry about the content of the signals in the D channel, for example. The major switch vendors originally implemented slightly different formats for their U interfaces. This meant that terminal adapters would work with only one brand of switch.

Consequently, the devices that used ISDN chip technology to distribute data from a T-1 node were data termination units and not ISDN terminal adapters. The difference mattered not at all when a DTU was connected to the matching T-1 node.

In early versions of DTUs integrated into T-1 nodes, the B channels were used for data (not voice). The D channel was the supervisory channel for the network management system of the T-1 backbone, and is not available to the end user directly. Thus one DTU on one pair of wires carried two data channels at up to 64 kbit/s each. In some circumstances it is possible to create a single channel of 128 kbit/s from the two B channels, but that is the limit of the "basic rate interface" (BRI) technology.

A microprocessor in the DTU makes it intelligent enough to rank as a node on the network, though a quite small node. Intelligence is applied to monitoring the DTU itself, generating alarms, and responding to central site commands to set up loopbacks and perform other diagnostic tests. It is possible to control, configure, and maintain the DTU from the central site.

The smart DTU also gathers statistics on usage, transmission quality, and other factors. This information is useful in analyzing performance and planning expansion of a network.

Compared to LDM units, a DTU can cost about the same ($450 list for a dual data port DTU vs. $400 for four RS-232

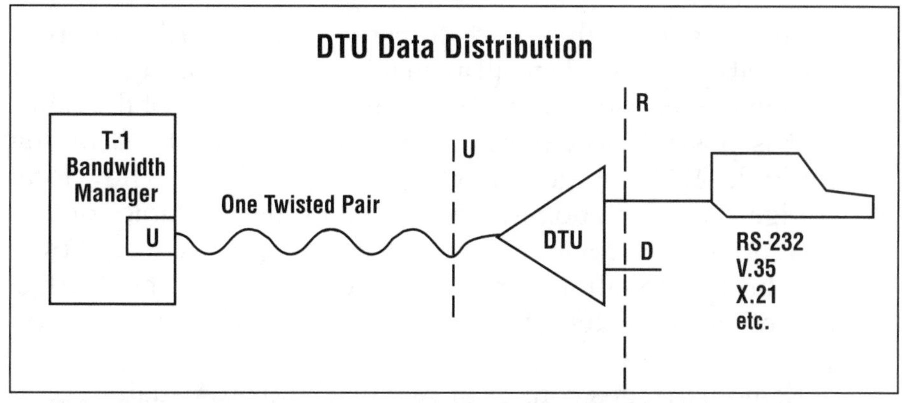

Figure 7-8 *As used for data distribution, ISDN basic rate technology connects a high-density data card in the T-1 node (up to 12 U interfaces) to remote data termination units. These appear as data ports to the central network management system as well as the end user's terminal or PC.*

LDMs). But the DTU uses one-quarter of the wires (2 vs. 8) and operates under the main network management system.

Designed for the local loop, these ISDN chips should drive information at least 18,000 feet on 26 gage wire. Because the reality of inside wiring includes horrors like bridged taps, changes in wire gage, and perhaps even sloppy punch-downs, the range on a campus or within a building probably won't exceed 12,000 feet. Practically speaking, however, that includes any building you are likely to encounter.

The data termination unit designed for premises distribution may lack the lightning protection needed when connected to aerial wire in Florida. Be sure any TA or DTU is rated for the your application, or add the necessary protection in the form of gas tubes or other surge suppressors.

As was inevitable, standard TA's have come to work with T-1 nodes and CO switches, with voice as well as data ports.

Cable-Based LAN

Ethernet, Token Ring, StarLAN, and other types of local area networks connect dozens to hundreds of terminals or computer ports together on one local cable systems. In most cases the length of the cable is limited to a few thousand meters.

Within that distance, however, the LAN offers flexible connectivity and high speed transmission; 10, 100, or 1000 megabits per second are all available. Practical throughput per station will be much less due to the shared nature of the cable.

It is possible to extend the reach of a LAN by bridging multiple LANs together, daisy-chain fashion. The "local bridge" is a black box into which you plug the cables of two or more LANs. Messages generated on one LAN pass over the bridge to all other LANs plugged into the bridge. Essentially repeaters, LAN bridges work with many protocols at one time.

Remote bridges consist of two black boxes, usually at different sites. Between these two boxes you connect a transmission facility. The LANs at both ends constitute a single segment; that is, all stations see all traffic, unless the bridge "filters" traffic based on learning user addresses.

The speed of the facility will depend on the amount of traffic between the sites and the transmission delay you can tolerate. The facility is often part of a wide area network, like a channel on a T-1 network (Fig. 7-9). It also may be a dedicated line.

Filtering bridges block messages from LANs and transmission lines that do not lead to the station addressed in the frame. Filtering is based on the MAC address, the permanently assigned address of an Ethernet adapter. This limits the volume of unnecessary traffic.

Routers operate at a higher protocol level (level 3 in the OSI model), and therefore are sensitive to protocols. They operate only with specific protocols like TCP/IP, AppleTalk, IPX, etc. Routers remove the MAC address, saving bandwidth.

Some routers include the functions of a bridge to pass those protocols they cannot route.

Routers generally perform several networking functions:

- set up routes between LAN segments automatically, so LANs can communicate with each other without manual configuration;
- find alternate routes if the primary is lost;
- allow an operator to manage the network by assigning priorities, passwords, permissions, etc.

Remote bridges and routers can be attached to a wide area network, occupying the position of the terminal adapters in Fig. 7-9. Often a bridge, and especially a router, will have multiple aggregates to transport on the WAN. The more sophisticated routers support multiple logical connections per router, growing into complex mesh networks. The goal of many T-1 vendors is to integrate bridges and routers into the WAN. The software will be implemented in stages.

Initially management will be put on a single workstation, followed by progressively tighter integration until LAN and WAN provide seamless end to end connection management. The first components of this line of development are seen in ATM-based extensions to backbone networks for LAN emulation.

One of the workstations on a LAN can be connected to a data port on a T-1 mux. Then, at least electrically, every terminal device on the LAN is connected to the mux. Whether the mux port can logically address each terminal on a LAN depends on what kind of a LAN port the mux connects to. You can look for three options:

1. A low-speed (RS-232) data port on a terminal server or terminal adapter which is on the LAN cable (Fig. 7-9). The mux port appears to the LAN to be any low speed data terminal. Any device on the LAN can address the multiplexer. If the LAN port can accept connection requests at low speeds (e.g., 9600 bit/s), then the mux (really the user at the other end of the T-1 line) can set up a logical connection to any one terminal on the LAN.

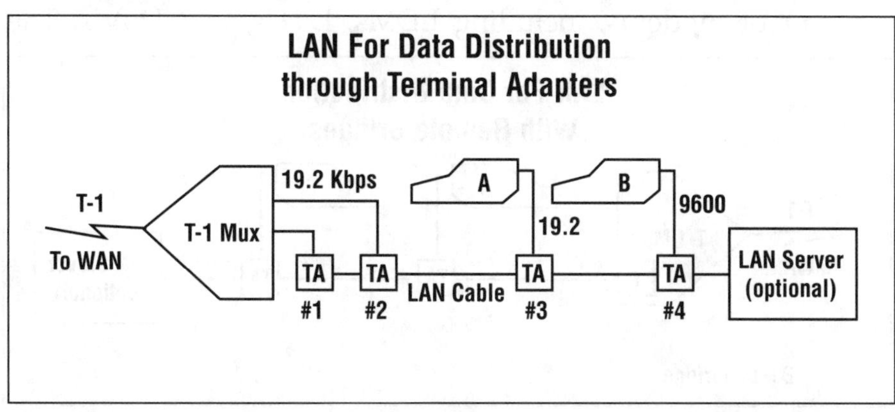

Figure 7-9 *Local data distribution over a LAN can be as simple as treating multiplexer ports like terminals on the LAN. Users at any location on the wide area network then appear locally connected to the LAN. Terminals A and B can set up connections through terminal adapters 1 and 2 to gain access to the WAN.*

Multiple data ports on the T-1 mux could be connected to an equal number of ports on LAN terminal servers.

2. A LAN bridge or router, separate from the multiplexer, occupies the same position as the TA in Fig. 7-9, but the cable labeled 19.2 kbit/s is an aggregate, not a single logical connection.
3. An integrated bridge or router in the mux has a high-speed connection directly to the LAN cable (Fig. 7-10). In this case the T-1 mux looks to the LAN like another segment of the LAN. Messages from the mux to the LAN must be in the LAN format, at the full LAN speed. Only one such connection would exist between a T-1 node and any LAN segment, but the one physical connection could carry many logical connections.

Beyond this level of description, the questions of LAN throughput, bridging between LANs, and LAN design for higher level protocols become quite specialized and beyond the scope of this book.

Statistical Multiplexer

Where many async data circuits tail out of a T-1 node, statistical multiplexers increase the number of logical channels handled by a T-1 mux port. In the same way, a stat mux improves the data handling capability of a LAN port or a pair of modems, including LDMs, DTUs, and DoV's. The

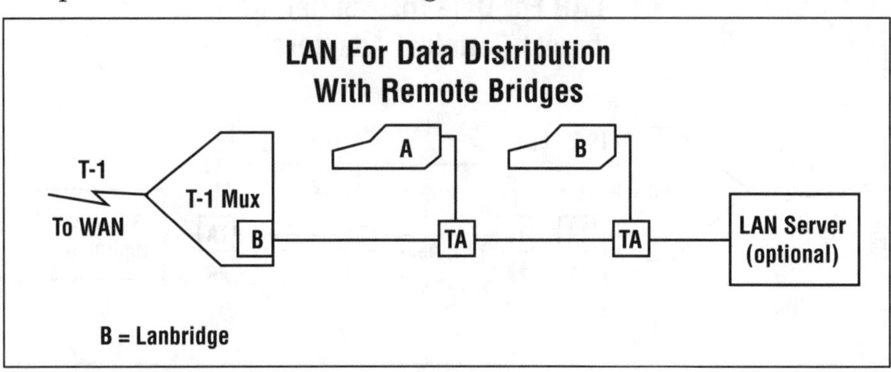

Figure 7-10 *Direct LAN connections makes the multiplexer appear to be a bridge or terminal adapter on the LAN. With similar LAN interfaces in other multiplexers on the WAN, the WAN could become a remote bridge, perhaps multiple bridges or even a distributed bridge.*

goal is to minimize the number of LDMs in a network.

By a judicious choice of locations for stat muxes (in communications closets around a building, for example), you can avoid LDMs between those points and end users (Fig. 7-11). That is, you can keep the RS-232 cable length for PCs or terminals under the practical limit of several hundred feet.

This means that only the aggregates from the stat muxes will require LDMs (or LAN ports, etc.). Concentration can reduce the number of LDMs by a factor as great as 48:1.

Most stat muxes provide in-band flow control (Xon and Xoff). It is possible to eliminate control leads and carry far fewer than the full 25 conductors in RS-232 cables that complete the circuit. This saves much of the cost (and complexity).

Going beyond simple port expansion, some stat muxes offer additional functions:

- switching functions exactly like a data PBX (see below);
- multiple aggregates from each stat mux chassis, for redundancy or drop-and-insert networking locally, and wide area networking over transmission paths on the T-1 nodes;
- speed and parity conversions between async devices;
- remote configuration, control, and diagnostics (though

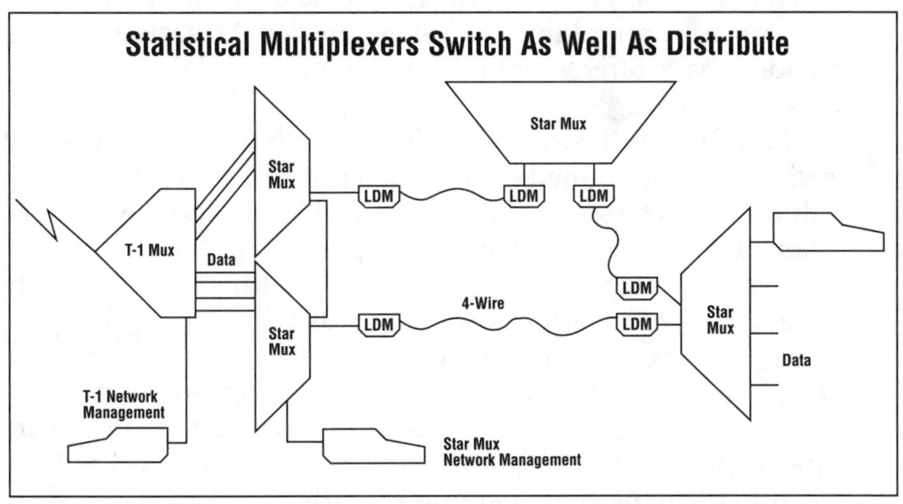

Figure 7-11 *Statistical multiplexers increase the number of data channels carried by a pair of LDMs within a building. They offer their own networking features, including contention for resources, user-controlled switching, and remote management and diagnostics (via a separate network management system).*

seldom by the same network management system as the T-1 nodes).

Any limitations imposed by stat muxes are not of distance, but of time. To share the aggregate line, individual user ports may have to wait in queue. Data sent by a terminal often will be buffered or stored temporarily until the aggregate is free. The time spent in storage shows up as delay or slowed response from the host.

Data PBX

The functions of a data PBX fall across those of a LAN and a stat mux. That is, a data PBX provides:

- Local connectivity among PCs, hosts, printers, terminals, etc. in a way very similar to a LAN. The bit rate is lower.
- Contention for resources, which offers the same benefits as stat mux concentration in reducing LDMs and ports on the T-1 nodes;
- Speed and parity conversions between async devices to simplify network set up and management.
- Password control per port on the data PBX.

This wide range of benefits keeps the data PBX functionality useful as an adjunct to T-1 networks. At various times, vendors have offered both T-1 nodes and data PBX's, more or less integrated.

Data Switch Basics. When first installed, the data PBX or switch is programmed so that each port has a number, name (alpha characters), or both as the address of the device permanently attached to that switch port. To keep it short, the address might be only "3-12," the twelfth port on switch three. For easy addressing it could be "JOES_PC" or "HQ_PRINTER." This procedure parallels how you attach a telephone to a voice PBX and assign an extension number.

The data switch is configured to match the device in speed, parity, stop bits, etc. Ports on the switch do not all need to have the same parameter settings.

A terminal or PC sets up a connection by telling the data PBX which ports to connect together. It does not matter if the switch ports are configured identically: the switch will convert

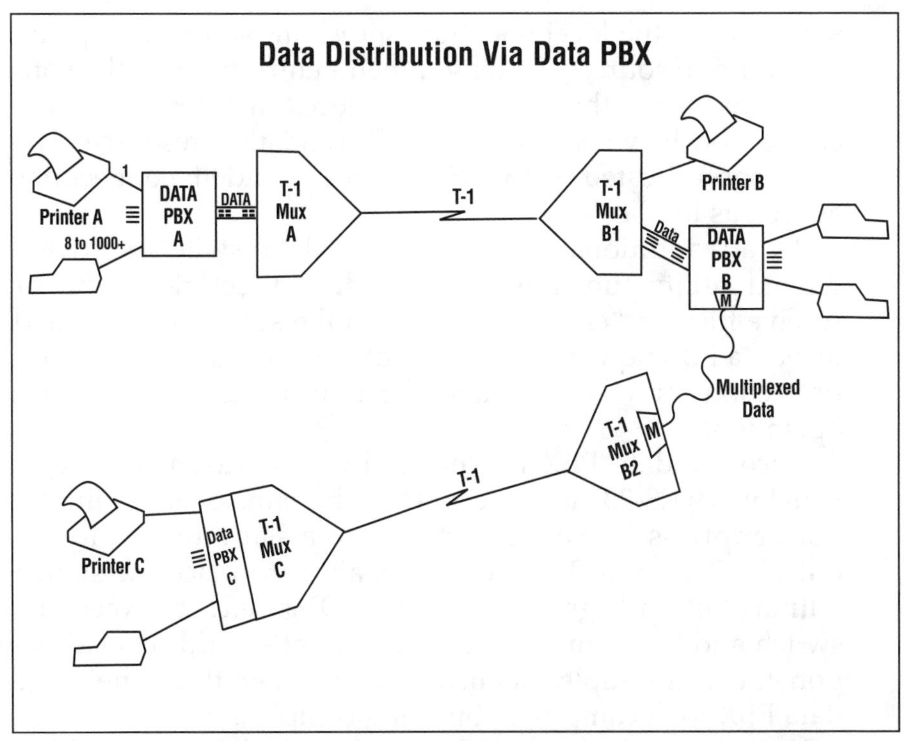

Figure 7-12 *Data PBX ports into a T-1 node offer a gateway between local devices and the wide area network. A data switch also provides local connectivity, like a LAN, and speed and parity conversion. PC software to control the data PBX makes networking of tandem switches transparent to the end user.*

speed and parity internally. For example, a higher speed port sending to a lower speed port will experience flow control.

Part of the Network. Some of the ports on a switch may connect to ports on a T-1 node (Fig. 7-12). Each WAN port could connect either to a remote device (printer B) or to another switch (B). The printer will appear to be a local resource of switch A even when the connection is extended over the WAN.

For a PC on switch A to reach a device attached to switch B requires the set up of two connections, one in each switch. It is possible to do this manually, but it requires a knowledge of the network topology and the address of each switch and device in the circuit.

Some data switch makers offer software for the PC that automates call set up. Rather than issue commands to each switch, even the local one, the user of this software pops up a menu of resources. Picking a menu entry triggers the software to send the commands necessary to make the connection between the calling PC and that resource. The more sophisticated software will set up tandem connections as well as local calls.

As an additional benefit, some data switch software offered utility functions: browse disk directories, send or receive files, or "chat" by typing on the screen. If interested in a data PBX network, look for enhancements like password protection for your files and file transfer as a background operation.

Since the data PBX is almost always an async device, it usually has RS-232 interfaces. Thus the same distance limitations apply as for connections out of a stat mux or directly out of a T-1 node. The strategy again is to place the switch within the work group to avoid LDMs, etc., between the switch and the terminals. In many situations this consideration favors multiple small switches rather than one large data PBX for a campus or building complex.

Network design. How does data switching function of T-1 fit into the T-1 backbone? A data PBX was sold as a feature option of a large T-1 node (B2 and C in Fig. 7-12). The switch was under the main network management system. Other T-1 vendors offered separate data PBX's, not fully integrated into the T-1 nodes (A in Fig. 7-12).

The trend is toward integration, but the data PBX appears to have given way to the router.

Network Management And Control

▶ All the hardware in the world does not a network make. Software is just as important, and taking an ever-larger role in supporting connectivity. As the size of a network grows beyond 10 to 20 nodes, the questions of control, continuing management, maintenance , and growth planning — all functions that depend on software—become so important they help determine basic long-term reliability of a network.

Just as the meaning "T-1" has changed with the technology (see Chapter 2), the understanding of "network management" has too. Originally influenced by early hardware designs, network management was later affected by LAN strategies, particularly the Simple Network Manatgement Protocol (SNMP).

In the jargon of the T-1 industry common up to 1987, the network management system (NMS) was an option, something added to the functionality of the nodes themselves. As we shall see, this definition of management resulted from an early emphasis on distributed network control.

The latest way to look at NMS centers on a workstation where application software manages the network. This view grew out of routers and the Internet, where autonomous systems (based on routers, a form of packet switch) had to be managed across vendors and geographic areas.

Management and Control Software Functions

Control	Management	Design
Call Set up	Trouble Ticket	Traffic Analysis
Call tear down	Inventory	Topology design
AAR	Vendor files	Tariff Files
Bumping	User records	Node design
Configuration	Moves & Changes	
Diagnostics	Historical statistics	
Alarm reporting		

Table 8-1

To make the discussion as clear as possible, we will divide NMS software into three components (Table 8-1):

Control, the normal operation functions that must be "on line" at all times. This segment includes connection set up and tear down, automatic alternate routing (AAR), alarm reporting, diagnostics, and all matters related to network topology and connectivity. Typically these functions are hardware specific, based on software that is proprietary to each vendor.

Management, the support functions that make life easier for the network operator. These can be "off line" in that they are not required for basic connectivity. Familiar functions include trouble ticketing, vendor liaison, user records, moves and changes, etc. Several software vendors offer generic products to deal with these aspects for any make of hardware.

Design Tools, which help in setting up the initial network and in evaluating proposed changes. Tariff files concerned with public network offerings and facilities, traffic analyzers, topology aids, and so on, are generally generic. Node configuraters, which produce a parts list for a specific way to deal with the topology at one location, fall in the "Tools" category even though proprietary to one hardware vendor.

Control, then, has been part of the nodes. Until 1988 it was not considered separately from the networking hardware. This function is now starting to stand alone, on dedicated processors, either within a node or on an external computer. *Management* continues to be the add-on functions of something outside the nodes. *Design tools* for T-1 networks appeared later than the other two forms of software. Early tools stood alone, but later versions are integrated into the control and/or management products from specific vendors.

In the SNMP world, control resides in the nodes, be they routers, LAN switches, or anything else. Management is performed centrally, or regionally, in NMS workstations. Each node contains software, the SNMP agent, that reports errors to the manager. The manager polls the agents to confirm they are "alive" and gather statistics. Commands from the manager to the agent can configure the node, reset statistics counters, and perform other functions.

The SNMP agent is the equivalent of the proprietary NMS software in each node that allows control and configuration from a central site.

Network Control

The first networking T-1 node was made intelligent enough to manage the network by itself. That is, the nodes in a network could stand alone, without an additional network management system, and deal with any anticipated event during the life of that network. As a first approximation to the ultimate need, it served well.

Unfortunately, those pioneers were not able to anticipate how the practice of networking was to develop years later. In particular, the need for large networks has grown beyond anything that early forecasters would dare predict. However, a failure to foresee 10,000 nodes in a network, and to allow for only 200 or so, should be considered wisdom compared to not seeing beyond point-to-point applications.

Now the need for larger networks is clear (many companies have 500 to 1000 locations, and may need multiple nodes within large sites). Distributed control architectures cannot deal with that many locations. The stage is set for a repeat of the debate that packet switched networks went through a decade ago: Should network control be distributed or centralized?

Packet networks, like T-1 networks, started with distributed control. When ARPANET, the original packet net, grew beyond 50 nodes, it crashed several times. Later the control architecture was modified to become more centralized, but not totally so. Since then it grew to many hundreds of nodes. Control of the Internet, which grew out of ARPANET, is distributed, but directory services are entirely centralized — there is only one name service authority at this time.

T-1 equipment seems to be following a similar path. The original designs had distributed control. They struggled with (and conquered) problems in clocking, alternate routing, and other networking features as networks grew to 40, 80, and 150 nodes.

The second wave of imitators and competitors again emphasized the benefits of distributed control. And again they had stability problems around 50 nodes. Only in 1987 did vendors begin to ship new products with centralized control architectures.

The reasons for the change deserve some study. We will examine the benefits of both, the trade offs between centralized and distributed, and a compromise approach called center-weighted control. The three are summarized in Table 8-2.

Distributed Control

Early T-1 node designs helped create a market, rather than respond to an existing demand. They did not have the benefit of hindsight. Designers had to extrapolate from data networks based on statistical multiplexers and low speed time division multiplexers. (All the early T-1 nodes came from companies with a history in data communications, rather than telephony.) Control of private data networks, at that time, was distributed.

That is, the nodes in the network performed all the functions of the network — there was no additional control processor.

The new T-1 products followed the established pattern. They used distributed control too. Every node had to know everything about the network: routing, topology, connectivity, priorities, resources, and so on.

Inherent Redundancy

There were good reasons, as a distributed control architecture has many strong points:

- The network is autonomous, and can be operated from relatively simple terminals (often a dumb ASCII device) (Fig. 8-1).
- There is nothing else to buy after the nodes; it's all there on the initial installation.
- Distributed intelligence tolerates quite well the loss of a single element, either a line or a node. Inherent redundancy has been the strongest selling point.

Types of Network Control and Management

Feature	Distributed	Center-Weighted	Centralized
Maximum number of nodes in net.	32-256	1000-10,000	Unlimited
Cost per port	High	Medium	Low
Time for AAR to reconnect full T-1	2-15 minutes	10-20 seconds	Manual or may be automated
Location of control intelligence	Equally in every node	Some in nodes; most centralized	All centralized
Control Redundancy	Inherent in nodes	Backup processors as desired at any locations	Backup processors as desired at any locations
Node complexity	High	Medium	Low
Migration to large node size	Hard	Smooth	Simple
Size of Control processor	Limited	Unlimited	
Upgrades require:	Software change in every node	Mostly change in central software (sometimes in nodes)	Change in central software

Table 8-2

The distributed control architecture works well in small networks. Its drawbacks are felt only as the network grows.

Practical Limitations

In recent years one problem for large networks in particular has been the time to recover from the loss of a T-1 line or node. Distributed control requires that the nodes themselves find alternate paths. Typically they negotiate one "hop" or node pair at a time, from node to node, for each logical circuit. Some versions attempt to handle many circuits simultaneously. Still, the process can take tens of minutes to find alternate paths, allocate bandwidth, and reconnect all the logical channels that were on a T-1 line that failed.

Obviously if every node must contain a map of the entire network, then for each topology change you must update that map (a set of routing tables in every node). Some vendors require that you update the map (tables) manually. Others provide a software utility to update the maps automatically. But it must happen in every node.

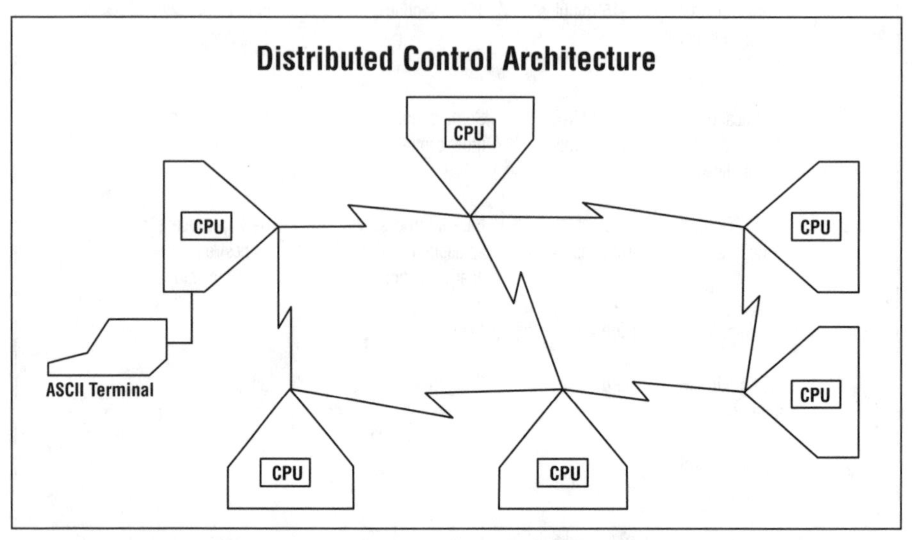

Figure 8-1 *Autonomous nodes include the computer processing power necessary to run a network. The nodes, among themselves, choose routes, alternate routes, deal with priorities, find backup clock sources, etc. The only management tool absolutely necessary is a dumb terminal. The nodes present the operator with either a command line or (highly preferable) a menu of selections and prompts.*

Then there is the question of enhancing the control software, or removing bugs. If all the nodes are controllers, then that's where the software must be upgraded. Typically the change has been a physical replacement of a PROM set; FLASH memory now allows downloads over the network. To give all nodes the new software, compatible with each other, every node needs new PROM's, or a software download. Some early users changed all their PROM's four, five, and six times. Sometimes they had to shut down the network, or at least one or two nodes at a time.

Administering a network takes processing power. Distributing control means that every node needs significant processing capacity dedicated to this job. How much can you get into a node? How much is practical to put into nodes that may never be part of a large network? How much will it cost? The answers always lead to a trade-off solution that limits the number of nodes in a logical network.

Distributed nodes shipping in 1989 offered different maximum network sizes: 32 nodes, 64, 96, 160. These number changed for some vendors of distributed architectures to 256 nodes.

Centralized Control

The telephone company approach to managing leased line networks resulted in a centralized architecture (Fig. 8-2). It requires very little of a node.

The prime example from current technology is a DACS. This switch does almost nothing unless commanded to by a human operator or computerized controller (as in CCR). Commands reach the DACS over a separate packet network, Common Channel Interoffice Signaling (CCIS). (Signaling System #7 is the successor to CCIS.)

A DACS will generate alarms and recover from a power loss (by restoring connections that existed earlier).

Less Intelligent Nodes

Nodes under centralized control perform only local processes. That is, the node "does its thing" without know-

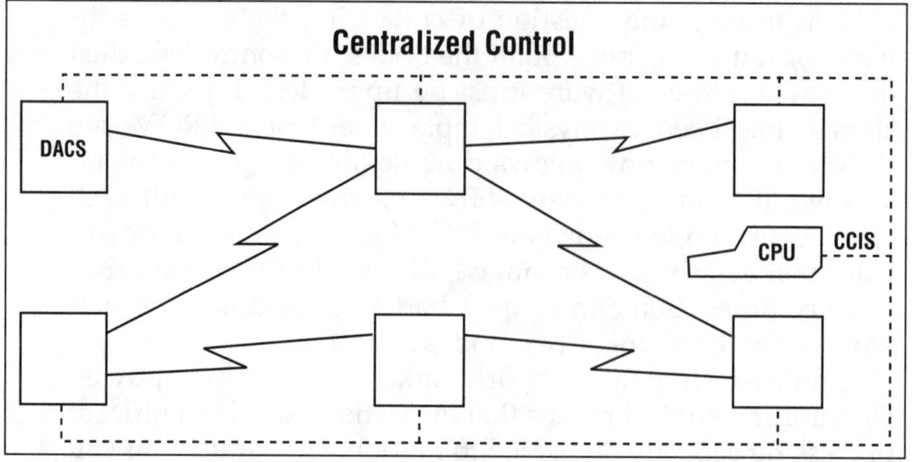

Figure 8-2 Centralized control asks very little of a node, only local processing. Decisions about network-wide matters (routing, rerouting, priorities, backup clocks, etc.) come from the central processor. With all networking functions relying on that central computer, it should be redundant for any sizable network.

ing anything about the topology of the network, routing alternatives, or connectivity. It has no need to know because it gets complete instructions from the central site.

Lacking a control processor, this type of node avoids that considerable expense. It seems clear that a DACS is much less expensive than a digital central office switch. If the 5ESS switch didn't cost far more than a DACS, no one would deploy a DACS.

Continuing deployment of large numbers of DACS in the public network also shows that centralized control imposes no limit on the number of nodes under one management scheme. This applies equally in private networks. At least one vendor controls and manages about 10,000 nodes in one logical network.

Networks under central control are relatively easy to expand or modify. When a new DACS or central office switch is added in New York, a DACS in San Diego needn't be reprogrammed.

Powerful Central Processor

The greatest benefit to centralization may be the ease of enhancing the networking functions. To enhance, debug, or add a feature, you change only the software at the central processor. The fundamental operations in the DACS nodes (e.g., making cross connections) do not change. Therefore you need not change software in every node.

A carrier like AT&T will develop new service offerings based on software in the CCR control center. Changes in the nodes will be rare. Private networks of nodes under central control can add software to implement better routing algorithms, to partition control among end user departments, and to integrate more management functions based on control information.

With so much responsibility on a central processor, it must be redundant. The simplest form is a complete second processor. If the second is identical to the first, control redundancy can seem expensive. This expense is very visible as it takes the form of a definite work station. However, the more accurate cost comparison is between complete networks, not parts of them.

Ideally, it should be possible to have multiple back up processors, all at different geographical locations. Each back up processor must be kept up to date by the on-line machine. When the secondary processor takes over, its data base must reflect the real network. There should be no delay to poll the nodes or "learn" the network.

All major carriers have dispersed control centers now, and private networks are acquiring them as fast as practical. Any one of them must be able to assume control, either on failure of the master controller or on command. A worldwide corporate network is usually managed from several sites during the day, following the sun around the globe. This kind of flexibility is almost inherent in distributed architectures, but should be included in centralized control stations as well.

Center-Weighted Control (CWC)

The goal of a center-weighted architecture is to combine the advantages and best features of both distributed and centralized control. Therefore the network assumes aspects of both approaches. The responsibility for controlling the network is divided between the nodes and a central processor (CP). Nodes and CP share control by dividing functions:

- The nodes do as much as they can while having only local knowledge. Nodes deal only with what is connected directly to them. Nodal functions include digital encoding and compression of voice inputs, making digital circuit cross-connections, generating statistics and alarms, and invoking protection switching for circuits and modules.
- A central processor handles those functions that involve network topology: circuit set up end-to-end, automatic alternate routing, prioritizing activities, and anything that extends beyond a single node.

In other words, network control is weighted toward the central processor. But the nodes provide a high level of distributed control over local operations.

Local Signal Processing

As a result, center-weighted control relies heavily on functions built into the nodes:

Alarms. Rather than tell the central site only that there is a problem, the node under center-weighted control should identify the exact nature of the problem. This way the central processor can act immediately. It will not have to query the node first to get the details.

Protection Switching. The node can switch from a failed circuit to a back up circuit. A node can also fall back from a failed hardware module to a redundant spare when installed in the node. Rather than protect only full T-1 lines or complete modules, center-weighted control should also protect individual logical circuits or channels. This may involve trunk conditioning that is transmitted throughout a network.

Trunk Conditioning. The node should condition all channels affected by a failure. Conditioning imposes specific bit patterns (like all 1's) or a specific signaling state on every channel disconnected by a failure. Conditioning spreads immediately through the network after it is applied by the nodes.

Conditioning by nodes on either side of a failed T-1 line allows end nodes of a multi-hop logical circuit to know when a failure occurs (Fig. 8-3). This information allows the end nodes to invoke protection switching (if a back up circuit is provided) without knowing anything about the topology of the network.

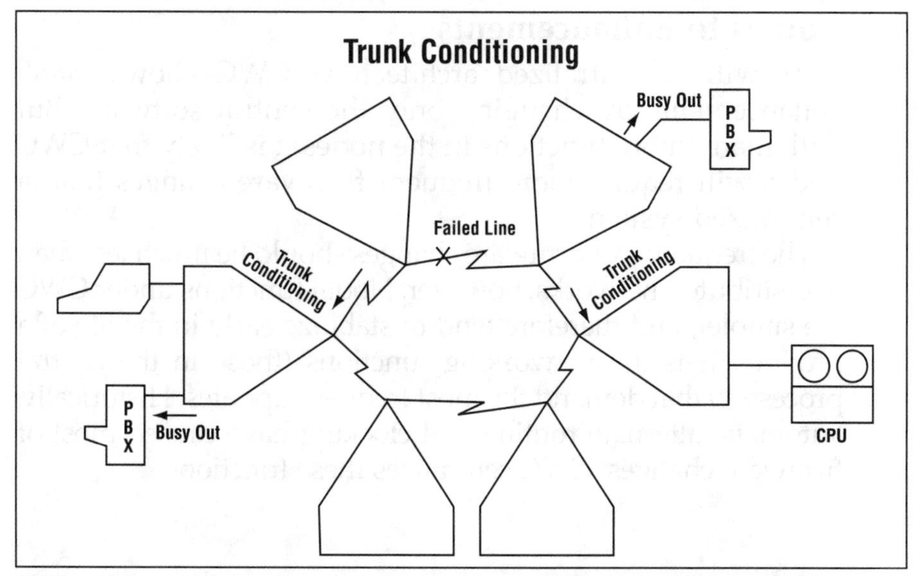

Figure 8-3 *In addition to local processing of voice (compression) and data (multi-drop bridging), nodes must also condition the trunks (DS-0s or sub-rate channels) affected by the loss of a T-1 line. This allows end points of logical circuits to invoke rapid protection switching to alternate paths. Conditioning also serves to busy out PBX trunks, preventing high and dry situations.*

For another example, look at any large mesh network without circuits available for protection switching. The end points then must know to busy out a voice interface port when the connection is broken by a failed line in the middle of the network. The end points, however, have not lost sync on any line connected to them. But if the nodes on both sides of the broken line recognize the loss, they can "condition" the channels on the lost line. By setting signaling bits to specific states, a message is carried end-to-end on that channel. Therefore the end nodes for that broken connection know to take some (local) action. They can busy out the voice ports to a PBX.

Statistics Gathering. The local operations in a node are of great interest to the network manager. Utilization levels affect advance planning. A node can track how busy each link is.

Many operators want to bill back the cost of actual usage to each department. A node can record the times of connections in a format that the central processor can convert to invoices.

Routes to Enhancements

As with a centralized architecture, CWC allows many enhancements by changing only the central software. But with many more functions in the nodes it is likely that CWC nodes will require more frequent firmware changes than a centralized system.

The frequency of firmware changes should be much less than in distributed networks, however. Nodal functions under CWC are simpler, and therefore tend to stabilize early in the life of a product. It is the networking functions (those in the central processor) that demand the most frequent updates. Historically, automatic alternate routing and clocking have caused most of firmware changes. CWC centralizes these functions.

Control Functions

Regardless of how you control a network, there are certain operations that are essential (Table 8-3).

Making Connections

Say this terminal wants a line to that computer. How is the call or connection set up? What does the network operator do? Current T-1 bandwidth managers offer at least four alternatives.

Physical Routing would be used for drop and insert nodes. When cables are required to establish a connection an operator must make physical changes at each node. The very earliest T-1 multiplexers (from around 1976) and the least expensive current models work this way. In the worst case the operator calculates TDM frames and sets switches and straps at each node.

Manual Routing, from the central site, involves assigning a path by specifying each time slot on each link, and how the channel is handled within each node. This is typical of a DACS, older matrix switches, and applies particularly where products of different manufacturers are mixed in one network. It requires an exhaustive knowledge of the network's topology, the command language at each node, and the ter-

Shared Features Among Control Types

Distributed — Center-Weighted — Centralized

Functions of the node	Functions of the central processor
Protection Switching to spare circuits	Automatic alternate routing
Fallback to spare modules	Circuit prioritizing
Select alternate clock sources	Mapping network topology
Invoke alternate node configuration (time of day, etc.)	Hold connectivity tables
	Call setup

Table 8-3

minal characteristics at both ends. Control is usually from a terminal or a personal computer. The PC may run either terminal emulation or some software designed for a particular type of equipment. Perfectly acceptable in small networks, manual routing becomes difficult for more than approximately a dozen T-1 nodes.

Determinant routing, performed by the network, was the philosophy of the Bell System for many years. All central offices, starting with the local serving office and working up to the regional centers, were arranged in a strict hierarchy. Once the call was dialed, routing was handled automatically. Calls that couldn't be completed in a central office were passed up to the next higher level for routing to completion.

Gradually, as programmable switches were introduced into the PSTN, ad hoc arrangements were added. New trunk lines, outside the hierarchy, were dedicated to frequently used routes between end offices. These shortcuts reduced the load on higher levels and made "a real mesh" of the neat hierarchy; but they also increased efficiency.

AT&T made the shift official in 1985, with an announcement that ATT-Com (AT&T Communications) had abandoned the strict hierarchical philosophy (required by inflexible mechanical switches) for a mesh plan. They programmed switches and installed trunks to reflect traffic patterns, rather than strictly a tree design.

A network with either distributed or centralized control may use determinant routing: the user specifies the end points and the network finds the route. Distributed nodes exchange messages to establish the network topology automatically and store it in look-up tables at every node. A central processor calculates the optimal new path from information in the central data base, then issues commands to the nodes to set up the connection.

However, the user may not control the routing any more than a caller can control the routing of a phone connection. Several vendors do give the operator an option to put "class of service;" restrictions on a connection request. That is, a high quality voice circuit might be marked "not satellite" to avoid the delay. If only satellite is available when bandwidth is requested, the call will be blocked. The alternative class "prefer not satellite" would let the connection be made. "Prefer," "must have," and "don't care" complete the range of options.

One problem does arise in some forms of determinant routing. The most troublesome case is when the network always picks a new path each time a connection is set up, for example when a voice call is placed. The network keeps no record of the routing, so the operator cannot determine what that path is. If a user reports intermittent trouble, there is no way to associate the fault with a specific line or node. This makes diagnostics nearly impossible.

Automatic Routing with Manual Override acts like a determinant system when the operator wants simply to specify the end points of a connection. That is, the network finds the path. For example, an algorithm might minimize the number of links or nodes in the path.

There is additional flexibility in some systems that lets the operator:

- alter the routing tables in each node.
- explicitly set a path from a central control point.

This opens the possibility of assigning different routes to different ports, applications, or categories of data. It might be desirable, say, to route the president's voice tie lines via terrestrial facilities rather than satellite circuits, even if the satellite route normally is preferred. Remote printers for stock picking tickets at the warehouse could be allowed to take any path, or directed to the noisy lines (where reduced throughput from retransmissions isn't so noticeable).

For critical traffic, the operator might be prudent to specify different paths for portions of the same trunk groups. Diverse routing improves reliability and availability.

Automatic Alternate Routing (AAR)

When the network has the intelligence to route connections, the network can repair itself to a certain extent. That is, the nodes or the central processor can find alternate routes for connections (voice or data) when the path in use "goes away." Whether the cause of the path loss is a missing circuit or a missing node makes no difference. Automatic alternate routing, which depends on having data bypass in those nodes, protects against either (or both).

In AAR under distributed control, loss of a connection causes the node that originated a call to attempt to re-establish it. Since the first path listed on that node's routing table is not working, it tries the second choice. Most major T-1 nodes offer multiple routing options, for example seven alternate paths out of a node in addition to the ideal one. The originating node negotiates bandwidth on the alternate link with the next node, then passes the connection request. The second node looks in its routing tables for a way to reach the requested destination, and the process repeats. This link-by-link reconnection continues until the call is blocked or successfully reconnected.

Under centralized and center-weighted control, the master processor performs AAR. First it examines its data base of network topology to determine which connections were broken. It can then sort them by priority and look for alternate paths, again within the data base on the processor. When a path is calculated, the processor sends commands to the affected nodes, instructing them on how to set up new cross connections.

In general, AAR processing (in nodes or central station)

takes place after a fault is detected. For T-1 lines, this may mean waiting more than 2.5 seconds (an Accunet requirement) before declaring the line down. This avoids unnecessary AAR due to short "error bursts," but delays the start of AAR and the completion of alternate connections.

A centralized control architecture can offer something faster by taking advantage of the fact that it is self-contained. Rather than wait 2.5 or 3 seconds to start calculating, the central processor can begin immediately (say 50 ms after loss of sync). No AAR commands are issued however, until the waiting period is up. If the line clears within 2.5 seconds, then the new commands are discarded.

Going beyond standard AAR, a centralized processor, working with local protection switching in the nodes, can establish alternate routes before a line fails. The process relies on the common practice of not filling T-1 lines to capacity. In the portion of bandwidth left unused on each line for future growth, new applications, and AAR, the central processor reserves bandwidth for an alternate path.

When a new connection is established, the central control asks for a priority rating on the connection. If the operator assigns a priority high enough, the processor then seeks an alternate path in spare bandwidth. Ideally the alternate avoids all the links and tandem nodes of the primary path. Only the end nodes are common to both paths (Fig. 8-4).

It is up to those end nodes to act as A-B protection switches. Upon loss of connectivity over the first path, the end nodes switch to the second. Note that the nodes themselves know nothing about the alternate path, or network topology. They simply sense a break and fall back to the spare path. Since A-B switching can be very rapid, this scheme restores protected channels much faster than AAR.

Following Priorities

As the AAR feature seeks paths to re-establish connections, the bandwidth available on the remaining T-1 circuits may not be sufficient to accommodate every request. There are three options open:

Random reconnection — make connections in any order without regard for the type of traffic or the application carried by the channel. The possibility that the president's or customer's call might be dropped (while the remote order printer continues) makes this unacceptable. Every multiplexer and bandwidth manager with a claim to networking ability has surpassed this level of performance, but can be programmed to work this way.

Priority assignment — each connection priorities is assigned a number (1-10, say). When AAR restores broken connections, those with the highest priorities are handled first. Calls are completed until all are done or available bandwidth is exhausted. For practical reasons, this is the most commonly applied technique.

Priority with bumping — the same as 2, but existing connections with lower priorities are interrupted to free bandwidth for higher-rated calls and data connections. While several multiplexers have this ability, one maker reports that no customer has been able to designate a significant amount of bandwidth (read "significant number of users") as expendable. Implementation of bumping can be a sensitive political issue within the user organization. The likeliest candidates for bumping are voice tie lines whose traffic can spill over onto the PSTN.

Remote Reconfiguration

Anything the nodes can do automatically, the operator should be able to do manually — from the central site. The most desirable benefit of this feature is the ability to reconfigure the network. Shifting requirements mean the operator often must reassign bandwidth, add or drop terminals, and make other changes. If this can be done entirely by "soft" commands, there is no need to send a technician to make physical adjustments. Moving jumpers on a printed circuit board, throwing DIP switches, or entering codes at front panel switches requires trained personnel on site. It is also time-consuming and expensive.

Most sophisticated bandwidth managers can perform a complete reconfiguration automatically, a useful trick. The user can store an extra set of tables (or many sets) in the network management equipment at the central site. On a single command, every node can be given a new routing and connection table, in standby memory. Then on another command, which may be a time of day signal from the system clock or each node's local clock, all nodes start operating from the alternate tables. The entire network is reconfigured on one or two commands, perhaps without operator intervention.

Applications for alternate configurations include changing assignments of bandwidth from night to day conditions. Normally more voice tie lines are needed during the day than at night. Then computer back-up might need many

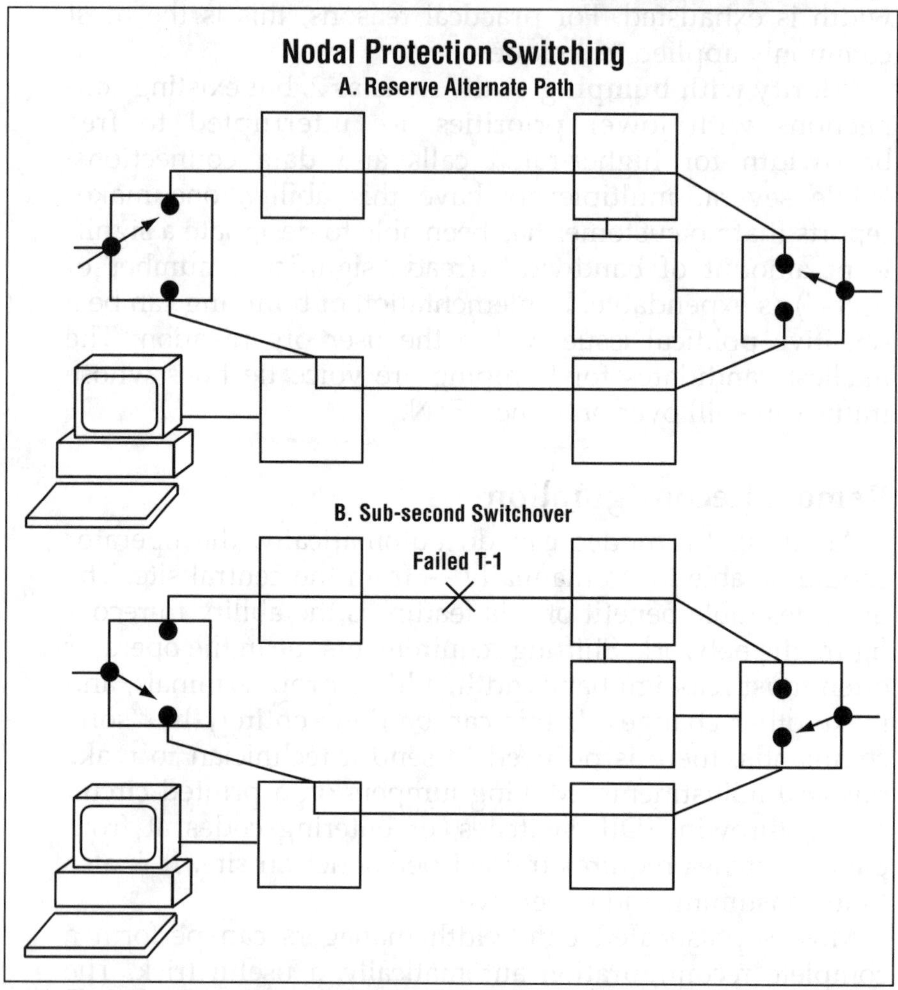

Figure 8-4 *Protection switching to backup circuits does not require that the nodes "know" the network topology. Even relatively simple nodes can sense the loss of connectivity on the primary path, for example from trunk conditioning. The nodes can then switch the local port to a preassigned standby circuit.*

wideband channels. With reconfiguration so easy to do, it is practical to reassign bandwidth on short notice.

For example, the number of voice tie lines might be reduced sharply at lunch time, permitting a mid-day file transfer (the morning's sales, for example). There could be an alternate configuration that on demand frees a large channel for video teleconferencing.

Experience indicates that a network operator eventually uses all the configuration flexibility available. In an emergency, the more options the better. As a network grows to near full capacity, the ability to juggle connections can extend the time before additional circuits are needed.

Transfer of Control

Must the central site remain manned all the time? Three shifts are possible, but often difficult to staff. On global or multinational networks there will be other sites in different time zones that are manned when the central site closes. The single point of control should be transferable to "follow the sun." In case of disaster — power loss at the main site, say — having a backup control point at a distant location can keep the remainder of the network operating.

When the power comes back, it should not be necessary to reprogram the affected nodes. Battery backup of memory (or some form of static memory) can preserve all settings. The more subtle question is whether the node can restart automatically. Some can, even to the extent of re-establishing the connections that were broken by the disaster.

Diagnostics

Any fault in a working or spare module must be reported so repair can be arranged. The first step in arranging repair is learning about it.

A key point is that all reports—errors, statistics, whatever — from any part of the network, should come to one point: the central control station. Here will be found the most able operators, the people who can best deal with a problem.

Alarms should be able to trigger something dramatic. People must be alerted, something must grab their attention.

- Visual indication (flashing or colorful display on a CRT);
- Audible signal like a bell or horn (activated by a dry contact closure);
- Permanent record on a line printer.

Supervisory Port

The "soup port" is the physical connector on the node where the operator attaches a terminal, or perhaps a computer, to collect alarms or fault reports (and statistics, more of which later). The same interface gives instructions to the network nodes. But how does the operator know something is wrong if the supervisory port fails? Not easily, which is why this critical element in the network certainly must be redundant, too. In addition, supervisory messages had better be protected by some error checking protocol to prevent garbled instructions from reconfiguring the network in a random way.

In addition to the configuration control covered earlier, the sup. port is the diagnostic window on the network. It should provide the means to exercise the equipment and conduct various tests.

Testing

To keep a network operating requires an ability to identify, and then isolate, faults.

Self checks. Every node and component in the network should be able to test itself on command, at power-up, and continuously in background. Any failure of a test (barring a failure in its communications sections) must generate an alarm or report.

Another form of checking is simply to determine the status of a port or line. Is it connected? Set for what speed? Arranged as DTE or DCE? Much can be learned from these simple facts, available as part of the self check in most intelligent nodes.

Fault isolation. Working with the results of self checks, and additional exercises, the operator should be able to isolate most faults to the board level. In some designs, failures can be traced to specific chips.

After self-checks by the nodes, the principal requirement for actively testing the network is flexible test access. Essentially access allows an operator at the central site to split a channel at any point; that is, to break a logical channel and route a connection from either side of the break back to the test position (Fig. 8-5).

Access allows the remote operator to inject test signals into segments of line that have a loopback at the other end. Comparing sent and received data blocks over progressive line sections will isolate the area at fault. One possible procedure would be to set up a loopback at both ends, then work from one end to the other, splitting the channel at every access point and testing each section from either end.

By sending known bit patterns through successively larger loops in the network, the fault can be identified as lying between two loopback points: the last one that returned a correct signal, and the first one that didn't (Fig. 8-6).

Therefore, loopback should be built into every port of every component in the network. Ideally these loop backs, at least on T-1 ports, should be selectable by DS-0, by logical channel, and for the full T-1.

CSUs provide loopbacks in response to commands on the line. Multiplexers may also do loopbacks, usually when

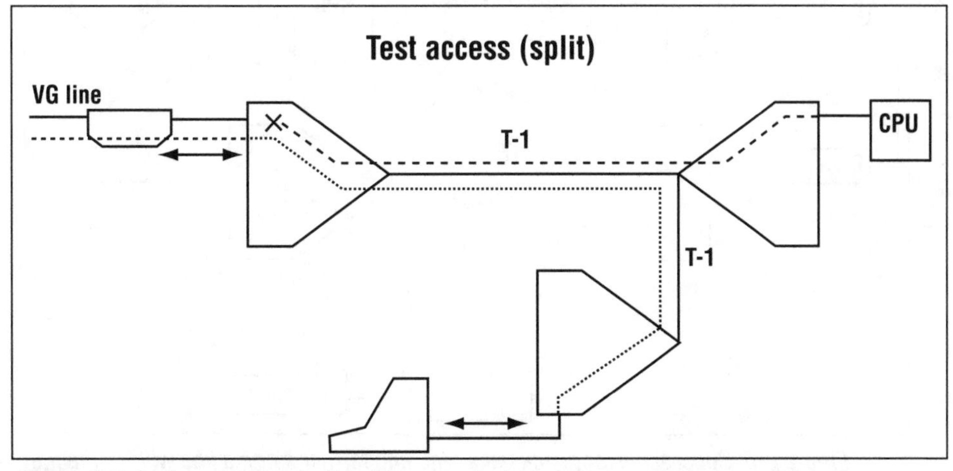

Figure 8-5 *Split access breaks the channel under test at the remote access point, and connects the tester's instruments to either side of the break. It is then possible to inject test messages, or perform loopbacks to help isolate line or equipment faults.*

commanded via their supervisory ports or network management system.

The necessary test pattern generator and comparator are built into some major multiplexers. More often a network has stand-alone testers of greater sophistication.

Advanced Diagnostic Features

If the fault is like most, it "fails to fail" whenever an operator attempts to look at it (The Disappearing Pain Syndrome when one gets to the dentist's office). The network should help in these cases by watching itself all the time.

Keeping Statistics. Data multiplexers and packet devices routinely count the number of frames received in error and retransmitted. This count can indicate a degrading line. But on T-1 circuits, the multiplexing is time division, and frames are not retransmitted. Still, a T-1 mux can count errors like loss of sync occurrences. With ESF service, the CSU or multiplexer itself records CRC errors in the framing bits. This is one of the principal benefits of ESF service—continuous measurement of error level.

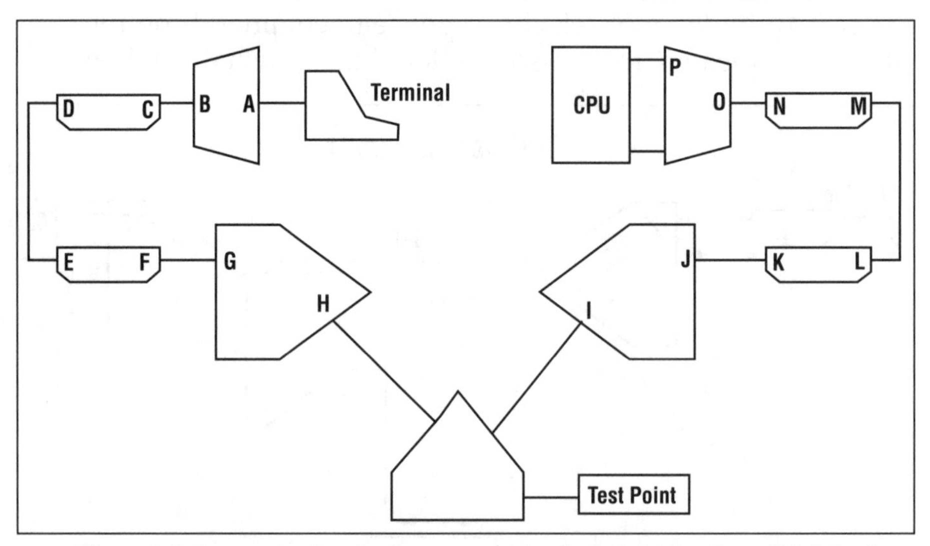

Figure 8-6 *Successive loopback tests reaching farther along a line will fail the first time the loop includes the faulty component or line section. The connection between the terminal and the CPU could be looped back toward the test point at any of the locations A to P.*

Statistics of normal activities like connections made and broken can also help. Any of these statistics contribute to diagnostics and overall network management. The more of them offered, the better.

Problems can exist even when the network itself is operating flawlessly. The nature of the traffic may give some indication, if the operator can observe it. Here a monitor feature is a blessing.

Monitor mode. allows the operator to tap into a channel, usually at one of the end ports. In effect, the connection is bridged digitally. The same signals that enter and leave the port are available at the central site. There the operator can attach a datascope or protocol analyzer to test for abnormal signals entering or leaving the network, possibly the fault of computers or terminal equipment. This access feature should also relay the status of EIA signal leads. In many instances, monitoring has shown the problem was simply a terminal not turned on. Some nodes allow a remote test center to tap into a channel at any site (Fig. 8-7).

Line monitoring on the T-1 now benefits from newer, more compact (and more cost effective) test equipment. Not only can portable test instruments look at the D4 or ESF framing patterns, they can do far more:

- examine CRC bits in ESF framed circuits for errors;
- confirm conformance to basic standards (like the ones density spelled out in Tech. Pub. 62411);

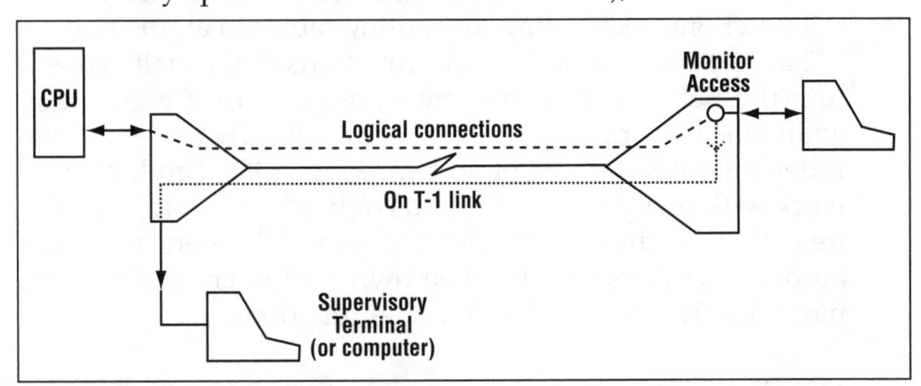

Figure 8-7 Monitor mode of test access allows the technician to intercept traffic on an individual channel at a remote location. It is highly useful in examining protocols and operator performance.

- decode PCM bit streams in a particular DS-0 and play the voice signal through a speaker;
- inject T-1 bit streams with various known error rates, to see how the nodes respond;
- generate alarm conditions (all 1's, yellow, etc.) to measure response and recovery times in the nodes;
- send zeros in strings of increasing length to test for proper clock recovery;
- alter the frequency of a nominal T-1 signal to evaluate response to jitter and wander.

Network Management

Because all early T-1 nodes contained the network control function, "network management" came to mean something additional. In early cases it was simply a friendlier user interface. Early nodes presented the operator with a command line that demanded extensive knowledge of the command syntax. The early network management systems converted these arcane symbols to English language menus.

Once the processing power of the central controller became available, equipment designers added more features. The PC was programmed to track the topology of the network, and the exact configuration of each port and component. Utilities were added to take over formerly manual functions, like setting up routing tables in all the nodes.

There has developed a set of functions that usually constitute the network management system. Most are generally available or promised by most hardware vendors. Independent suppliers of software also offer products that work with many different brands of hardware, but generally treat that hardware in a generic way. Therefore what the hardware makers offer for their own equipment will often be more tightly integrated with nodal functions.

Configuration Setup

The drudgery of early network installations is disappearing into automated utilities within network management systems. Former time consumers like routing table set up, configuration of the supervisory channels, and so on no longer require direct operator input.

After installation, the same utilities deal with moves and changes, network expansion, and emergency reconfigurations much more rapidly than humanly possible.

Inventory

It would take thousands of man-hours and reams of paper to track the physical connections, installed hardware, and software settings in even a medium size corporate network. Fortunately for users, many of the major T-1 vendors now can query nodes to determine their exact configurations. In some cases the reports at the network management system, often presented graphically, include:

- topology of lines between nodes;
- cards and modules installed in each shelf;
- hardware revisions of each module;
- software settings for each port;
- firmware versions installed in each module;
- cost;
- source (manufacturer at least, and distributor if applicable).

This level of detail most often comes from the vendor's own NMS software. However, it is very valuable to be able to include all the equipment in the network, from every vendor.

Trouble Ticket

Every alarm that requires individual attention, and every problem reported by a user, should be tracked until the case is closed satisfactorily. After a time interval that may depend on the nature of the problem, attention should escalate upward in the management structure. To make sure it happens, software can create reminders at key times, to ensure that problems are not overlooked.

In addition to versions integrated into vendor's NMS, trouble ticketing is available from many sources as a stand alone application.

Moves and Changes

A special type of "problem" that never goes away is the constant request for new services, changes to old services, and disconnections in a private network. If the end user community is more than a few hundred, some formal system to track requests is essential.

Not only is an automated system needed to ensure completion of requested work, many organizations require:

- management approval before work proceeds;
- clear tracking of work for bill back to departments;
- justification of work done.

Automation of provisioning (moves and changes) in the network management system allows new connections to be entered now for set up at a later time or date. This means an operator can work to a more convenient schedule, and need not be on-line at the exact time a new connection goes into service.

User Records

Most end users on private networks are not sophisticated in communications, nor even power users of PCs. They often forget, if they ever knew, what equipment they have on their desks or how it is set up. Network operators find that it pays to keep this information themselves.

A record of an individual's history involving communications equipment is often a valuable tool when trouble shooting the network.

Vendor Files

Is that component under warranty? Who should we call for maintenance service? When and where did we buy it?

The ideal is an integrated package that cross references the files for inventory. Records should give the operator instant access to names and phone numbers of the people who can get fast action direct from the vendor, as well as the distributor or reseller who actually delivered the equipment.

Historical Statistics

Nodes have limited capacity for statistics. ESF CSUs, for example, save only one day's Error history of errors, over 15 minute intervals. Over the years the number of quarter hours, times the number of T-1 lines, times two ends, will add up. Watching trends in these error statistics (preferably in graphs) can warn of deteriorating line quality, and therefore can be important.

The only place to collect and analyze this volume of data is in a central network management system.

Network Design Tools

A relatively new class of software available to the private network operator, design tools have evolved from larger scale programs designed to model the public phone network. New experience with private networks has prompted consultants to put down some of their knowledge in the form of application software.

Traffic Analysis

The oldest discipline in design, long a staple of telephone consulting, traffic analysis establishes the number of circuits needed to provide a given quality of service for a certain number of users. Assumptions are made about the nature of the traffic from the type of work handled, geographical distribution, and other inputs.

Tariff Files

Carriers have always published the costs of their facilities, but the paper document was the only form available at a reasonable cost to most end users. The increasing number of private networks has created a market for computerized tariff files, needed to optimize configurations and to simply check the monthly bill.

Several publishers offer tariff services (frequent changes to tariffs make any given version obsolete very rapidly). They can be on-line, or on diskettes mailed at regular intervals. Costs vary widely also.

This information is essential for the next design tool, the topology designer.

Topology Design

With the results of the traffic analysis, and drawing on the tariff files, the topology design software decides the most economical choices for locations of nodes, types of service, and routes for the backbone circuits.

These tools are relatively new and, with the necessary cost of the tariff service, far from cheap. However the best of them do more than configure a network from scratch. They can play "what if" games with the network, off-line:

- Hold the tariffs constant, but adjust the traffic analysis to reflect anticipated growth. The simulation can point out shortages of bandwidth, congestion points, and remedial action.
- Simulate a major failure by "taking a line out of service." The topology design tool should be able to apply the same algorithm as your network to show what you would get from automatic alternate routing under real conditions.
- Plug in anticipated tariff trends to see how they affect different topologies; what might future costs be?

Most creators of topology design tools seemed to write for a generic audience (that is, not for a specific brand of hardware). The smarter programs, however, incorporate at least some accounting for the nature of the T-1 nodes to be used.

Node Design

The flexibility and power of T-1 nodes is increasing dramatically, but so is their complexity. Even after the topology is roughed out for minimum line costs, there remains the job of specifying the exact node configurations. You can't order it unless somebody writes down the quantities and part numbers.

Vendors perform this job as a service to prospects and to customers expanding existing networks. Without automated assistance it can be a very large job just to keep up with proposal requests. Therefore several vendors put the job of node configuration on PCs. The next logical step is to make the tool available to customers.

As software tools evolve, look for greater integration

among these design packages. For example, a user with an established network would plan out the next stage in expansion by building on the installed base (current inventory). Future tariffs could be anticipated, as well as traffic volume. Folding in plans for new building sites, new staff, and perhaps mergers should also be possible.

The result will be greater control of costs, of course. But mostly the network manager will benefit from delivering better service to end users on the network.

T-1 Case Examples

- **Cost savings.**
- **Flexibility.**
- **Availability/Reliability.**

▶ T-1: reasons to install

These are the three main reasons companies install T-1 lines and networks. Here are some examples of actual networks and the thinking that went into them.

From these practical answers to questions of network design we hope you can find something to apply to your own network. Just remember that your design will depend on the actual locations of sites and the tariffs.

Pricing of T-1, as for other forms of lines, has been variable and promises to continue to change. It is impossible lay down hard rules under such circumstances, but experience indicates several things to look for:

- Intra-LATA T-1 may be less expensive than inter-LATA.
- Look for ways to build intra-LATA subnets that can be joined as efficiently as possible via inter-LATA T-1 lines.
- Check pricing on various circuit routes: the standard practice is to choose the closest serving office, but it may not result in the shortest route — and you pay by distance.
- Compare costs of local access lines provided by AT&T with the same service from the operating company (or a

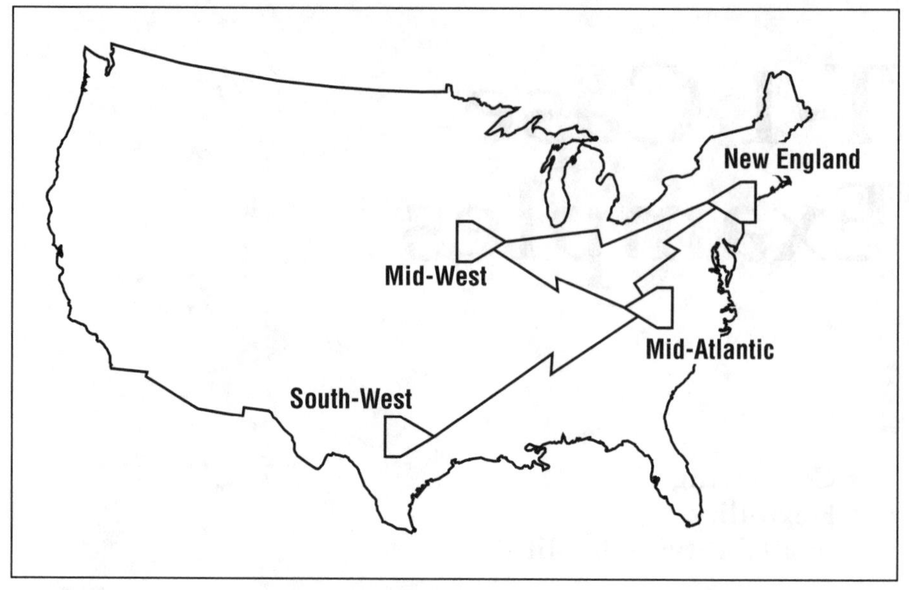

Figure A-1. Intra-company communications prompted a move to a T-1 network using compressed voice. Reduction in the number of leased voice grade tie lines saved $2 million per year before the network was fully implemented. Data channels ride almost free.

bypass route).

- Network integration, combining voice with data and other forms of information on a single digital network, typically produces the largest savings.
- Last, determine well in advance who will maintain the network, and exactly how.

Manufacturer I

In most corporations, voice dominates communications. Data takes a secondary role, in terms of sheer traffic volume.

A large corporation with many plants and offices in five clusters found that leased voice tie lines among its many PBX's cost about $10 million per year (1984). T-1 lines over large distances (Fig. A-1) could link the clusters, but with standard PCM encoding for the voice lines, the cost would be about the same as for multiple voice grade lines. Voice compression in the nodes made T-1 much more appealing.

Originally this corporation used CVSD, largely because that was the encoding technique then available. CVSD also allowed the data rate per channel to be adjusted above or below the standard (for CVSD) of 32 kbit/s. Because the network handles mostly employee to employee conversations, a slight reduction in quality was acceptable on some legs to save bandwidth. A relatively small number of data channels piggyback on the voice network.

On the other hand, within clusters where links were not yet fully occupied (the Midwest in Fig. A-1), the data rate on voice conversations could be increased toward 64 kbit/s to improve perceived quality.

Those legs dominated by voice tend to employ a single T-1 link. Each can fall back to the public network if necessary (in case of an emergency or the need for instant heavy bandwidth). Routes carrying heavier data traffic (as in the mid-Atlantic area) duplicate intra-cluster circuits for reliability as there is no public network fallback facility. The private network must supply alternate routing if the primary route fails.

Overall savings on line charges exceeded $2 million per year to start. The T-1 network expanded further, and likewise the savings.

By 1987 the traffic volume had almost filled the inter-cluster T-1 lines. Not only did this situation halt further expansion, it posed a problem in finding alternate routes during failures.

The customer's response was to go further in compressing voice—to 16 kbit/s per channel. This step freed enough bandwidth to again allow alternate routing. It also avoided the need to lease additional T-1 lines, a very significant dollar savings.

Manufacturer II

The following is excerpted from an internal memo written by a user corporation. The purpose of the memo was to explain to non-telecommunications company executives the reasons for recommending a T-1 network for that division in 1986. Only the company's name and its location has been changed.

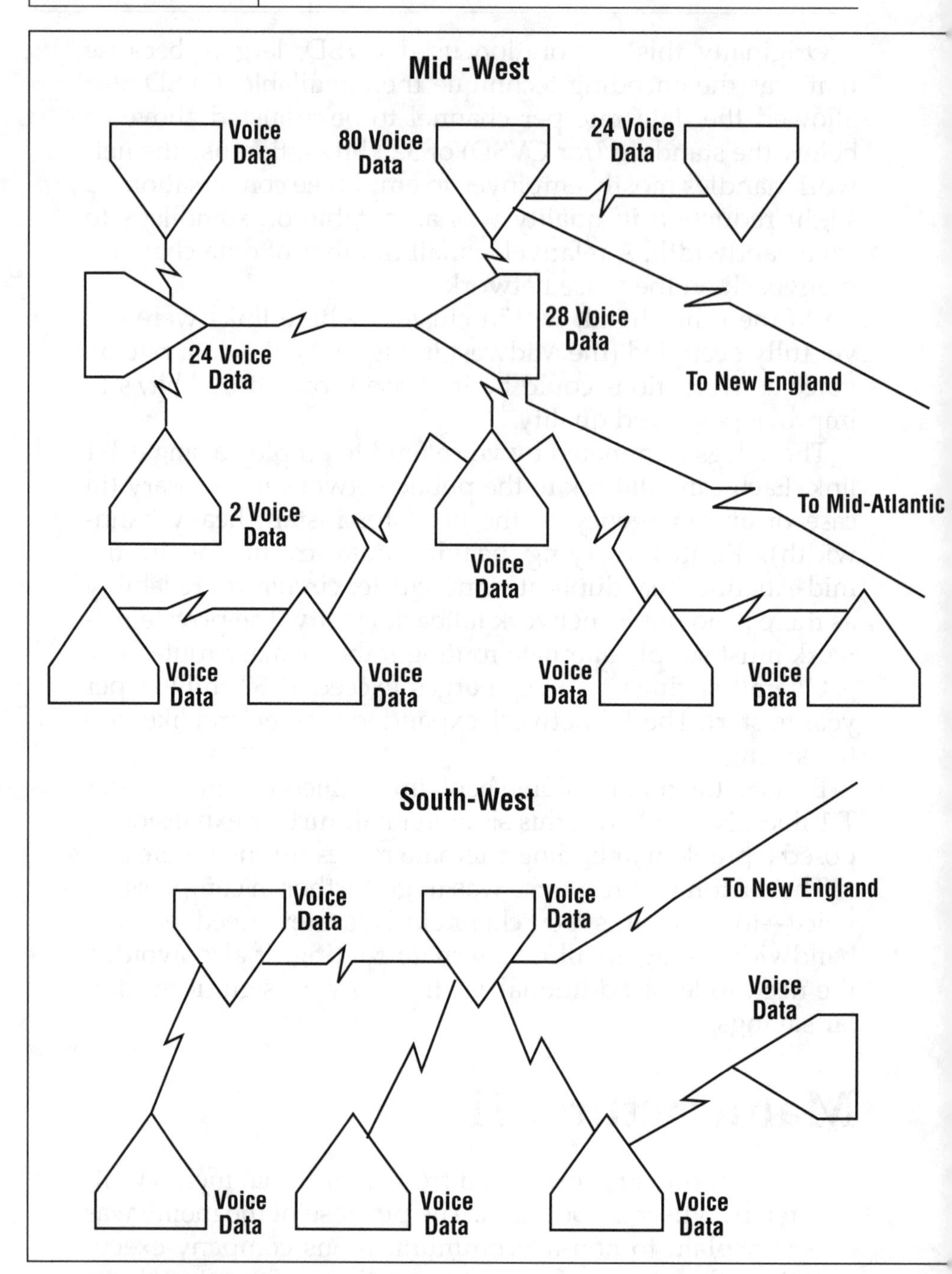

Figure A-2. *The four regions of the corporate voice network consist of nodes serving PBXs in clusters of offices and plants. Direct access at each location to the public switched telephone network forms the backup to the single T-1 circuit*

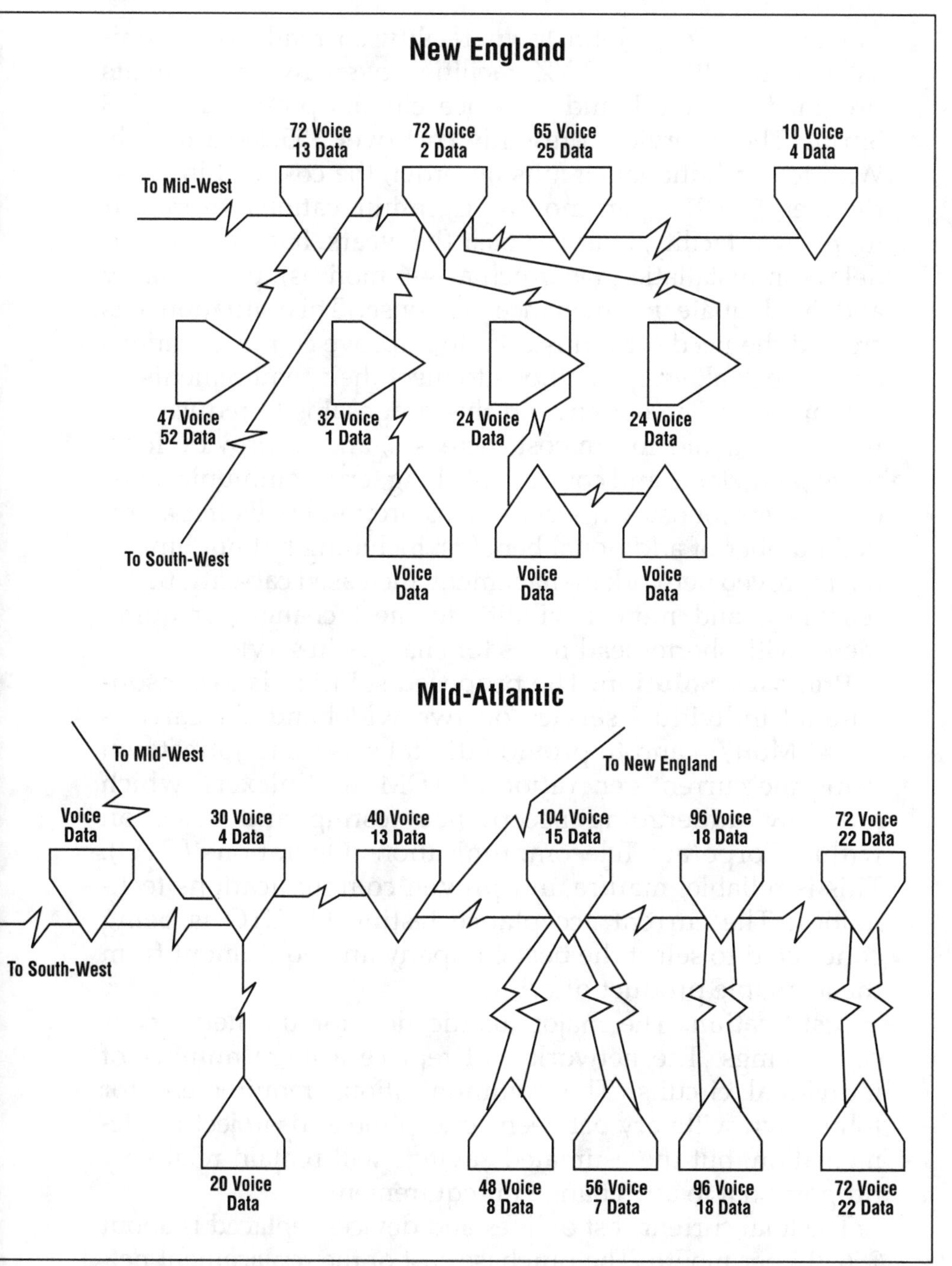

serving most locations. Concentrations of data ports, lacking a fallback to a public network, often employ dual T-1 links.

The XYZ Corp. Johnstown facility currently communicates with all other XYZ facilities over 16 data circuits provided by AT&T and 54 voice circuits provided by US Sprint. These services presently cost over $55,000 a month. With four additional circuits on order, the cost will increase to over $56,914 per month. Communications from our Johnstown facility, over the past five years, has experienced delays in installation (as much as 4-6 months), poor quality and inadequate maintenance response. This situation has created the need for Johnstown to improve communications and to seek alternate methods to meet their requirements.

Objective: The objective in this proposal is to reduce current costs, avoid future cost increases, and to provide for a more predictable and controllable long term communications expense for Johnstown service. This proposal will offer a limited number of additional benefits, including but not limited to: improved network management, increased capacity, better reliability, and more flexibility to meet changing requirements with shorter lead times for changes in service.

Proposed Solution: The proposed solution is to consolidate all individual service on two wideband T-1 carriers (1.544 Mbit/s) and to provide digital service to Johnstown using the current generation of TDM multiplexers, which are now undergoing acceptance testing and selection within Corporate Telecommunications Operations (CTO). This is reliable, mature and proven communications technology. The current acceptance testing by CTO is being conducted to select the best company and equipment from the available product offerings.

Justification: The major justification for the network is cost savings. The network will replace a large number of individual circuits. The communications requirement for Johnstown will vary between conception and project implementation, but the estimated savings will remain relatively proportional to any change in requirement.

The total current cost of lines and devices replaced is about $56,914 per month. The purchase cost of the replacement network circuits and equipment, expressed on a monthly basis is $35,740 per month with ABC Equipment Co. and $36,854 with

DEF Equipment Co. These are the two vendors undergoing CTO acceptance testing. Additional cost savings are possible as well. For example, the additional unused capacity on the replacement network is worth $9,126 per month at current market rates. A considerable cost benefit should also be realized by the XYZ Corporation with an improvement in lead times to acquire additional service to Johnstown, and an increase in network quality and availability.

Network Management: The proposed network will have a central network control terminal which will provide reliability measurements on all trunk lines, reporting on alarm conditions in the network, and which will allow the traffic on the network to be rerouted for load balancing and recovery. Also, it will be possible from this central point to monitor any individual data subchannel traffic for diagnostic reasons. Built in redundancy and automatic fall back on common equipment will minimize the impact of component failures, and allow repair and replacement activities to proceed without impacting the operation of the network.

Security: Central management of the facility, and the reduction in the number and type of circuits and devices in the network, will facilitate the National Bureau of Standards Data Encryption Standard (DES) for future encryption of all traffic on the network. It is anticipated that this requirement will be levied on all DDD contract holders by the National Security Administration as a contract term for protection of Unclassified National Security Related (UNSR) information.

Increased Capacity: Economies of scale inherent in large bandwidth telecommunications circuits and large scale network management equipment means the carrying capacity of the proposed network exceeds the carrying capacity of the facilities it replaces. This additional capacity will make it possible to accommodate new demand without the long lead times that are typical today, and without incurring additional outside costs.

Reliability: Each multiplexer will contain redundant backup modules, providing automatic fall back on all common equipment, with central notification and alarms. The network topology provides multiple paths to all sites. The reliability of T-1 service is enhanced by built-in interoffice redundancy.

Coupled with the fact that the network has unused capacity, means that alternate paths can be defined if a primary path fails. Finally, the network can eventually interface with High Speed Switched Digital Systems (ACCUNET RESERVE 1.544 T-1) as backup mode if a dedicated circuit fails.

Cost Analysis: The total current cost of lines and devices replaced is about $56,914 per month. The cost of the replacement network circuits and equipment expressed on a monthly basis is $35,740 per month with ABC Equipment Co. and $35,740 with DEF Equipment Co. A break even cost analysis based on different traffic volumes indicates that the replacement network will continue to be cost effective at 89% of current traffic volume for ABC Equipment Co. and 78% for DEF Equipment Co. Additional unallocated initial traffic capacity on the network will be worth $9,126 monthly at current commercial rates and usage patterns.

Three Nodes Intra-State

An intra-state company with three offices in Memphis, Chattanooga, and Knoxville, Tenn. Between each pair of offices (Fig. A-3) there are:

- 12 voice tie lines between PBX's
- 4 DDS lines at 56K
- 8 DDS lines at 9.6K

Should this company build a new digital backbone network based on T-1 circuits? Separate lines are working satisfactorily, so the appeal of a new network depends on a significant dollar saving to justify the effort involved in a change. Table A-1 shows the expense per month for lines.

Looking at the bandwidth requirements for the existing lines:

Voice	12	at	32K	=	384.0 kbit/s
DDS	4	at	56K	=	224.0 kbit/s
DDS	8	at	9.6K	=	76.8 kbit/s

digital bandwidth needed per leg: 684.8 kbit/s

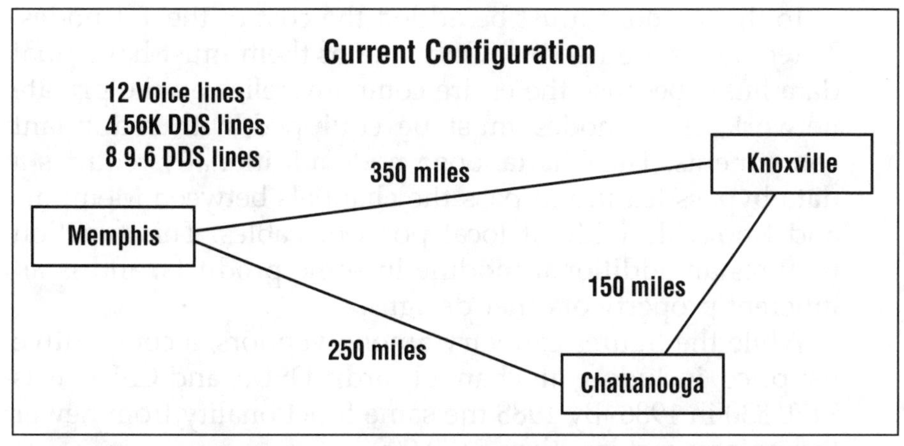

Figure A-3. *A company with three locations within a state and modest communications volume (12 voice tie lines, four 56K DDS, and eight 9.6K DDS between each city pair) is a candidate for a private T-1 backbone network.*

Since this is less than half a T-1, a linear topology (Fig. A-4) is practical. It requires only two circuits:

T-1 Leg:	Cost, 1986	Cost, 1989
Memphis to Chattanooga	$9,198.	$3,900.
Chattanooga to Knoxville	6,198.	3,150.
Total Monthly T-1 charges	$15,396.	$7,050.

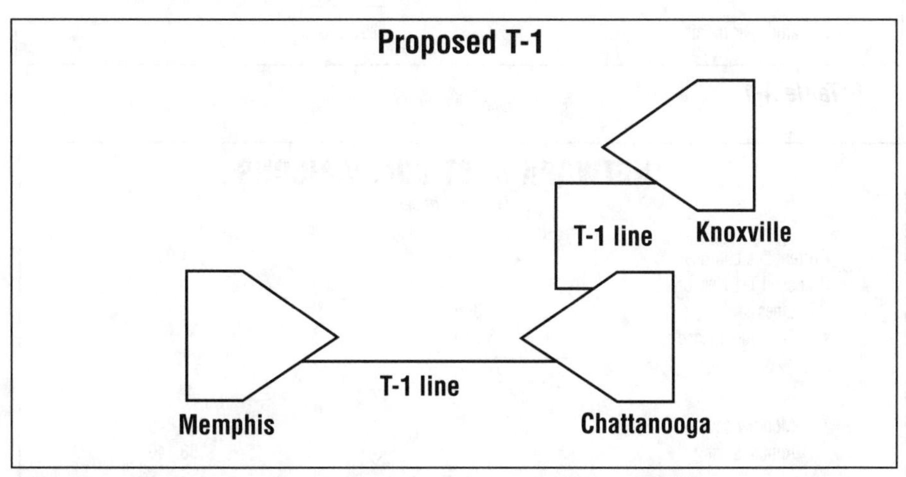

***Figure A-4**. The proposed T-1 configuration, a linear topology, will do the job and save over $29,000 per month. Adding a third T-1 from Memphis to Knoxville, to make a ring, despite the higher costs, still saves about $15,000 per month over the present non-T-1 solution.*

To this amount must be added the cost of the T-1 nodes. Three nodes are needed. Only one of them must have dual data links. Because the entire company relies on the private network, these nodes must be equipped with redundant components. The Chattanooga node additionally requires a data bypass feature to pass the channels between Memphis and Knoxville without local ports or cables. This function requires an additional module in some products, and is an inherent property of other designs.

While the figures can vary among vendors, a competitive list price, including all channel cards, DSUs, and CSUs, was $170,830 in 1986. By 1988 the same functionality from newer technology cost less than $150,000.

LINE CHARGES PER MONTH

For three nodes in Tenn.

City Pairs:	**Memphis – Chattanooga**	**Chattanooga – Knoxville**	**Knoxville – Memphis**
Distance:	250 Miles	150 Miles	350 Miles
12 voice lines	$5,208.	$4,080.	$6,336.
4 56K DDS lines	8,600.	5,780.	10,480.
8 9.6K DDS lines	3,528.	2,776.	4,272.
	$17,336.	$12,636.	$21,088.
Total charges per month		$51,060.	

Table A-1

NETWORK COST COMPARISONS

For three nodes in Tenn.

Current line charges		$51,060.
Proposed T-1 Network		
Lines	$15,396.	
Equipment lease	5,819.	
	21,215.	
Monthly saving		$29,845.
Annual saving		$358,140.
Payback period	12 (170/358) = 5.7 months	

Table A-2

Equipment like this may be purchased, but most often it is leased. The monthly rate for this example would be about $5,819 over three years. It will fluctuate with the current interest rate, and the expense taken on owned equipment will depend on the depreciation method, but this figure is representative. The comparison then, is in Table A-2:

An objection to the suggested solution in this simplified case might be that it loads both the T-1 lines near capacity. Let's look at adding a third T-1 circuit, to create a triangle. That third T-1 circuit cost $12,200 per month in 1986, less after 1987. Points:

- Each link then would be less than half full;
- The T-1 ring would offer alternate routing and thus increased reliability and availability;

...and the configuration, would still save money.

Additional T-1 cards to handle the third data link are the only additional hardware cost for the ring configuration. They reduce the savings slightly. However, the T-1 lines continue to cost less, compared to DDS and analog circuits, so the trend in savings is still up. While saving less than the linear (point A to point B) network, the triangle/ring design still pays back in well under a year.

Basic Multiplexing

 There are three aspects of a T-1 multiplexer that are decided early in its design.

- Multiplexing techniques: FDM, TDM, packetized
- Interleaving: bit, byte, or packet
- Framing: fixed or flexible

The ways the multiplexer work are fixed and can't be changed by the buyer:

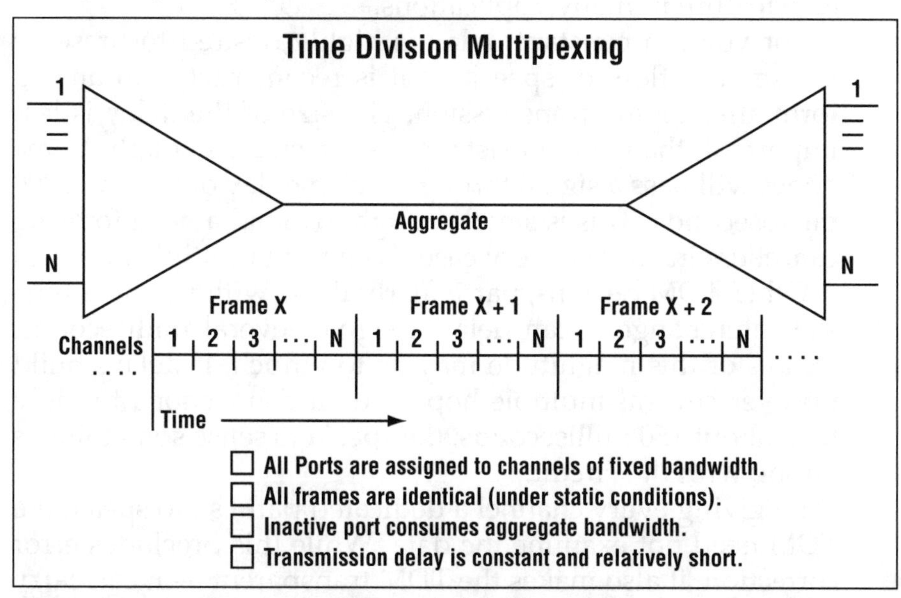

Figure B-1 *Time Division Multiplexing assigns each channel to a fixed time slot on the aggregate line. The bandwidth is reserved for each connection and not interpreted or examined by the multiplexer. Therefore each channel has fixed delay and can carry synchronous data transparently.*

FDM, TDM, or STDM/Packet?

The techniques available for a multiplexer are frequency division multiplexing, time division multiplexing, and packet switching (which is essentially the same technique as statistical time division multiplexing, also called statistical multiplexing).

Time Division Multiplexing

TDM is digital in and out. It is a good match on this basis for data traffic and digitized voice. TDM methods assign fixed amounts of bandwidth to each connection, making them independent. (Fig. B-1) However, TDM's can assign different amounts of bandwidth to each channel, as needed.

As a result, the start-up delay introduced by a TDM network (the time between first receipt of a bit and the time it is sent out) is relatively small. Most of the delay in practical networks of TDM's can be attributed to the propagation time across the lines. Total transit delay is also constant, a desirable feature in many applications.

For voice, a constant delay is highly desired to preserve the smooth flow of speech as it is reconstructed in analog form after digital transmission. The size of the delay is less important than its consistency — if small enough. Some nodes will pass a signal through with a delay of less than 300 microseconds. This is small enough to avoid a need for echo cancellers in almost every case, even with multiple hops.

Other TDM designs, particularly those with multiple time slot interchangers, can delay a signal several milliseconds. Delays of this magnitude may be too much. If delays build up over several multiple-hop tandem connections to more than about 150 milliseconds, the speakers sense something is wrong with the circuit.

By giving every channel a dedicated path, so to speak, the TDM need not examine the data. While this precludes error correction, it also makes the TDM transparent — it can carry any synchronous bit pattern.

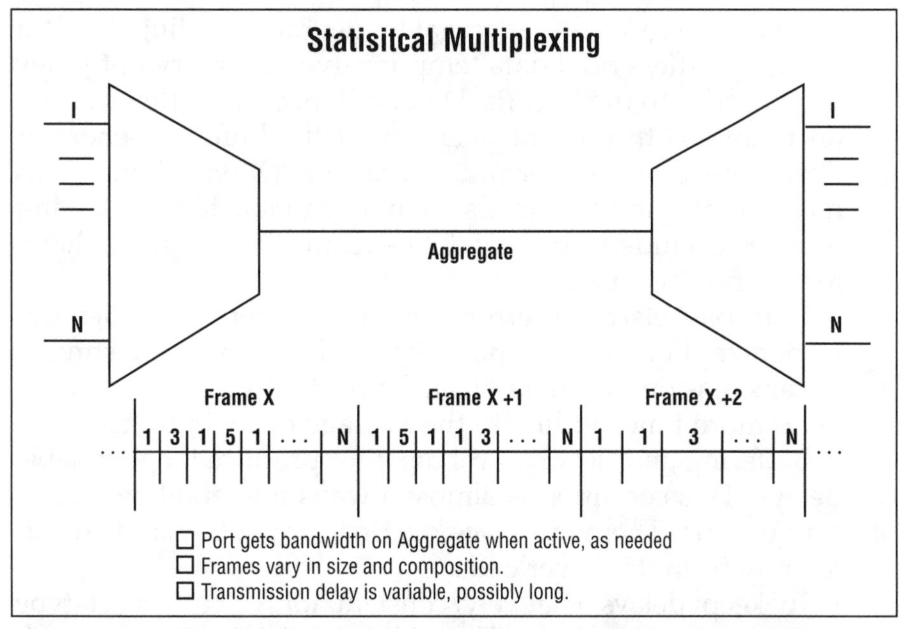

Figure B-2. *Packetized information in digital form ignores idle channels. The aggregate sees only active traffic. Statistically, the chances are that only one fourth or less of the end users will need access to the aggregate at any given moment. This allows significant improvements in bandwidth utilization compared to TDM. Variable, and possibly long, delays require special handling for voice .*

Packet Switching

The same principle as statistical multiplexing, packet switching assigns bandwidth only when there is traffic to send. (Fig. B-2) The control module responds to interrupts from the I/O ports. "Statistical" in this context refers to probability; namely, that not all the users will be active at the same time. Taking advantage of the statistical fluctuations in data terminal usage, a stat mux or packet switch allows connection of an aggregate input greater than the line capacity. Similar statistics apply to voice conversations: usually only one end speaks at a time and there are pauses when neither party speaks.

Suppressing idle periods results in a positive concentration ratio, for both voice and data, of 4:1 typically. It can go much higher for low speed async data.

If many terminal users try to send data simultaneously,

they can exceed the aggregate line rate, leading to data backup. Buffers hold data temporarily in memory until they can be fed into the line. Packet handlers enforce flow control on terminals to prevent overflow of the buffers. Generally, flow control uses in-band characters (X-on, X-off). This means that some characters are not available to the user and must be excluded from the data stream. The stat mux, therefore, is not transparent.

Data packets carry error checks, so receiving nodes can recognize line errors and correct them by requesting a retransmission of a data block found faulty. Retransmission takes more time, adding to the average delay in transit.

Buffering and flow control together produce long, variable delays. This condition is almost always intolerable for voice connections. Therefore voice overload must be handled differently from data overloads.

To keep delays relatively constant for voice, packet type nodes give voice priority. The additional delay in the data traffic is usually less objectionable. During very high instantaneous traffic loads, some voice packets may be discarded. This may be done gracefully, with little impact on voice quality.

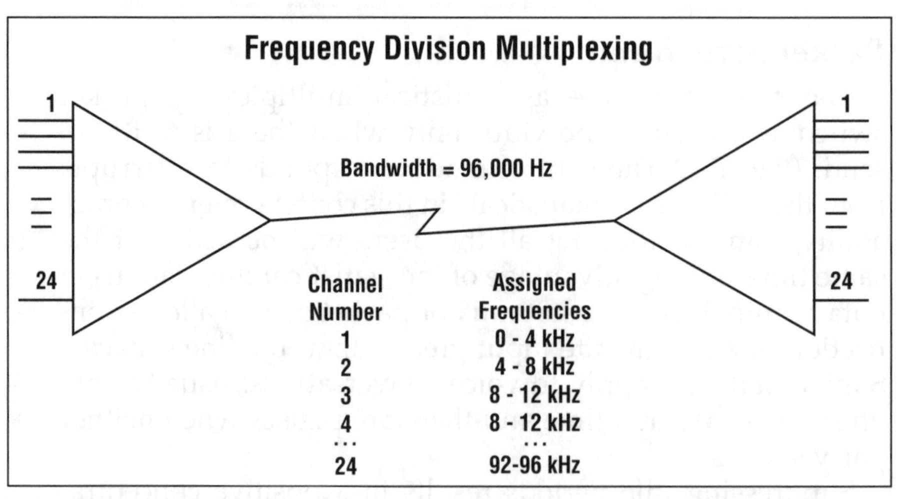

Figure B-3. *Frequency Division Multiplexing converts a voice channel to a higher frequency of the same range. That is, a 0 to 4,000 Hz voice grade input will be translated to the same bandwidth at a higher frequency; for example, 40,000 to 44,000 HZ. Twenty four inputs occupy 96,000 Hz of bandwidth, which can be carried by two twisted pair.*

For this reason, and until recently the lack of microprocessors that could handle packets at T-1 rates, this technique has not been widely adopted for T-1 nodes. With faster processors and some specialized chips available, several vendors are developing fast packet switches for voice. By 1990, they had not delivered full voice and data switching of individual packets. Rather, to preserve time ordering of information (particularly voice) and to simplify switching at tandem nodes, these packet nodes employ connection oriented virtual circuits rather than dynamic packet switching. That is, the route for a connection is determined at call set up, and does not change for the duration of the call.

Frequency Division Multiplexing

FDM has been used almost since vacuum tubes appeared early in this century. Much equipment of this type continues in operation but very little is manufactured and installed new.

In the FDM equivalent to T-1, each analog input (nominally 4,000 Hz bandwidth) is assigned a 4 kHz segment in a broadband circuit. (Fig. B-3) For example, channel 10 might occupy 40-44 kHz. The translation between the audible 300-3300 Hz signal and the assigned frequency slot is done electronically. Adding 24 such channels produces a single composite signal of 96 kHz bandwidth which can be transmitted over two twisted pairs of wires.

Analog signals are subject to degradation by the noise in each line section. The signal to noise ratio gets worse in each amplification stage. FDM is an analog technique, suitable for voice (and modem signals). Despite small delays, FDM is not at all a good match for digital signals.

Bit vs. Byte Interleaving

In assigning fixed bandwidth to multiple channels, TDM's work in one of three ways:

- Bit interleaving; the central control module constructs a data frame by taking one bit at a time from each input port, and sending it out on the T-1 line. That is, the trans-

Framing Summary

Feature	Fixed Frame	Flexible in D4	Flexible Frame
Frame length	fixed	fixed	variable
Time slot size	fixed	fixed	variable
Time slots/frame	fixed	variable	variable
Data efficiency	poor	good/excellent	excellent
Voice flexibility	none	good	good
Voice quality	high	variable	variable
Typical in:	channel bank	T-1 mux, fast-packet switch	T-1 mux, stat mux packet switch

Table B-1

mitted frame consists of a single bit from each channel.

- Byte interleaving; data is taken from each port a byte at a time, making the TDM frame a collection of bytes from the input channels.
- Packetized nodes accumulates 1 to 256 (or even 1024) characters from each input, though 192 bits is popular because it matches a D4 frame. The node then assembles each group of characters into a packet. The node sends packets in the order they are filled or ready to transmit. Packet nodes interleave entire packets.

How does this affect a network? There is an impact on perceived delays.

Start Up Delay

The time between the first bit of a transmission entering one port of the network and the time it reaches the exit or destination port consists of three components. It is independent of message length.

1. input/output processing delay,
2. transit time through intermediate nodes, and
3. propagation delay in transmission lines.

Input/output processing time for a bit interleaved mux is about 1.5 bit times at the low-speed channel or I/O rate. For a 9600 bit/s channel this is about 150 microseconds. This period, about the time of a T-1 frame, is the time a single bit waits in a buffer, on average, before it is picked up by the control module and multiplexed on the T-1 line. Byte interleaved data waits longer, until a full byte is accumulated in the buffer. On average, a byte waits 1.5 byte times or about 1 millisecond for 9600 bit/s. The start delay for a packet is the time it takes to receive the full packet (or less, the system may be set up to send packets after a fixed delay, full or not).

Transit time is the period between a message entering and leaving a node, when that message is just passing through. That is, it's the delay for digital bypass in the node. For the fastest processors, the time gets down to about 2 frames of the D4 format (250 microseconds). This element of delay can't be reduced much as the T-1 line signal must be buffered at least once.

Watch out for nodes with large transit delays. They can accumulate to the point where you will need echo cancellers for relatively short hops even in smaller networks.

Propagation delay depends on the speed of electrical or optical waves in the transmission medium. Signals travel at nearly the speed of light (186,000 miles per second) over microwave links. Over older wire cable the signal may slow to as little as 100,000 miles/s. Delay is fixed by the distance between entry and exit ports. At the full speed of light it would be just over 5 microseconds per mile. Under most conditions, delay is more than 10 microseconds per mile. For 1000 miles, the time is several milliseconds. In reality, with repeater delay, the transit time increases to tens of milliseconds.

Propagation delay of 10 milliseconds completely swamps the processing time difference between bit and byte interleaving. Packets may or may not add significantly to delay, depending on the size of the packets.

Fixed Vs. Flexible Framing

The frame on the data link defines how a block of bits is organized so the receiver can understand them. Framing can be fixed or variable, depending on the type of multiplexer.

Fixed implies that the same organization is used with any mix of data and voice channels. A channel bank, for example, is a fixed frame device. It assigns the same 64,000 bit/s time slot to any connection.

A statistical multiplexer or packet switch responds to varying demands from I/O ports. They must assemble data in different ways and usually frame flexibly. (By restricting the form of the packets this type of node can also use a fixed frame.)

Certain TDM's also frame flexibly, assigning bandwidth (that is, the size of a channel's time slot) in proportion to the data volume in that channel. Flexible frame in TDM's help optimize bandwidth utilization under two fairly common conditions:

- many data channels at speeds lower than DS-0, like 9600 bit/s, or
- reduced bit rate voice encoding, perhaps at 32,000 bit/s.

In the past the trade off between the two types of framing was rather dramatic (Table B-1). Fixed framing was inefficient, but compatible with telco services. Flexible framing was very efficient, but had to be treated by a telco as a "big pipe." It was compatible just enough for transmission, but not enough to draw on Telco services.

An emerging approach to flexible framing is a middle path between rigid formats like D3/D4 and completely flexible proprietary techniques. It is possible to

- observe DS-0 boundaries within the D4 frame;
- frame flexibly within the DS-0, and
- assemble super-rate channels from multiple DS-0's.

This approach gives up little in bandwidth efficiency on subrate data channels and compressed voice connections. At the same time it remains compatible with telco standards, like D4 format and CCR switching.

Conclusion

The standard today in T-1 nodal processors is byte interleaved, time division multiplexing. This technology can provide a rich selection of features while staying fully compatible with carrier services. Flexible framing, an early choice for its high efficiency on the data link, is being joined by a variant of framing that is flexible within the D4 format.

Global Framing Formats

ITU Format For 2.048 Mbit/s

More and more frequently, a major private network will extend internationally. Outside of the United States and Canada the "T-1" or DS-1 line bit rate for "primary rate" service is usually 2.048 Mbit/s in the CEPT format, now defined by ITU recommendations G.703 and G.704. Japan, France, and West Germany impose slight variations that make their formats unique, though based on either T-1 or CEPT standards.

Only one element remains constant: the DS-0 (Fig-C1). The 64 kbit/s channel is universal. Most often it represents a PCM voice signal sampled 8000 time per second. However, the form of PCM encoding differs between T-1 (mu-law) and E-1 (A-law companding). The differences are not so great that a multiplexer cannot convert between them (CO switches routinely convert them). Conversion of E-1 to T-1 involves both the compression law and the signaling format.

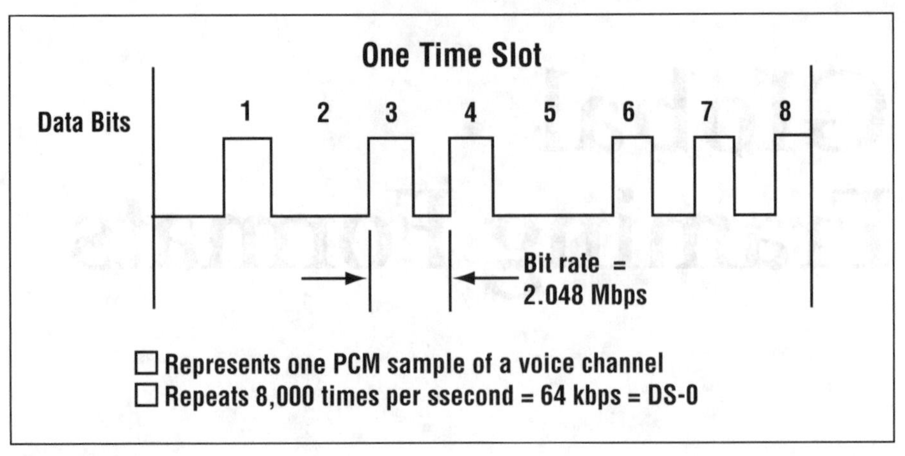

Figure C-2.

At the higher bit rate of 2.048 Mbit/s, 32 time slots are defined at the G.703 interface. But the different form of the encoding, and especially the different way to handle signaling, means that not only is the frame bigger, the superframe also looks very different from the T-1 frame.

HDB-3 Coding

The E-1 line encoding or signal is similar to the Alternate Mark Inversion of T-1, but with a limit of only three zeros in a

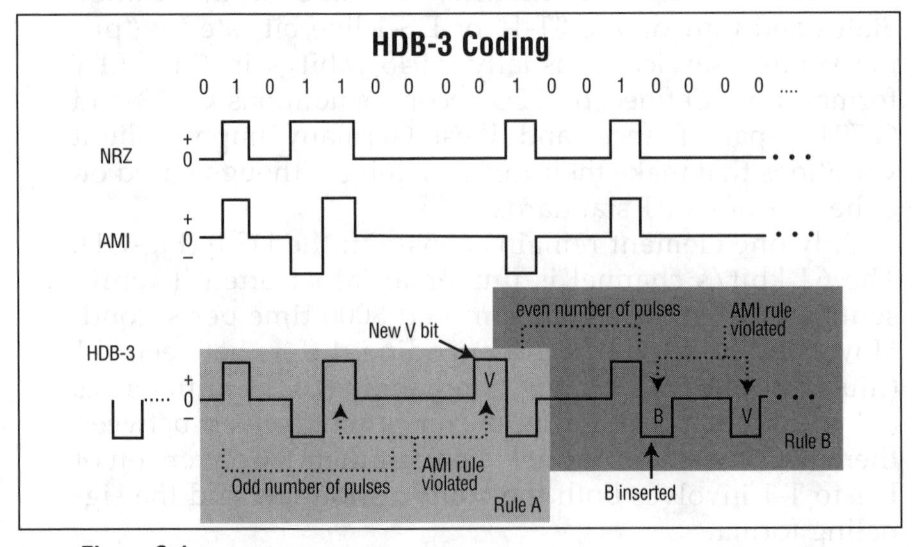

Figure C-1.

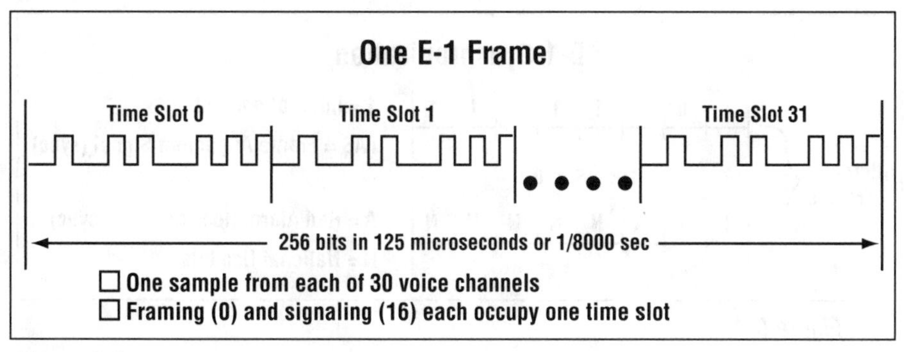

Figure C-3.

row. If a string of four zeros appears, the AMI rule is violated intentionally (bipolar violation) to insert pulses in a way that is recognized by the receiver as standing for zeros (Fig. C-2).

Every time slot in an E-1 is a 'clear channel.' Zeros are transmitted transparently, and no bits are robbed from the voice channel for signaling.

Frame

As in T-1, it is necessary to identify the DS-0's for the receiver. CEPT format uses TDM framing (Fig. C-3) for this function, the same as T-1 does. That is, there are 8000 frames per second, each frame containing one sample from each time slot, numbered zero to 31.

However the frame synchronization (Figs. C-4) uses half of time slot 0:

- a fixed 7-bit pattern of 0011011 in bit positions 2 to 8,
- a single 1 bit in position 2, in alternate frames.

There is no extra 'framing bit' inserted between frames. All E-1 frames are exactly 32 octets.

In the UK, the national use bits are available to end users as a Facilities Data Link. The N bits are more functional than the FDL within the framing bits of the ESF T-1 format because they are carried end to end (CPE to CPE). The first N bit (in position 4, also known as the Sa4 bit) is used to indicate an error rate above the alarm threshold. Setting Sa4 = 0 indicates that there are too many CRC errors in the received signal. Some multiplexers put the inter-node supervisory channel in the N bits.

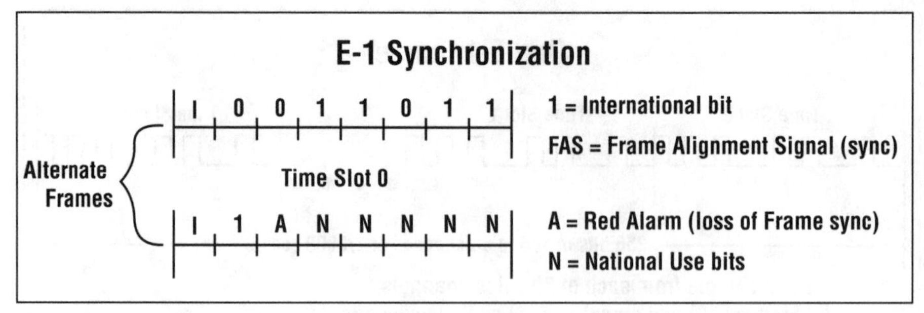

Figure C-4.

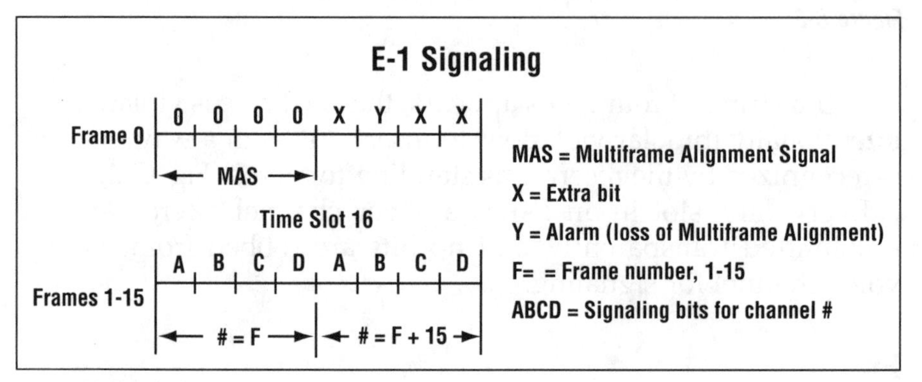

Figure C-5.

Time Slot

While the time slot (Fig. C-1) is the same size in CEPT and T-1, the CEPT format separates signaling from the voice information. Signaling is carried separately, in time slot 16 (actually the 17th because counting starts from 0, Fig. C-5). There are no robbed bits in the voice channel.

This also means that data can run at the full 64 kbit/s rate. There is no need to avoid signaling bits. And since ones density is handled automatically, by the line coding, there is no restriction on the user's data. The DS-0 in a CEPT environment is a clear channel.

Multiframe

Just as T-1 has a superframe, E-1 has a multiframe. Time slot zero marks each frame. Signaling (Fig. C-5) occupies time slot number 16 (in the 0-31 sequence). After reserving 0 and 16, there are 30 channels left for user information.

Time slot 16 is time division multiplexed. Each of the 30 voice channels needs half a byte in a superframe for its ABCD signaling bits for channel associated signaling. The 30 user channels then take 15 frames to cycle through all the signaling bits. One additional frame is needed to synchronize the receiver to the signaling channel, or to establish 'multiframe alignment.' So the full multiframe has 16 frames (Fig. C-6).

The Multiframe Alignment Signal consists of an all-zeros nibble (4 bits). The remainder of that byte carries an alarm bit and three extra bits.

Possible Variations

Both France and West Germany have floated proposals for primary rate service at 1.92 Mbit/s. This allows the user access to 30 of the 32 time slots. The carrier-owned CSU would add time slot zero, containing synchronization.

The difference will mean another variant for multiplexer makers. It also means that customers (and their hardware nodes) are denied the national use bits. Any network management and control information therefore will have to pass through a DS-0.

As carriers move toward ISDN technology, they move away from bit-oriented signaling to message-oriented signaling. The best known message signaling standards are Signaling System #7, the CCITT recommendation for use inside a carrier network, and Digital Subscriber Signaling System #1, the user form of signaling for ISDN.

Before the standard was defined, some carriers implemented their own versions. British Telecom offers DPNSS on E-1 lines

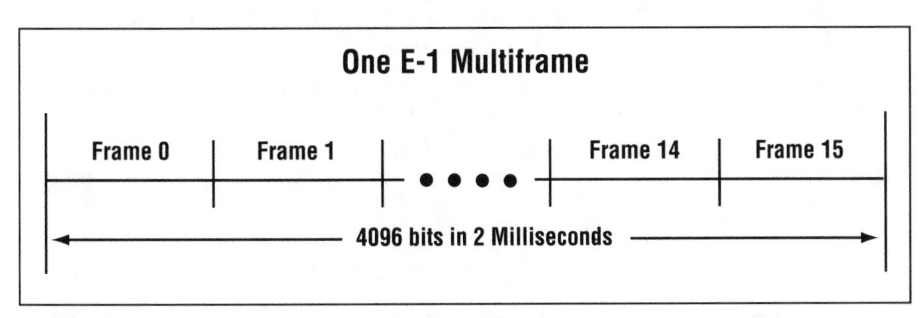

Figure C-6.

connected to PBX's. AT&T offered something close to but not quite ISDN signaling on a PRI service, but that format is migrating to comply with the National ISDN standard (in the US).

T-1 in Japan

The basic format for 1.5 Mbit/s transmission facilities in Japan is similar to ANSI standards. There are the familiar 24 channels of 64 kbit/s each (DS-0's). The 193-bit frames repeat at 8000 per second (Fig. C-7). The demark is a DB-15 connector, though with a different pin out.

However, the CMI line coding; used in Japan differs distinctly from that used in North America. Rather than run 50% duty cycle (half width) "mark" pulses, Japan transmits continuous marks, inverting the voltage after each mark (Fig. C-8). The spaces also have a transition, but in the middle of the pulse period. Thus there are always sufficient transitions to hold synchronization, whether the signal is all

Framing Bits in Japanese T-1

Frame Number	F Bit	Function
1	CRV	Coding Rule Violation, frame and superframe synchronization
even	D	Data link, 4 kbit/s, CPE to CPE or CPE to Network
17	DNR	DCE Not Ready, DCE —> DTE, DSU indicates to terminal that circuit is out of order, communications is not guaranteed by the carrier when DNR = 1.
19	UNR	Uncontrolled Not-Ready, DCE —> DTE, local DSU indicates far-end DSU lost sync with DTE or sees no signal from terminal.
21	S	Status of the circuit, sent by each DTE to the other, 1 = in use, 0 = out of use. Communications not guaranteed when S = 0.
23	SEND	Clear to Send, DCE —> DTE, normally SEND = 0, DSU sets SEND = 1 when it loses signal or sync with DTE. Communications not guaranteed when SEND = 1.
others	-	Don't care, ignored by the public network

Table C-1.

zeroes or all ones. DTE must be loop timed, but there is plenty of signal to sync to.

The distinctive coding line, may be summarized as:

0 = positive going transition at mid point;

1 = constant during interval (transition in either direction at edge of pulse period).

A clear rule also allows for a clear coding rule violation

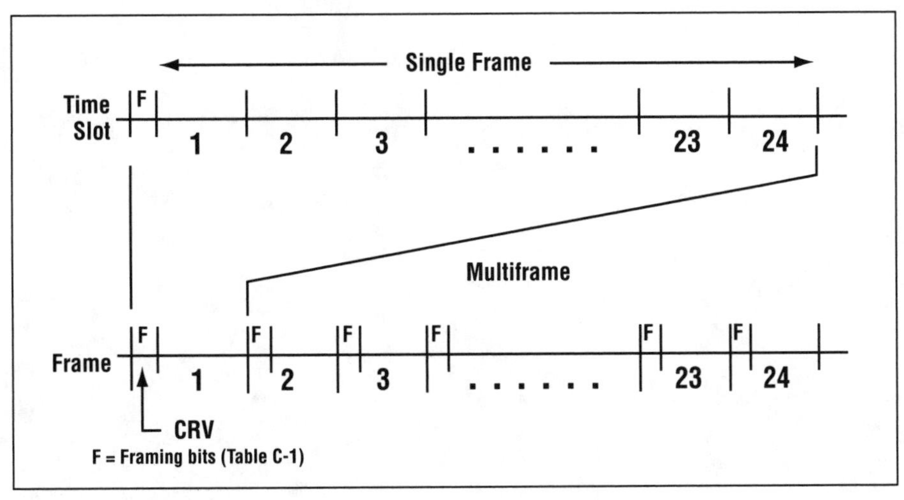

Figure C-7.

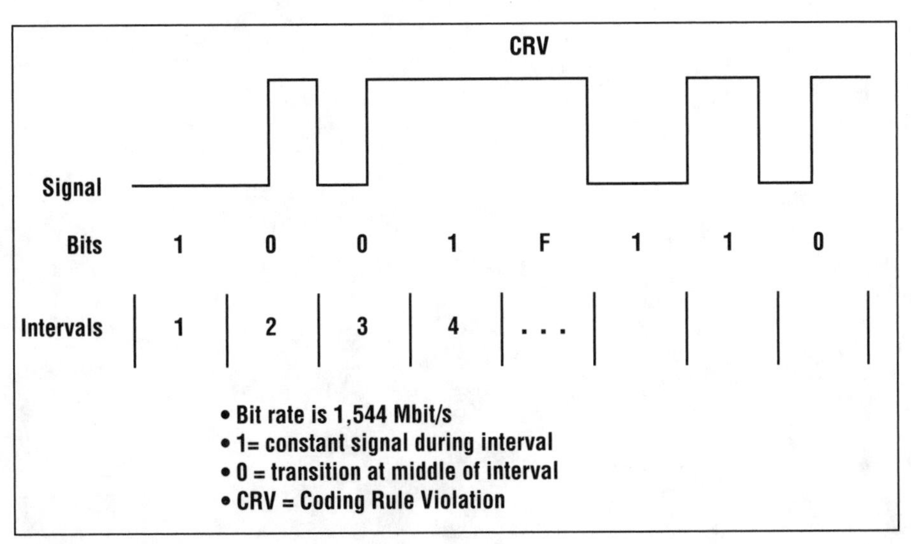

Figure C-8.

(CRV): a negative going transition at the midpoint is distinct because it violates the rules above. This CRV is all that is necessary to synchronize an entire extended superframe (24 frames). Thus the other 23 F bits may be used for other purposes (Table C-1).

Networking Acronyms

This glossary tries to define acronyms in telecom that are common enough for a network manager to come across in connection with the next generation of equipment as well as historical documents. Among the true acronyms listed here in all capitals are some abbreviations which appear, as they normally do, in lower case letters. Numerical items are listed last.

Note that the Index doesn't cover this glossary.

Many items carry a source reference in parentheses. (802.x) = IEEE LAN standard; (Tel) = telephone terminology; (SS7) = signaling system 7; (Layer x) = in OSI model; (A.NNN) = ITU-T recommendation; other references are acronyms, listed here.

A

A Ampere, unit of electrical current.

AAL ATM Adaptation Layer, two sublayers concerned with segmenting large PDUs into ATM cells; type 1 = CBR, 2 = VBR, 3/4 = connectionless services, 5 = for LAN and FR frames, 6 = MPEG video. See also SAR, CPCS.

AAP Alternate Access Provider, carrier other than local telco that can provide local loop into IXC or LEC; CAP.

AAR Automatic Alternate Routing, failure recovery.

ABAM Order code for the 22 AWG shielded twisted pair cable used for manual cross connects of DS-1 through DS-2 signals.

ABCD Signaling bits, for robbed bit signaling with ESF; only A and B are available with SF (Tel).

ABM Asynchronous Balanced Mode (Layer 2).

ABR Available Bit Rate, 'best effort' BoD (ATM) with flow control and 'CIR.'

ABS Alternate Billing Service; credit card, 3rd party, etc. (SS7).

a.c. Alternating Current, the form of analog voice signals, ringing, and power lines; also 'ac'.

ACARS Aircraft Communications and Reporting System, VHF network to extend ADNS.

ACCTS Aviation Coordinating Committee for Telecommunications Services, part of ARINC.

ACD Automatic Call Distributor, PBX function or machine to spread calls among phones.

ACELP Algebraic CELP, form of voice compression (G.729, etc.).

ACF Access Control Field, first byte in ATM header (802.6).

ACF Access Coordination Function, tariffed service where AT&T obtains local loops between customer premises and AT&T serving office.
ACF Advanced Communications Function, SNA software.
ACK Positive Acknowledgment, message or control bytes in a protocol; report a frame was received OK.
ACL Applications Connectivity Link, Siemens' PHI.
ACM Address Complete Message, signaling packet equivalent to ring-back tone or answer (SS7).
ACR Actual (Average) Cell Rate, ATM interface parameter.
ACSE Association Control Service Element (OSI).
ACT ACTivate, BRI control bit, turns on NT.
ACT Applied Computer Telephony, Hewlett-Packard's PHI.
A/D Analog to Digital, usually a conversion of voice to digital format.
ADCCP Advanced Data Communications Control Procedure, ANSI counterpart to HDLC.
ADCR Alternate Destination Call Redirection, service diverts calls to second site (AT&T).
ADCU Association of Data Communications Users.
AEPA ATM Endpoint Address, globally unique user address on ATM network.
ADM Add/Drop Multiplexer, node with 2 aggregates that supports data pass-through.
ADM Asynchronous Disconnect Mode (Layer 2).
ADNS ARINC Data Network Service, a packet network.
ADPCM Adaptive Differential Pulse Code Modulation, a form of voice compression that typically uses 32 kbit/s.
ADSL Asymmetric Digital Subscriber Line, local loop technology for unequal transmission speeds in CO—>CP and CP—>CO directions (HDSL).
AESA ATM End System Address.
AF ATM Forum.
AFI Authority and Format Identifier, part of network level (NSAP) address header (MAP, ATM).
AFIPS American Federation of Information Processing Societies.
AFR Alternate Facility Restriction.
AI Alarm Indication.
AIB Alarm Indication Bit, BRI control bit.
AIM ATM Inverse Multiplexing.
AIN Advanced Intelligent Network, carrier offering more than 'pipes' to users.
AIS Alarm Indication Signal, unframed all 1's (Blue Alarm) sent downstream (to user) from fault site. Also used as T-1 keep-alive signal.
ALS Active Line State, possible status of FDDI optical fiber.
AMI Alternate Mark Inversion, line coding for T-1 spans where 0 (space) is no voltage and successive 1s (marks) are pulses of opposite polarity. See also DMC, NRZ, 4B/5B.
AMIS Audio Messaging Interchange Specification, for voice mail.
AMPS Advanced Mobile Phone Services (System), analog cellular in N.A.
AMS Audio/visual Multimedia Services (ATM).
AN Access Node.
ANI Automatic Number Indication, display of calling number on called phone.
ANM ANswer Message, signaling packet returned to caller indicating called party is connected (SS7).
AN Access Network, OSP like local loops, etc.
ANP AAL-CU Negotiation Procedure, sets up ATM call.
ANR Automatic Network Routing (APPN).
ANSI American National Standards Institute, the US member of the ISO.
ANT Alternate Number Translation, ability to reroute 1-800 calls on NCP failure.
AOS Alternate Operator Service, non-telco firm responding to "dial zero."
AP Action point, SDN switch located closest to customer site.

APDU Application PDU (OSI).
API Application Programming Interface, software module or commands to separate OS or network from application.
APP ATM Peer-Peer connection, between adjacent devices; half duplex.
APPC Advanced Program to Program (Peer-to-Peer) Communications, session level programming interface (APPN).
APPI Advanced Peer to Peer Internetworking, cisco version of APPN that encapsulates in IP.
APPN Advanced Peer-to-Peer Networking, IBM networking architecture.
APS Automatic Protection Switch.
AR Access Rate, speed of a channel into a backbone network.
ARAP AppleTalk Remote Access Protocol.
ARD Automatic Ring Down, lifting your phone rings the fixed far end without dialing.
ARINC Aeronautical Radio, Inc., operator of private airline networks.
ARP Address Resolution Protocol, a way for routers to translate between different forms of protocol addresses or domains.
ARPA Advanced Research Projects Agency, created Arpanet packet network, first X.25 net; folded into NSFnet in 1990.
ARQ Automatic Repeat reQuest, for retransmission; an error correction scheme for data links, used with a CRC.
ASAI Adjunct Switch Applications Interface, AT&T's PHI.
ASCII American Standard Code for Information Interchange, based on 7 bits plus parity.
ASDS Accunet Spectrum of Digital Services, AT&T fractional T-1.
ASE Applications Service Element, protocol at upper layer 7 (SS7, OSI).
ASIC Application Specific Integrated Circuit, custom chip.
ASN.1 Abstract Syntax Notation #1, language to manage network elements.
ASR Automatic Send/Receive, a printer with keyboard or a Teletype machine.
ASTN Alternate Signaling Transfer Network, a CCS6 that backs up CCS7.
ATAP All Things to All People, the mythical perfect product.
ATD Asynchronous Time Division, ETSI proposal for pure cell relay, without SONET or other framing.
ATM Asynchronous Transfer Mode, a type of cell transfer protocol or packet framing.
ATMF ATM Forum; also AF.
ATMM ATM Management entity, function in ATM device.
ATN Aeronautical Telecommunication Network, as in ATN Protocol Architecture used by ARINC.
AU Administrative Unit, payload plus pointers (SDH).
AUU ATM User-User connection, end to end.
AUG AU Group, one or more AUs to fill an STM (SDH).
AUI Attachment Unit (Universal) Interface, standard connector between MAU and PLS (802.x Ethernet).
Autovon Automatic Voice network, a U.S. military net.
AVS Available Seconds, when BER of line has been less than 10-3 for 10 consecutive seconds until UAS start.
AWG American Wire Gauge, conventional designator of wire size.

B

B Bearer channel, a DS*f*0 for user traffic (ISDN).
B Beginning, bit in header of VoFR subframe, for data fragmentation.
B1 SOH byte carrying BIP-8 parity check (SONET).
B2 LOH byte carrying BIP parity check.
B3 POH byte carrying BIP parity check.
B3ZS Binary 3-Zero Substitution, line coding for DS-3 signal substitutes a 'coding violation for 4 consecutive zeros.

B8ZS Binary 8-Zero Suppression, substitutes 000+-0-+ for 00000000 to maintain ones density on T-1 line.

BAN Boundry Access Node, edge device in APPN.

BAsize Buffer Allocation; number of octets in L3-PDU from DA to Info, +CRC if present (SMDS, ATM).

BATT Battery, the -48 (-40 to -52) V d.c. supply in the CO.

Bc Committed Burst, amount of data allowed in time T=Bc/CIR without being marked DE.

BCC Bellcore Client Company, one of 7 RBOCs who owned Bellcore until 1997.

BCC Block Check Code, a CRC or similarly calculated number to find transmission errors.

BCD Binary Coded Decimal, 4-bit expression for 0 (0000) to 9 (1001).

BCM Bit Compression Mux, same as M44 for ADPCM.

BCN Beacon, frames sent downstream by station on token ring when upstream input is lost (802.5).

B-DCS Broadband Digital Cross-connect System, DACS OC-1, STS-1, DS-3 and higher rates only (see W-DCS).

BDLC Burroughs Data Link Control, layer 2 in Burroughs Network Architecture.

Be Excess Burst, transient capacity above CIR and Bc in FR net.

BECN Backward Explicit Congestion Notification, signaling bit in frame relay header.

BER Bit Error Ratio (Rate), errored bits over total bits; should be <ˇ107 for transmission lines.

BERT Bit Error Rate Test(er).

BES Bursty Errored Second, from 1 to 319 CRC errors in ESF framing, ESB.

BEtag Beginning/End tag; same sequence number put at head and tail of L3-PDU.

BGF Basic Global Functions, requirements for ISDN.

BIB Backward Indicator Bit, field in SUs (SS7).

B-ICI Broadband ICI, interface between public ATM networks.

BIOS Basic Input/Output System, part of OS; may include communications functions.

BIP-x Bit Interleaved Parity, error checking method where each of x bits is parity of every xth bit in data block (x=8 in SONET, 16 in ATM).

B-ISDN Broadband ISDN, generally ATM access at more than 100 Mbit/s.

B-ISSI Broadband Inter-Switching System Interface, e.g., between ATM nodes.

BISYNC Binary Synchronous communications, a protocol; also BSC.

bit One eighth of something, e.g., half the $ cost of 'shave & a haircut.'

BITS Building Integrated Timing Supply, stratum 1 clock in CO.

BIU Basic Information Unit, up to 256 bytes of user data (RU) with RH and TH headers(SNA).

BLERT BLock Error Rate Test.

BLU Basic Link Unit, data link level frame (SNA).

BMS Bandwidth Management Service, AT&T offering like a private network with equipment in CO.

BN Bridge Number, device identifier in LAN for routing/bridging.

BNA Burroughs Network Architecture, comparable to SNA.

BNS Broadband Network Switch, usually ATM or packet based, DS-3 and faster.

BOC Bell Operating Company, a local telephone company.

BoD Bandwidth on Demand, dynamic allocation of line capacity to active users.

BOM Beginning of Message, type of segment (cell) that starts a new MAC frame, before COM and EOM (SMDS).

BOM Bill Of Materials, list of all parts in an assembly.

BONDING Bandwidth On Demand INteroperability Group, makers of inverse muxes and standard they adopted.

BOOTP Boot Protocol, lets station get IP address from server using UDP/IP; enhanced RARP.

BORSHT Battery feed, Over-voltage protection, Ringing, Signaling (Supervision), Hybrid, Test; classic functions of analog interface.

bps Bits per second, serial digital stream data rate, now bit/s.

BP/S Burroughs Poll/Select, legacy data protocol.

BPV Bipolar Violation, two pulses of the same polarity in a row.
BR Bureau of Radiocommunications, part of ITU that allocates international spectrum.
BRA Basic Rate Access, ISDN 2B+D loop.
BRI Basic Rate Interface, 2B+D on one local loop.
BRT Broadband Remote Terminal, node with DS-3, OC-1, or faster access into ATM, etc.
BSC Binary Synchronous Communications, a half-duplex legacy data protocol.
BSN Backward Sequence Number, sequence number of packet (SU) expected next (SS7).
BSP Bell System Practice, document associated with equipment used by telcos for I&M, etc.
BSRF Basic System Reference Frequency, formerly Bell SRF; Stratum 1 clock source of 8 kHz.
BSS Broadband Switching System, cell-based CO switch for B-ISDN.
BT British Telecom, primary phone company in the United Kingdom.
BTag Beginning Tag, field in header of frame whose value should match ETag.
BTAM Basic Telecommunications Access Method, older IBM mainframe comm software.
BTU Basic Transmission Unit, LU data frame of RU with RH (SNA).
BUS Broadcast and Unknown Server, last resort to map MAC to ATM address.
BWB Bandwidth Balancing, method to reduce a station's access to a transmission bus, to improve fairness (802.6).
BX.25 see SX.25.

C

C Capacitance, the property (measured in farads) of a device (capacitor) that holds an electrical charge.
C-Plane Control Plane, out of band signaling system for U-Plane.
CA*net Canadian Academic Network.
CAC Connection (Call) Admission Control, process to limit new calls to preserve QoS (ATM).
CAD Computer Aided Design, drafting on computers.
CAE Computer Aided Engineering.
CALM Connection Associated Layer Management, F5 flows (ATM),
CAM Computer Aided Manufacturing.
CAP Carrierless Amplitude and Phase modulation, a modem technique applied at up to to 50 Mbit/s in LANs and HDSL.
CAP Competitive Access Provider, alternative to LEC for local loop to IXC or for dial tone.
CAS Channel Associated Signaling, bits like ABCD tied to specific voice channel by TDM.
CASE Common Application Service Elements, application protocol (MAP).
CAT Category, often with a number (CAT-3) to indicate grade, as of UTP wiring.
CATV Community Antenna Television, cable TV.
CB Channel Bank, 24-port voice multiplexer, to T-1 interface.
CB 7
CBEMA Computer and Business Equipment Manufacturers Association.
CBR Constant (Continuous) Bit Rate, channel or service in ATM network for PCM voice or sync data in a steady flow with low variation in cell delay; emulates TDM channel.
CBX Computerized Branch eXchange, PABX.
CC Call Control.
CC Continuity check, OA&M cell (ATM).
CC Cluster Controller, for group of dumb terminals (SNA).
CCBS Completion of Calls to Busy Subscribers, supplementary service defined for ISDN.
cch Connections per Circuit-hour, in Hundreds.
CCIS Common Channel Inter-office Signaling.
CCITT Comite Consultatif Internationale de Telegraphique et Telephonique, The International Telegraph and Telephone Consultative Committee, part of ITU that was merged with CCIR in 1993 to form TSS.
CCIR International Radio Consultative Committee, sister group to CCITT, became part of TSS (1993).
CCR Commitment, Concurrency, and Recovery (OSI).

CCR Current Cell Rate (ATM).
CCR Customer Controlled Reconfiguration, of T-1 lines via DACS switching.
CCS Common Channel Signaling.
CCS Common Communication Subsystem, level 7 applications services (SNA).
CCSA Common Control Switching Arrangement.
CCSS CCS System, usually with a number.
CCS6 CCS system 6, first out of band signaling system in N.A. (CCIS).
CD Carrier Detect, digital output from modem when it receives analog modem signal on phone line.
CD Compact Disk, as in CD-ROM data storage.
CD Count Down, a counter that holds the number of cells queued ahead of the local message segment (802.6).
CDMA Code Division Multiple Access, spread spectrum; broadcast frequency changes rapidly in pattern known to receiver.
CDPD Cellular Digital Packet Data.
CDT Cell Delay Tolerance, ATM parameter.
CDV Cell Delay Variation, ATM UNI traffic parameter.
CE Circuit Emulation, ATM function to carry TDM circuits (T-1, etc.) on cell stream.
CE Connection Element (ATM, LAP-D).
CEI Connection Endpoint Identifier (ATM, LAP-D).
CELP Code-Excited Linear Predictive coding, a voice compression algorithm used at 16 kbit/s, 8 kbit/s, and slower rates.
CEN Committee for European Standardization.
CENELEC Committee for European Electrotechnical Standardization.
CEP Connection End Point (ATM).
CEPT Conference on European Posts & Telecommunications (Conference of European Postal and Telecommunications administrations), a body that formerly set policy for services and interfaces in 26 countries.
CERN Nuclear research facility (particle physics) in Geneva, Switzerland.
CES Circuit Emulation Service, carries TDM channel on cells (ATM).
CES Connection Endpoint Suffix, number added by TE to SAPI to make address for connection; mapped to TEI by L2.
CFA Carrier Failure Alarm, detection of red (local) or yellow (remote) alarm on T-1.
CGA Carrier Group Alarm, trunk conditioning applied during CFA.
CHAP Challenge Handshake Authentication Protocol, log-in security procedure for dial-in access.
CI Congestion Indication.
CI Connection Identifier, frame or cell address.
CI Continuation Indicator (ATM).
CI Customer Installation, all the phone equipment and wiring attached to the PSTN; see CPE.
CIB CRC Indication Bit, 1 if the CRC is present, 0 if it is not used (SMDS).
CICS Customer Information Control System, IBM mainframe comm software with data base.
CID Channel Identifier, subframe address (VoFR).
CIDR Classless Inter-Domain Routing (IP).
CIF Cells In Flight, number of cells sent before first cell reaches far end.
CIF Cell Information Field, 48 byte payload in each cell (ATM).
CIR Committed Information Rate, minimum throughput guaranteed by FR carrier.
CIT Computer Integrated Telephony, DEC's PHI.
CL Common Language, Bellcore codes to identify equipment, locations, etc.
CL ConnectionLess.
CLASS Custom(er, ized) Local Area Signaling Services; ANI, call waiting, call forwarding, trace, etc.
CLEC Competitive Local Exchange Carrier.

CLEI Common Language Equipment Identifier, unique code assigned by Bellcore for label on each CO device.
CLID Calling Line IDentification, ANI.
CLIP Calling Line Identity Presentation, ISDN-UP service to support ANI.
CLIR Calling Line Identity Restriction, feature where caller prevents ANI (ISDN-UP).
CLIST Command List, similar to .BAT file (SNA).
CLLM Consolidated Link Layer Management (802).
CLN ConnectionLess Network, packet address is unique and network routes all traffic on any path(s); e.g., most LANs like IP.
CLNAP CLN Access Protocol.
CLNIP CLN Interface Protocol.
CLNP ConnectionLess mode Network (layer) Protocol; see CLN.
CLNS ConnectionLess mode Network (layer) Service, ULP (SNA).
CLP Cell Loss Priority, signaling bit in ATM cell (1=low).
CLR Cell Loss Ratio (ATM).
CLS Connectionless Service.
CLTS ConnectionLess Transport Service, OSI datagram protocol.
CMA Communications Managers Association.
CMD Circuit Mode Data, ISDN call type.
CMDR Command Reject, similar to FRMR (HDLC).
CMI Coded Mark Inversion, line signal for STS-3.
CMI Constant Mark, Inverted; line coding for T-1 local loop in Japan.
CMIP Common (network) Management Information Protocol, part of the OSI network management scheme, connection oriented.
CMIS Common (network) Management Information Service, runs on CMIP (OSI).
CMISE CMIS Element.
CMOL CMIP Over LLC, reduced NMS protocol stack.
CMOS Complementary Metal Oxide Semiconductor, low power method (lower than NMOS) to make ICs.
CMOT CMIP over TCP/IP.
CMT Connection Management, part of SMT that establishes physical link between adjacent stations (FDDI).
CND Calling Number Delivery, another name for CLID, one of the CLASS services.
CNF Confirmed (OSI).
CNIS Calling Number Identification Service, provide, screen, or deliver CPN or caller ID (ISDN).
CNLP Connectionless Protocol.
CNLS Connectionless Service.
CNM Communications Network Management (SNA).
CNM Customer Network Management (Bellcore).
CO Central Office, of a phone company, where the switch is located; the other end of the local loop opposite CP.
C-O Connection Oriented.
COC Central Office Connection, separately tariffed part of circuit within a CO.
COCF Connection Oriented Convergence Function, MAC-layer entity.
CODEC COder-DECoder, converts analog voice to digital, and back.
COFA Change of Frame Alignment, movement of SPE within STS frame.
CO-LAN Central Office Local Area Network, a data switching service based on a data PBX in a carrier's CO.
.com Commercial, first Internet address domain of businesses.
COM Continuation Of Message, type of segment between BOM and EOM (ATM, SMDS).
comm Communications.

CON Connection-Oriented Network, defines one path per logical connection (FR, etc.).
CONP Connection mode Network layer Protocol.
CONS Connection-Oriented Network Services, ULP (SNA).
COS Class Of Service.
COS Corporation for Open Systems, R&D consortium to promote OSI in the US; see SPAG.
COSINE Cooperation for Open Systems Interconnection Networking in Europe.
COT Central Office Terminal, equipment at CO end of multiplexed digital local loop or line.
COT Customer-Originated Trace, sends CPN to telco or police (CLASS).
CP Central Processor, CPU that runs network under center-weighted control.
CP Control Point, function in APPN node for routing, configuration, directory services.
CP Customer Premises, as opposed to CO.
CPAAL Common Part AAL, may be followed by a number to indicate type.
CPCS Common Protocol Convergence Sublayer, pads PDU to N x 48 bytes, maps control bits, adds FCS in preparation for SAR.
CPE Customer Premises Equipment, hardware in user's office.
CPI Computer-PBX Interface, a data interface between NTI and DEC.
CPIC Common Programming Interface for Communications, a software tool for using LU6.2 adopted by X/Open as a standard.
CPN Calling Party Number, DN of source of call (ISDN).
CPN Customer Premises Node (or Network), CPE.
CPNI Customer Proprietary Network Information, customer data held by telcos.
CPSS Control Packet Switching System, subnetwork of supervisory channels (Newbridge).
CPU Central Processor Unit, the computer.
CR Carriage Return, often combined with a line feed when sending to a printer.
CRC Cyclic Redundancy Check, an error detection scheme, for ARQ or frame/cell discard.
CRF Connection Related Function (ISDN).
CRIS Customer Records Information System, telco OSS.
CRT Cathode Ray Tube, simple computer terminal.
CRV Coding Rule Violation, unique bit signal for F bit in frame 1 of CMI.
CS Circuit Switched, uses TDM rather than packets.
CS Convergence Sublayer, where header and trailer are added before segmentation (ATM).
CSA Carrier Service Area, defined by a local loop length (<12,000 ft) from CO, or from remote switch unit or RT.
CSA Callpath Services Architecture, for PBX to IBM host interface.
CS-ACELP Conjugate Structure-ACELP, specifically G.729.
CSC Circuit-Switched Channel (Connection).
CSDC Circuit Switched Digital Capability, AT&T version of Sw56. 61330
CSMA Carrier Sense Multiple Access, a LAN transport method, usually with "/CD" for collision detection or "/CA" collision avoidance; LAN protocols at physical layer.
CSO Cold Start Only, BRI control bit.
CSPDN Circuit Switched Public Data Network.
CS-PDU Convergence Layer PDU, info plus new header and trailer to make packet that is segmented into cells or SUs.
CSTA Computer Supported Telephony Application, PHI from ECMA.
CSU Channel Service Unit, the interface to the T-1 line that terminates the local loop.
CT2 Cordless Telephone, second version; digital wireless telephone service defined by ETSI.
CTR Common Technical Requirements, European standards.
CTS Clear To Send, lead on interface indicating DCE is ready to receive data.
CU Channel Unit, plug in module for channel bank.
CU Composit User, form of AAL for multimedia traffic (ATM).
CUG Closed User Group.

CV Coding Violation, transmission error in SONET section.
CVSD Continuously Variable Slope Delta modulation, a voice encoding technique offering variable compression.
CWC Center-Weighted Control, A central processor runs a network-wide functions while nodes do local tasks.

D

D Delta (or Data) channel, 16 kbit/s in BRI, 64 kbit/s in PRI, used for signaling (and perhaps some packet data).
D3 Third generation channel bank, 24 channels on one T-1.
D4 Fourth generation digital channel bank, up to 48 voice channels on two T-1's or one T-1C.
D5 Fifth generation channel bank with ESF.
DA Destination Address, field in frame header (802).
D/A Digital to Analog, decoding of voice signal.
D/A Drop and Add, similar to drop and insert.
DACS Digital Access and Cross-connect System, a digital switching device for routing T-1 lines, and DS-0 portions of lines, among multiple T-1 ports.
DAMA Demand Assigned Multiple Access, multiplexing technique to share satellite channels (Vsat).
DAMPS Digital AMPS.
DARA Dynamic Alternate Routing Algorithm.
DARPA Defense ARPA, formerly just ARPA.
DAS Dual-Attached (Access) Station, device on main dual FO rings, 4 fibers (FDDI).
DASD Direct Access Storage Device (SNA).
DASS Digital Access Signaling System, protocol for ISDN D channel in U.K.
DAVIC Digital Audio Video Interoperability Council.
dB Decibel, 1/10 of a bel; 10 log (x/y) where x/y is a ratio of like quantities (power).
dBm Power level referenced to 1 mW at 1004 Hz into 600 ohms impedance.
dBm0 Power that would be at zero TLP reference level, = measurement - (TLP at that point).
dBrn Power level relative to noise, dBm + 90.
dBrnC dBrn through a C-weighted audio filter (matches ear's response).
DB-25 25-pin connector specified for RS-232 I/F.
d.c. Direct Current, used for some signaling forms; type of power in CO.
DCA Distributed Communications Architecture, networking scheme of Sperry Univac.
DCC Data Communications Channel, overhead connection in D bytes for SONET management.
DCC Digital Cross Connect, generic DACS.
DCC Direct Connect Card, data interface module on a T-1 bandwidth manager.
DCE Data Circuit-terminating Equipment, see next DCE.
DCE Data Communications Equipment, 'gender' of interface on modem or CSU; see DTE
DCS Digital Cross-connect System, DACS.
DDCMP Digital Data Communications Message Protocol.
DDD Direct Distance Dialing, refers to PSTN.
DDS Digital Data System, network that supports DATAPHONE Digital Service.
DDSD Delay Dial Start Dial, a start-stop protocol for dialing into a CO switch.
DE Discard Eligibility, bit in FR header denoting lower priority; as when exceeding CIR or Bc.
DEA DEActivate, BRI control bit.
DECmcc Digital Equipment Corp. Management Control Center, umbrella network management system.
DEO Digital End Office, class 5 CO or serving office.
DES Data Encryption Standard, moderately difficult to break.
DFC Data Flow Control, layer 5 of SNA.
DFN Deutsche Forschungsnetz Verein, German Research Network Association.
DGM Degraded Minute, time when BER is between 10^{-6} and 10^{-3}.
DHCP Dynamic Host Configuration Protocol, enhanced BOOTP, negotiates several parameters.
D/I Drop and Insert, a mux function or type.

DID Direct Inward Dial, CO directs call to specific extension on PBX, usually via DNIS.
DIP Dual In-line Package, for chips and switches.
DIS Draft International Standard, preliminary form of OSI standard.
DISA Direct Inward System Access, PBX feature that allows outside caller to use all features, like calling out again.
DISC Disconnect, command frame sent between LLC entities (Layer 2).
DL Data Link.
DLC Data Link Connection, one logical bit stream in LAPD (Layer 2).
DLC Data Link Control, level 2 control of trunk to adjacent node (SNA).
DLC Digital Loop Carrier, mux system to gather analog loops and carry them to CO.
DLCI Data Link Connection Identifier, address in a frame (I.122); LAPD address consisting of SAPI and TEI.
DLE Data Link Escape, ESC.
DLL Data Link Layer, layer 2 (OSI).
DLS Data Link Switching, IBM way to carry Netbios and spoofed SDLC over TCP/IP.
DM Disconnected Mode, LLC frame to reject a connection request (Layer 2).
DMC Differential Manchester Code, pulse pattern that puts transition at center of each bit time for clocking, transition [none] at start of period for 0 [1] (802.5).
DME Distributed Management Environment, OSF's network management architecture.
DMI Digitally Multiplexed Interface, AT&T interface for 23 64 kbit/s channels and a 24th for signaling; precursor to PRI.
DMPDU Derived MAC Protocol Data Unit, a 44- octet segment of upper layer packet plus cell header/trailer (802.6); see L2PDU.
DMS Digital Multiplex System.
DMT Discrete MultiTone, form of signal encoding for ADSL.
DN Directory Number, network address used to reach called party (POTS, ISDN).
DNA Digital Network Architecture, DEC's networking scheme.
DNIC Data Network Identification Code, assigned like an area code to public data networks.
DNIC Data Network Interface Circuit, 2B+D ISDN U interface, from before 2B1Q.
DNIS Dialed Number Identification Service, where carrier delivers number of called extension after PBX acknowledges call.
DNR DCE Not Ready, signaling bit in CMI.
DoD Dept. of Defense.
DOD Direct Outward Dialing.
DoV Data over Voice, modems combine voice and data on one twisted pair.
DP Draft Proposal, of an ISO standard.
DP Dial Pulse, rotary dialing rather than DTMF.
DPBX Data PBX, a switch under control of end users at terminals.
DPC Destination signal transfer Point Code, level 3 address in SU of STP (SS7).
DPCM Differential Pulse Code Modulation, voice compression algorithm used in ADPCM.
DPGS Digital Pair Gain System, multiplexer and line driver to put 2+ channels over 1 or 2 pair of wires.
DPNSS Digital Private Network Signaling System, PBX interface for common channel signaling.
DPO Dial Pulse Originate, a form of channel bank plug-in that accepts dial pulses.
DPT Dial Pulse Terminate, a channel bank plug that outputs pulses.
DQDB Distributed Queue Dual Bus, an IEEE 802.6 protocol to access MAN's, typically at T-1, T-3, or faster.
DS-0 Digital Signal level 0, 64,000 bit/s, the worldwide standard speed for PCM digitized voice channels.
DS-0A Digital Signal level 0 with a single rate adapted channel.
DS-0B Digital Signal level 0 with multiple channels sub-rate multiplexed in DDS format.

DS-1 Digital Signal level 1, 1.544 Mbit/s in North America, 2.048 Mbit/s in CCITT countries.
DS-1A Proposed designation for 2.048 Mbit/s in North America.
DS-1C Two T-1's, used mostly by Telcos internally.
DS-2 Four T-1's, little used in US, common in Japan.
DS-3 Digital Signal level 3, 44.736 Mbit/s, carrying 28 T-1's.
DSAP Destination Service Access Point, address field in header of LLC frame to identify a user within a station address (Layer 2).
DSG Default Slot Generator, the function in a station that marks time slots on the bus (802.6).
DSI Digital Speech Interpolation, a voice compression technique that relies on the statistics of voice traffic over many channels.
DSL Digital Subscriber Line, any of various ways to send fast data on copper loops (ISDN, HDSL, xDSL, etc).
DSLAM DSL Access Multiplexer, concentrator for xDSL in central office.
DSP Digital Signal Processor, specialized chip optimized for fast numerical computations.
DSP Display System Protocol, protocol for faster bisync traffic over packet nets.
DSR Data Set Ready, signal indicating DCE and line ready to receive data.
DSS1 Digital Subscriber Signaling system 1, access protocol for switched connection signaling from NT to ISDN switch (Q.931 & ANSI T1S1/90-214).
DSU Digital (Data) Service Unit, converts RS-232 or other terminal interface to line coding for local loop transmission.
DSX-1 Digital Signal cross connect, level 1; part of the DS-1 specification, T-1 or E-1.
DSX-z Digital Signal cross connect where 'z' may be 0A, 0B, 1, 1C, 2, 3, etc. to indicate the level.
DT Data Transfer, type of TPDU in ISDN.
DTAU Digital Test Access Unit, CO equipment on T-1 line.
DTE Data Terminal Equipment, 'gender' of interface on terminal or PC; see DCE.
DTI Digital Trunk Interface, T-1 port on Northern Telecom PBX.
DTMF Dual Tone Multi-Frequency, TOUCHTONE dialing, as opposed to DP.
DTP Data Transfer Protocol.
DTR Data Terminal Ready, signal that terminal is ready to receive data from DCE.
DTU Data Termination Unit, Newbridge TA for data.
DVD Digital Versatile Disk, CD format for sound, video, and data.
DVI Digital Video Interactive, applications with large, bursty bandwidth.
DWDM Dense Wave Division Multiplexing, using multiple colors of light to make many channels on optical fiber.
DWMT Discrete Wavelet MultiTone (DMT).
DX Duplex, a 2-way audio channel bank plug without signaling.
DXC Digital Cross Connect, DACS.
DXI Data Exchange Interface, serial protocol for SNMP for any speed.

E

E 'Ear' lead on VG switch, receives signaling.
E Ending, indicator bit in header of subframe carrying frame fragment.
E-1 European digital signal level 1, 2.048 Mbit/s.
E-ADPCM Embedded ADPCM, packetized voice with "core" and "enhancement" portion to each frame.
EA Extended Address, or address extension bit, =1 in last byte of frame header.
EASInet Network for European Academic Supercomputer Initiative.
E&M Earth and Magneto, signaling leads on a voice tie line, also known as Ear and Mouth.
EBCDIC Extended Binary Coded Decimal Interchange Code, extended character set on IBM hosts.
EC Echo Canceller.
EC Error Correction, process to check packets for errors and send again if needed.
EC Enterprise Controller, new terminal cluster controller, e.g. 3174.
EC European Community, covered by common telecom standards.
ECC Error Checking Code, 2 bytes (usually) in frame or packet derived from data to let

receiver test for transmission errors.

ECHO European Clearing House Organization, foreign exchange settlement network, run by SWIFT.
ECL Emitter Coupled Logic, transistor circuit type optimized for high speed.
ECMA European Computer Manufacturers Association.
ECN Explicit Congestion Notification, network warns terminals of congestion by setting bits in frame header (I.122).
ECO Engineering Change Order, document from designer ordering change in product.
ECTRA European Committee on (Telecom) Regulatory Affairs, created out of CEPT in 1990 to be regulatory half, as opposed to operational part, of carriers and PTTs.
ECSA Exchange Carrier Standards Association.
ED Ending Delimiter, unique symbol to mark end of LAN frame (TT in FDDI, HDLC flag, etc.).
EDI Electronic Document (Data) Interchange, transfer of business information (P.O., invoice, etc.) in defined formats.
EDSX Electronic DSX, usually followed by a "- N" for signal level.
EETDN End to End Transit Delay Negotiation, part of call setup via X.25 (ISDN).
EFCI Explicit Forward Congestion Indication, flow control method in ATM networks.
EFCN Explicit Forward Congestion Notification (ATM). See also, FECN.
EFI Errored Frame Indicator (ATM on fiber channel).
EFT Electronic Funds Transfer.
EGP Exterior Gateway Protocol, used between an autonomous user network and the Internet.
EIA Electronic Industries Association, publisher of standards (e.g. RS-232).
EISA Extended ISA, 32-bit PC bus compatible with AT style PCs.
EKTS Electronic Key Telephone Service, ISDN terminal mode.
ELAN Emulated LAN, logical net based on LANE (ATM).
E&M Earth and Magneto, signaling leads on a voice tie line, also known as Ear and Mouth.
EMA Enterprise Management Architecture, DEC's umbrella network management system.
EMI ElectroMagnetic Interference.
EMS Element Management System, usually a vendor-specific NMS for a hardware domain (OSI).
EN End Node, limited capability access device (APPN).
ENQ Enquiry, control byte that requests a repeat transmission or control of line.
ENTELEC Energy Telecommunications and Electrical Association.
EOC Embedded Operation Channel, D bytes devoted to alarms, supervision, and provisioning (SONET); control field in M channel of BRI.
EOM End of Message, cell type carrying last segment of frame.
EOT End Of Transmission, control byte; preceded by DLE indicates switched station going on hook.
EPSCS Enhanced Private Switched Communications Service.
ERL Echo Return Loss.
ERLE ERL Enhancement, reduction in echo level produced by echo canceller.
ERS Errored Second, a 1 sec. interval containing 1 or more transmission errors.
ES Errored Second, original term for 1 s period with 1 or more errors.
ESB Errored Second type B, new name for bursty ES.
ESC Escape, an ASCII character.
ESD ElectroStatic Discharge, electrical "shock" from person or other source that can destroy semiconductors.
ESF Extended Super Frame, formerly called Fe.
ESP Enhanced Service Provider, a firm that delivers its product over the phone.
ESS Electronic Switching System, a CO switch.
ET Exchange Termination, standards talk for ISDN switching equipment in CO.
ETag End Tag, field in trailer of frame whose value should match that in BTag.
ETC End of Transmission Block, control byte in BSC.
ETN Electronic Tandem Network.

ETO Equalized Transmit Only, voice interface with compensation to correct for frequency response of the line.
ETR ETSI Technical Report.
ETS Electronic Tandem Switching.
ETS European Telecommunications Standard, published by ETSI.
ETSI European Telecommunications Standards Institute, coordinates telecommunications policies.
ETX End of Text, control byte.

F

F Farad, electrical unit of capacitance.
F Final, control bit in frame header (Layer 2).
F Framing, bit position in TDM frame where known pattern repeats; first bit in T-1 frame.
F1 Flow (of OA&M cells) at level 1, over a SONET section (ATM).
F2 Flow of OA&M cells over a line.
F3 Flow of OA&M cells between PTEs.
F4 Flow of OA&M cells for metasignaling and VP management.
F5 Flow of OA&M cells specific to a logical connection on one VPI/VCI.
FACS Facility Assignment Control System, for telco to manage outside plant (local loops).
FAD Factory Authorized Dealer.
FADU File Access Data Unit (OSI).
FAS Facility Associated Signaling, D channel is on same interface as controlled B channels (ISDN).
FAS Frame Alignment Signal, bit or byte used by receiver to locate TDM channels.
FASTAR Fast Automatic Restoration, of DS-3s via DACS switching (AT&T).
FAX Facsimile.
FB Framing Bit.
FC Frame Control, field to define type of frame (FDDI).
FCC Federal Communication Commission, regulates communications in US; also FastComm Communication Corp.
FCOT Fiber optic Central Office Terminal.
FCS Frame Check Sequence, error checking code like CRC (Layer 2).
FDDI Fiber Distributed Data Interface, 100 Mbit/s FO standard for a LAN or MAN.
FDL Facility Data Link, channel in ESF framing bits for reporting error status, doing tests (available for user data in some cases).
FDM Frequency Division Multiplexer.
FDMA Frequency Division Multiple Access, wireless access where each channel has separate radio carrier frequency.
FDX Full DupleX, simultaneous transmission in both directions.
Fe Extended framing ("F sub e"), old name for ESF.
FEA Far End Alarm, repeating bit C-3 in DS-3 format identifies alarm or status.
FEBE Far End Block Error, alarm signal (count of BIP errors received; ATM uses Z2 byte in SONET LOH).
FEC Forward Error Correction, using redundancy in a signal to allow the receiver to correct transmission errors.
FECN Forward Explicit Congestion Notification, signaling bit in frame relay header.
FEP Front End Processor, peripheral computer to mainframe CPU, handles communications.
FERF Far End Receive Failure, alarm signal (ATM).
FEXT Far End Cross Talk.
FGA Feature Group A, set of signaling and other functions at LEC-IXC interface; also defined for FGB to FGD.
FH Frame Handler, term in standards for FRS or network.
FIB Forward Indicator Bit, field in SUs (SS7).
FID Format Identification, bit C-1 in DS-3 format shows if M13, M28, or Syntran signal.
FIFO First In First Out, buffer type that delays bit stream.

FIPS Federal Information Processing Standards, for networks.
FISU Fill-In Signaling Unit, 'idle' packet that carries ACKs as sequence numbers (SS7).
FITL Fiber In The Loop, optical technology from CO to customer premises.
FIX Federal Internet Exchange, point of interconnection for U.S. agency research networks.
FMBS Frame Mode Bearer Service, FR on ISDN.
FO Fiber Optic, based on optical cable.
FOTP Fiber Optic Test Procedure.
FOTS Fiber Optic Terminal System, mux or CO switch interface.
FPDU FTAM PDU.
FPDU Frame relay Protocol Data Unit (I.122).
FR Frame Relay, interface to simplified packetized switching network (I.122, T1.617).
FRAD Frame Relay Access Device (Assembler/ Disassembler), functions like a PAD or router for frame relay networks.
FRBS FR Bearer Service, newer name for FMBS.
FRF Frame Relay Forum.
FRF-TC FRF Technical Committee, writes implementation agreements.
FRS Frame Relay Switch or Service.
FRSE FR Service Emulation, as by IWF over ATM.
FRMR Frame Reject, LLC response to error that cannot be corrected by ARQ, may cause reset or disconnect (Layer 2).
FS Failed Second, now called UAS (see index).
FS Framing bit for Signaling, F bits that identify which frames carry RBS.
FSK Frequency Shift Keying, modem method where carrier shifts between two fixed frequencies in voice range.
FSN Forward Sequence Number, sent sequence number of this SU/packet (SS7).
FSS Failed Signal State, continuous SESs.
FT Framing bit for Terminal, F bits that identify frames and time slots.
FT-1 Fractional T-1, digital capacity of N x 64 kbit/s but usually less than 1/2 a T-1.
FTAM File Transfer, Access, and Management; an OSI layer-7 protocol for LAN interworking (802).
FTP File Transfer Protocol (TCP/IP).
FTTC Fiber To The Curb (Cabinet), local loop is fiber from CO to just outside CP, wire into CP.
FUNI Frame-based UNI, serial interface to ATM network.
FX Foreign Exchange, not the nearest CO. FX line goes from a CO or PBX to beyond its normal service area.
FXO Foreign Exchange, Office; an interface at the end of a private line connected to a switch.
FXS Foreign Exchange, Subscriber (or Station); an interface at the end of an FX line connected to a telephone, etc.

G

G Ground, lead at VG port.
G3 Group 3, analog facsimile standard at up to 9.6 kbit/s.
G4 Group 4, digital facsimile standard at 56/64 kbit/s.
GA Group Address, for multicasting.
Gbit/s Giga bits per second, billions (10^9) per second.
GCRA Generic Cell Rate Algorithm, how an ATM entity measures/controls negotiated service usage.
GFC Generic Flow Control, first half-byte in ATM header at UNI.
GOSIP Government OSI Profile, suite of protocols mandated for US Federal and U.K. contractors; -T = Transport model; -A = Application model.
.gov Government, Internet domain name.
GPS Global Positioning System, satellites that report exact time.
GS Ground start, analog phone interface.
GSM Group Special Mobile (Global System for Mobile communications), CEPT standard on digital cellular.

GSTM General Switched Telephone Network, CCITT term to replace PSTN after 1990's privatizations.
GUI Graphical User Interface.

H

H Halt, line state symbol (FDDI).
H Henry, unit of inductance.
Hx High-speed bearer channels (ISDN):
H0 384 kbit/s.
H1 Payload in a DS-1 channel.
H11 1.536 Mbit/s (N. Amer.).
H12 1.920 Mbit/s (CEPT areas).
H2 Payload of a DS-3 channel.
H21 32.768 Mbit/s (CEPT).
H22 43.008 Mbit/s, the payload of 28 T-1s (N. A.), to 41.160 Mbit/s, if including 18 more DS-0s from the DS-3 overhead (N. Amer.)
H3 Would have been 60-70 Mbit/s, but left undefined for lack of interest.
H4 135.168 Mbit/s (88 T-1s)
HCDS High Capacity Digital Service, Bellcore T-1 specification.
HCI Host Command Interface, Mitel's PHI.
HCM High Capacity Multiplexing, 6 channels of 9600 in a DS-0.
HCS Header Check Sequence, CRC on header fields only, not on info; HEC (ATM).
HCV High Capacity Voice, 8 or 16 kbit/s scheme.
HDB3 High Density Bipolar 3-zeros, line coding for 2 Mbit/s lines replaces 4 zeros with BPV (CEPT).
HDLC High-level Data Link Control, layer-2 full-duplex protocol.
HDSL High bit-rate Digital Subscriber Line, Bellcore Standard for way to carry DS-1 over local loops without repeaters.
HDT Host Digital Terminal, CO end of multiplexed local loop (see RDT).
HDTV High Definition Television, double resolution TV image and candidate application for broadband networks.
HE Header Extension, a 12-octet field for various information elements (SMDS).
HEC Header Error Control, ECC in ATM cell for header, not data. (See HCS)
HEL Header Extension Length, the number of 32-bit words in HE (802.6).
HEPNET High Energy Physics Network, international R&D net.
HFC Hybrid Fiber-Coax, fiber from CO to neighbor cabinet, then coax to CP.
HIPPI HIgh-speed Peripheral Parallel Interface, computer channel simplex interface clocked at 25 MHz; 800 Mbit/s when 32 bits wide, 1.6 Gbit/s when 64 bits.
HLM Heterogeneous LAN Management, OSI NMS protocol specification without layers 3-6, developed by IBM and 3Com to save memory in workstations.
HLPI High Layer Protocol Identifier; field in L3-PDU, included in SMDS to align with DQDB (802.6).
HOB Head Of Bus, station and function that generates cells or slots on a bus (DQDB).
HOPS Horizontally Oriented Protocol Structure, proposal for high performance interfaces at broadband rates.
HPR High Performance Routing, a form of dynamic call routing in the PSTN.
HQ Headquarters, network central site.
HRC Hybrid Ring Control, TDM sublayer at bottom of data link (2) that splits FDDI into packet- and circuit switched parts.
HSSI High Speed Serial Interface, of 600 or 1200 Mbit/s.
HSPS High Speed Peripheral Shelf.
HTML HyperText Mark-up Language, text commands to create WWWeb pages.
HTTP HyperText Transfer Protocol.
Hz Hertz, frequency, cycles/second.

I

I I class central office switch is not in HPR network.
I Idle, line state symbol (FDDI).
I Information, type of layer 2 frame that carries user data.
IA Implementation Agreement, based on a subset of CCITT or ISO standards without options to ensure interoperability.
Ia Interface point a, the network-side port of TE-1, NT-1, or NT-2 (ISDN PRI)
Ib Interface point b, the user-side port of TE-1, NT-1, or NT-2 (ISDN PRI).
IA5 International Alphabet #5, coding for signaling information (ISDN).
IAB Internet Activities Board, defines LAN standards like SNMP.
IACS Integrated Access and Cross-connect System, AT&T box with DACS and mux functions via packet switching fabric.
IAD Integrated Access Device, CPE that supports multiple services on public/private net.
IAM Initial Address Message, call request packet (SS7).
IBR Intermediate Bit Rate, between 64 and 1536 kbit/s; fractional T-1 rates.
IC Integrated Circuit.
ICA International Communications Association, a users group.
ICCF Industry Carriers Compatibility Forum.
ICIP InterCarrier Interface Protocol, connection between two public networks.
ICMP Internet Control Message Protocol, reports to a host errors detected in a router by IP.
ICR Initial Cell Rate.
IDF Intermediate Distribution Frame.
IDLC Integrated Digital Loop Carrier, combination of RDT (remote mux), transmission facility, and IDT to feed voice and data into a CO switch.
IDT Integrated Digital Terminal, M24 function in a CO switch to terminate a T-1 line from RDT.
IE Information Element, part of a message; e.g. status of one PVC in a report.
IEC Inter-Exchange Carrier, a long distance company, carries traffic between LATA's.
IEC International Electrotechnical Commission, standards body.
IEEE Institute of Electrical and Electronics Engineers, Inc.; engineering society; one of the groups which set standards for communications.
IETF Internet Engineering Task Force, adopts RFCs.
I/F Interface.
IG ISDN Gateway (AT&T).
IGOSS Industry/Government Open Systems Specification, broader GOSIP.
IGP Interior Gateway Protocol, IGRP.
IGRP Interior Gateway Routing Protocol, learns best routes through large LAN internet (TCP/IP).
ILMI Interim Local Management Interface, PVC management in ATM at UNI.
ILC Intra-LATA Channel, leased line between COs.
ILS Idle Line State, presence of idle codes on optical fiber line (FDDI).
IMD InterModulation Distortion.
IMPDU Initial MAC PDU, the SDU received from LLC with additional header/trailer to aid in segmentation and reassembly (802.6).
IN Intelligent Network.
INA Integrated Network Access, multiple services over one local loop.
IND Indication (OSI).
INE Intelligent Network Element,
I/O Input/Output.
IOC Inter-Office Channel, portion of T-1 or other line between COs of the IXC.
IOC Isdn Ordering Code, 1 or 2-letter code for a complete set of configuration parameters for ISDN BRI service.
IMPDU Initial MAC Protocol Data Unit.
IP Internet Protocol, connectionless datagram network layer (3) basis for TCP, UDP and the Internet.

IPX Internetwork Packet eXchange, Novell's networking protocol, based on XNS.
IR InfraRed, light with wave length longer than red, like 1300 nm used over fiber.
IS International Standard.
ISA Industry Standard Architecture, the personal computer design based on IBM's AT model.
ISDN Integrated Services Digital Network.
ISDN-UP ISDN User Part, protocol from layer 3 and up for signaling services for users, Q.761-Q.766 (SS7).
ISDU Isochronous Service Data Unit, upper layer packet from TDM or circuit-switched service (802.6).
ISI Inter-Symbol Interference, source of errors where pulses (symbols) spread and overlap due to dispersion.
ISO International Standards Organization, ANSI is US member.
ISR Intermediate Session Routing, performs address swapping and flow control (APPN).
ISSI Inter-Switching System Interface, between nodes in a public network, not available to CPE (e.g. SMDS to B-ISDN).
ISSIP ISSI Protocol.
ISUP ISDN User Part (SS7).
ITB End of Intermediate Transmission Block, control byte in BSC.
ITG Integrated Telemarketing Gateway.
ITU International Telecommunications Union, UN agency, parent of (former) CCITT, CCIR, etc.
ITUA Independent T1 Users Association (dissolved, 1995).
ITU-RS ITU-Radiocommunications Sector.
ITU-T Short for ITU-TSS
ITU-TSS ITU Telecommunications Standardization Sector, successor to CCITT.
IVR Interactive Voice Response.
IWF InterWorking Function, the conversation process between FR and X.25, FR and ATM, etc.
IWU InterWorking Unit, protocol converter between packet formats like FR and ATM.
IXC IntereXchange Carrier, a long distance phone company or IEC, as opposed to LEC.

J

J Non-data character for starting delimiter (11000) in 4B/5B coding (802.6).
JB7 Jam Bit 7, force bits in position 7 within a DS-0 to 1 for 1's density.
JPEG Joint Photographic Experts Group, part of ISO that defined digital storage format for still photos.
JTC1 Joint Technical Committee 1, of IEC and ISO.

K

K Non-data character for starting delimiter (10001) in 4B/5B coding (802.6).
k Kilo, prefix for 1000; 1000 bit/s; K = 1024 bytes when applied to RAM size.
K2 LOH byte (SONET).
kbit/s Thousands of bits per second
KDD Kokusai Denshin Denwa, Japan's international long distance carrier.
KG Key Generator (Krypto Gear), encryption equipment from NSA.
kHz Kilohertz, thousands of cycles per second

L

L2-PDU Layer 2 Protocol Data Unit, fixed length cell (SMDS).
L3-PDU Layer 3 Protocol Data Unit, a variable length packet at OSI level 3.
LADT Local Area Data Transport, telco circuit on copper pair.
LAN Local Area Network.
LANE LAN Emulation, ATM specification for logical linking of stations.
LAP Link Access Procedure (Protocol), layer 2 protocol for error correcting between master station and 1 or more slaves.
LAPB LAP Balanced, HDLC layer 2 protocol for data sent between 2 peer stations; used for X.25, SX.25, etc.

LAPD Variant of LAPB for ISDN D channels.
LAPD+ LAPD protocol for other than D channels, e.g. B channels.
LAPM LAP Modem, part of V.42 modem standard.
LAT Local Area Transport, DECnet protocol for terminals.
LATA Local Access and Transport Area, a geographic region. The LEC can carry all traffic within a LATA, but nothing between LATA's.
LBO Line Build Out, insertion of loss in a short transmission line to make it act like a longer line.
LBRV Low Bit Rate Voice, digital voice compressed encoded below 64 kbit/s.
LC Local Channel, the local loop.
LCD Liquid Crystal Display.
LCD Loss of Cell Delineation, receiver can't find cells (ATM).
LCI Logical Connection Identifier, short address in connection-oriented frame.
LCN Logical Channel Number, form of PVC address in an X.25 packet.
LCP Link Control Protocol, part of PPP.
LD-CELP Low Delay CELP, voice compression with small processing delay (G.728).
LDC LATA Distribution Channel, line between local CO and POP.
LDM Limited Distance Modem.
LE LAN Emulation (ATM).
LEC LAN Emulation Client, end station function maps MAC to ATM address.
LEC Local Exchange Carrier, a telco.
LECS LAN Emulation Configuration Server, part of LANE (ATM).
LED Light Emitting Diode, semiconductor used as light source in FO transmitters.
LEN Low Entry Networking, most basic subset of APPN.
LEN Local Exchange Node, CO switch of LEC.
LEO Low Earth Orbiting satellite, option for global 'cellular' phones.
LGE Loop- or Ground-start, Exchange; FXO analog voice interface.
LGS Loop- or Ground-start, Subscriber; FXS analog voice interface.
LI Length Indicator, field in VoFR subframe header.
LI Link Identifier, address consisting of VPI and VCI (ATM)
LIDB Line Identification Data Base (SS7).
LIV Link Integrity Verification (FR).
LLB Local Loop Back.
LLC Logical Link Control, the upper sublayer of the OSI data link layer (layer 2).
LLC1 Connection oriented LLC.
LLC2 Connectionless LLC.
LM Layer Management, control function for protocol.
LME Layer Management Entity, the process that controls configuration, etc. (802.6).
LMI Layer Management Interface, software at each OSI layer in SMDS, 802.6.
LMI Local Management Interface, transport specification for frame relay that sets way to report status of DLCIs.
LOF Loss Of Frame, condition where mux cannot find framing, OOF, for 2.5 sec.
LOFC Loss of Frame Count, number of LOFs.
LOP Loss of Pointer, SONET error condition, like LOF.
LOS Loss Of Signal, incoming signal not present (no received data).
LPC Linear Predictive Coding, voice encoding technique.
LPDA Link Problem Determination Aid, part of Netview NMS (SNA).
LS Loop Start, analog phone interface.
LSAP Link layer Service Access Point, logical address of boundary between layer 3 and LLC sublayer in 2 (802).
lsb Least Significant Bit, position in data field with smallest value.
LSSU Link Status Signaling Unit, control packet at layer 3 (SS7).

LSU Line State Unknown, possible report from FDDI line state monitor.
LT Loop (Line) Termination, in the CO on a BRI.
LTE Line Terminating Equipment, SONET nodes that switch, etc. and so create or take apart an SPE (SONET).
LU Logical Unit, upper level protocol in SNA.
LUNI LAN UNI, specific type of LAN emulation (ATM).
LU6.2 Set of services that support program to program communications.

M

M Maintenance, overhead bits in frames and superframes at BRI.
M Million when used as prefix to abbreviation: Mbit/s.
M 'Mouth' lead on VG switch, sends signaling.
m Milli (1/1000) when used as prefix: mm = millimeter
m Meter (39.37 inches).
M13 Multiplexer between DS-1 and DS-3 levels.
M24 Multiplexer function between 24 DS-0 channels and a T-1, a channel bank.
M28 Same as M13, but different format, not compatible.
M44 Multiplexer function to put 44 ADPCM channels into one T-1; four bundles, each of one common signaling channel with 11 voice channels; transcoder or BCM.
M48 Multiplexer function to put 48 ADPCM channels into one T-1; signaling in each voice channel.
M55 ADPCM multiplexer that puts 55 voice channels in five bundles on an E-1.
mA Milliampere, unit of electrical current, 1/1000 of an ampere.
MAAL AAL Management.
MAC Medium (Media) Access Control, the lower sublayer of the OSI data link layer.
MAN Metropolitan Area Network, typically 100 Mbit/s.
MAP Manufacturing Automation Protocol, for LAN's; closely related to TOP, and written MAP/TOP (802.4).
MAU Media Access Unit, device attached physically to Ethernet cable (802.3).
MAU Multiple (Multistation, Media) Access Unit, hub device in a TR LAN (802.5).
Mbit/s Megabit (1,000,000 bits) per second. .
MBS Maximum Burst Size, number of cells that may be sent at PCR without exceeding SCR (ATM).
MCC Master Control Center, part of DEC's umbrella network management system, EMA.
MCF MAC Convergence Function, how an SDU is framed into a packet (PDU), segmented, and loaded into cells (802.6).
MCP MAC Convergence Protocol, segmentation and reassembly procedure to put MSDUs into cells (802.6).
MCPC Multiple Channel Per Carrier, satellite connection method for point-point links.
MCR Minimum Cell Rate.
MDDB Multi-Drop Data Bridging, digital bridging of PCM encoded modem signals, equivalent to analog bridging.
MDF Main Distribution Frame, large CO wire rack for low speed data and voice cross connects.
MDI Medium Dependent Interface, link between MAU and cable (802 Layer 1).
MELP Mixed Excitation Linear Prediction, voice encoding method for LBRV; DoD standard.
MF Multi-Frequency, tone signaling on analog circuits between CO switches.
MFA MultiFrame Alignment, code in time slot 16 of E-1 to mark start of superframe.
MFJ Modified Final Judgment, court decision that split AT&T in 1984.
MHS Message Handling System, OSI store and forward protocol.
MHz Megahertz, million cycles per second.
MIB Management Information Base, OSI defined description of a network for management purposes (SNMP, IP).
MIC Media Interface Connector, dual-fiber equipment socket and cable plug (FDDI).

MID Message IDentification, a sequence number shared by all L2-PDUs holding segments of one L3-PDU (SMDS, 802.6) or all segments of same frame (AAL3/4).
MIDI Musical Instrument Digital Interface.
MIPS Millions of Instructions Per Second, speed rating for computer.
MIS Management Information Systems, dept. that runs the big computers.
MJU Multipoint Junction Unit, a digital data bridge for DDS (DS-0B or 56 kbit/s), often part of a DACS.
MLHG MultiLine Hunt Group, operation mode for ISDN terminal.
MLPPP MultiLink PPP, protocol to split data stream over multiple channels.
MMFS Manufacturing Messaging Format Standard, application protocol (MAP).
MML Man-Machine Language, commands and responses understandable by both human and device being controlled.
MNP Microcom Networking Protocol, error correcting protocol and compression in modems.
modem MOdulate/DEModulate, modulate analog signal from digital data and reverse.
MOS Mean Opinion Score, scale for voice quality from 5 (toll quality) to 1 (unusable) assigned by expert listeners.
MPA Manufacturing Program Analysis, evaluation standard for vendor plants (TR411).
MPEG Motion Picture Experts Group, part of ISO that defined digital video compression and file format.
MPDU MAC PDU (802.6).
MPL Maximum Packet Lifetime, number of hops allowed before packet is discarded.
MPMC Multi-Peer Multicast, "N-way" mutual broadcasting of information.
MPOA MultiProtocol Over ATM.
MRD Manual Ring Down, VG leased line where caller presses button to ring.
MS Management Service (SNA).
ms Millisecond, 1/1000 second.
M/S Master/Slave, relationship in a protocol where master always issues commands and slave only responds.
MSAP MAC Service Access Point, logical address (up to 60 bits) of boundary between MAC and LLC sublayers (802).
msb Most Significant Bit, high order bit in a data field.
MSDU MAC Service Data Unit, data packet in LAN format; may be long and variable length before segmentation into cells.
MSS MAN Switching System.
MSS Maximum Segment Size, limit on TCP frames sent.
MSU Message Signaling Unit, layer 3 packet (SS7).
MTA Metallic Test Access, service point on equipment in CO.
MTBF Mean Time Between Failures, average for one device.
MTBSO Mean Time Between Service Outages.
MTP Message Transfer Part, set of connectionless protocols at lower layer 3 and below (SS7); cf ISUP.
MTS Message Telephone (Toll) Service, normal dial up phone service.
MTSO Mobile Telephone Switching Office, CO that joins cellular and landline services.
MTTR Mean Time To Repair.
MTU Maximum Transmission Unit, largest PDU in IP.
MUX Multiplexer.
mW Milliwatt, unit of electrical power, 1/1000 watt.

N

N N class central office has tandem switch that participates in HPR.
N Digit from 2 to 9 inclusive.
n Nano, prefix meaning 10^{-9} of the unit as nm = 10^{-9} meter.
N.A. North America.

NAK Negative Acknowledgment, protocol control byte indicating error.
NANP North American Numbering Plan.
NAS Network Applications Solutions, set of DEC APIs for communication.
NAU Network Addressable Unit, addressable device or process running an SNA protocol.
NBS National Bureau of Standards, now NIST.
N.C. Normally Closed, switch contacts on at 'idle.'
NCB Network Control Block, command packet in SNMP.
NCB Network Control Block, transport protocol in LAN Manager (level 4).
NCC Network Control Center.
NCI Network Control Interface.
NCP Network Control Point, for SDN and AT&T switched network.
NCP Network Control Program, software for FEP in SNA; has FR interface after Ver. 7.1.
NCP Network Control Protocol, part of PPP.
NCTE Network Channel Terminating Equipment; first device at CP end of local loop; e.g., CSU.
NDF New Data Flag, inversion of some pointer bits to indicate change in SPE position in STS frame (SONET).
NDIS Network Driver Interface Specification.
NE Network Element, device or all similar devices in a network.
NEBS Network Equipment-Building System, Bellcore generic spec for CO equipment (TR63).
NECA National Exchange Carriers Association.
NET3 EC standard for BRI.
NET5 EC standard for PRI.
NET33 EC standard for ISDN telephones.
NEXT Near End Cross Talk, interference on 2-wire interfaces from sent signals leaking back into the receiver.
NFS Network File System, protocol for file transfers on a LAN.
NFS Network File Server, computer with shared storage, on a LAN.
NI Network Interface; demarcation point between PSTN and CPE/CI.
NIC Network Interface Card, add-in card for PC, etc. to connect to LAN.
NID Network IDentification, field in network level header (MAP).
NISDN Narrowband ISDN, access at T-1 or less.
NIST National Institute of Standards and Technology, name change for National Bureau of Standards.
NIUF North-american Isdn Users Forum, a group associated with NIST.
NLPID Network Level Protocol ID, control field in frame header identifying encapsulated protocol.
NM Network Management.
NME NM Element.
NMOS N-channel Metal Oxide Semiconductor, common IC type uses more power than CMOS.
NMP Network Management Protocol.
NMS Network Management System.
NMVT Network Management Vector Transport (SNA).
NNI Network-Network Interface, between two carriers or between carrier and private network (FR, ATM).
NNI Network-Node Interface, point to point interface between two switches for SDH, SONET, or B-ISDN network.
N.O. Normally Open, switch contacts off at 'idle.'
NOS Network Operating System.
NPA Numbering Plan Area, area code in phone number: NPA-NXX-5555.
NPDA Network Problem Determination Application, fault isolation software for IBM hosts, part of NetView.
NPDU Network PDU, layer 3 packet (OSI).
NPI Numbering Plan Indicator, field in message with DN to specify local, national, or international call (ISDN).

NPSI Network Packet Switching Interface, IBM software for packet connection to FEP (SNA).
NR Number Received, control field sequence, tells sender the NS that receiver expects in next frame (Layer 2).
NREN National Research and Education Network, U.S.
NRZ Non-Return to Zero, signal transitions from positive to negative without assuming 0 value. See also DMC, AMI.
NRZI NRZ Invert on ones, coding changes polarity to indicate '1' and remains unchanged for '0.'
ns Nanosecond, 10-9 second.
NS Network Supervision (ATM).
NS Number Sent, sequence number of frame in its control field; determined by sender.
NSA National Security Agency.
NSA Non-Service Affecting, fault that does not interrupt transmission.
NSAP Network Service Access Point, logical address of a 'user' within a protocol stack (ISDN).
NSC Network Service Center, for SDN.
NSDU Network Service Data Unit, basic packet passed by SCCP (SS7); also OSI.
NSFNET National Science Foundation Network.
NSP Network Services Part, reliable transport for signaling, MTP + SCCP.
NT-1 Network Termination 1, the first device on the CP end of the ISDN local loop (like the CSU).
NT-2 Network Termination 2, the second CP device, like the DSU (ISDN).
NTM NT Test Mode, BRI control bit.
NTN Network Terminal Number, address of terminal on data network, part of global address with DNIC (X.121).
NTSC National Television Standards Committee, group and format they defined for U.S. TV broadcasting.
NTT Nippon Telephone and Telegraph, the domestic phone company in Japan.
NUI Network/User Interface.
NV NetView, IBM's umbrella network management system.
NWT Network Technology, Bellcore group.
NXX Generic indication of exchange in phone number: NPA-NXX-5555.
NYSERnet New York State Education and Research Network, part of NSFnet.

O

OAI Open Applications Interface, Intecom's PHI.
OAM Operations, Administration, and Maintenance.
OAM&P Operations, Administration, Maintenance, & Provisioning, telco housekeeping.
OC-1 Optical Carrier level 1, SONET rate of 51.84 Mbit/s, matches STS-1.
OC-3 Optical Carrier level 3, SONET rate of 155.52 Mbit/s, matches STS-3.
OC-N Higher SONET levels, N times 51.84 Mbit/s.
OCD Out of Cell Delineation, receiver is searching for cell alignment (ATM).
OCR Office Channel Repeater, OCU.
OCU Office Channel Unit, "CSU" in the CO; also called OCR.
OCU-DP OCU-Data Port, channel bank plug I/O to 4-wire local loop and CSU on CP to provide DDS.
ODI Open Data-link Interface, driver interface, API for LAN cards.
OEM Original Equipment Manufacturer.
OF Optical Fiber.
OLTP On Line Transaction Processing.
OMAP Operations Maintenance and Administration Part, upper layer 7 protocol in SS7.
ONA Open Network Architecture, FCC plan for equal access to public networks.
ONI Operator Number Identification.
OOF Out Of Frame, mux is searching for framing bit pattern; LOF.
OOS Out Of Synchronization; multiplexers can't transmit data when OOS.
OPC Origination signal transfer Point Code, address in SU of source of packet (SS7).
OPR Optical Power Received, by a FO termination.

OPX Off-Premises Extension, line from PBX to another site.
OR Or, as in either/or, a logical device that outputs a 1 if any input is 1; a 0 only if all inputs are 0.
.org Organization, first Internet domain for non-profits, schools, etc.
ORL Optical Return Loss.
OS Operating System, main software to run a CPU.
OS Operations System, used by telco to provision, monitor, and maintain facilities.
OSF Open Software Foundation.
OSI Open Systems Interconnection, a 7-layer model for protocols defined by the ISO.
OSI/NMF OSI Network Management Forum, standards group for NM protocols.
OSIone Global organization to promote OSI standards.
OSI TP OSI Transaction Processing, a protocol.
OSPF Open Shortest Path First, standard routing protocol.
OSS Operations (or Operational) Support System, used by telco to provision, monitor, and maintain facilities.
OTC Operating Telephone Company, LEC.
OTC Overseas Telephone Company, international carrier in Australia.
OTDR Optical Time Domain Reflectometry (Reflectometer), method (tester) to locate breaks in OF.
OUI Organizationally Unique Identifier, code for administrator of PIDs.
OW Order Wire, DS-0 in overhead intended for voice path to support maintenance .

P

PA Preamble, a period of usually steady signal ahead of a LAN frame, to set timing, reserve the cable, etc.
PA Pre-Arbitrated, portion of traffic on DQDB MAN with assigned bandwidth, usually isochronous connections (802.6).
PABX Private Automated Branch eXchange, electronic PBX.
PAD Packet Assembler/Disassembler, device to convert between packets (X.25, etc.) and sync or async data.
PAL Programmable Array Logic, large semi-custom chip.
PAM Pulse Amplitude Modulation; used within older channel banks and at 2B1Q ISDN U interface.
PANS Peculiar and Novel Services, phone services that go beyond POTS: switched data, ISDN, etc.
PAP Password Authentication Protocol, encrypts passwords for security of dial-in access.
PARIS Packetized Automated Routing Integrated System, fast switch developed by IBM.
PBX Private Branch eXchange, small phone switch inside a company, manual or automatic.
PC Path Control, level 3 in SNA for network routing.
PC Personal Computer, often used as a data terminal.
PCB Printed Circuit Board.
PCC Page Counter Control (SMDS).
PCI Peripheral Component Interconnect, Intel's advanced bus for personal computers.
PCI Protocol Control Information (ATM).
PCM Page Counter Modulus, SMDS header field.
PCM Pulse Code Modulation, the standard digital voice format at 64 kbit/s.
PCN Personal Communications Network, second generation cellular system.
PCMCIA Personal Computer Memory Card International Association.
PCR Peak Cell Rate, traffic parameter applied per VC, VP, or channel (ATM).
PCR Preventive Cyclic Retransmission, error correction procedure that repeats packets whenever link bandwidth is available (SS7).
PCS Personal Communications Service, low-power portable phones based on dense public network of small cells.
PDG Packet Data Group, 12 octets in FDDI frame (outside of WBCs) not assignable to circuit-switched connections.
PDH Plesiochronous Digital Hierarchy, present multiplexing scheme from T-1 to T-3 and higher; contrast with SDH.

PDN Public Data Network; usually packetized.
PDS Premises Distribution System, the voice and data wiring inside a customer office.
PDU Protocol Data Unit, information packet (ADDR, CTRL, INFO) passed at one level between different protocol stacks (OSI).
pel Picture Element, the smallest portion of a graphic image encoded digitally.
P/F Poll/Final, bit in control field of LLC frames to indicate receiver must acknowledge (P) or this is last frame (F) (Layer 2).
PFT Power Failure Transfer, protection switch.
PHF Packet Handling Facility, packet switch for X.25 or FR service (ISDN).
PHI PBX-Host Interface, generic term for link between voice switch and computer, c.f., SCAI.
PHY PHYsical, layer 1 of the OSI model.
PI Primary In, FO port that receives light from main fiber ring (FDDI).
PIC Polyethylene Insulated Cable, modern phone wire.
PID Protocol ID, codes (some allotted by CCITT) to identify specific protocols.
PIN Positive-Intrinsic-Negative, type of semiconductor photo detector.
PIU Path Information Unit, BIU plus the transmission layer frame header (SNA).
PL Pad Length, number (0-3) of octets of 0s added to make Info field a multiple of 4 octets (802.6).
PL Payload Length, field in VoFR subframe header.
PL Physical Layer, level 1 in OSI model.
PL Private Line, a dedicated leased line, not switched.
PLAR Private Line Automatic Ring-down; see ARD.
PLB Performance Loop Back, LB done at point of ESF performance function in CPE.
PLCP Physical Layer Convergence Protocol (Procedure), part of PHY that adapts transmission medium to handle a given protocol sublayer (DQDB).
PLL Phase Locked Loop, electronic circuit that recovers clock timing from data.
PLP Packet Layer Protocol, at layer 3 like X.25.
PLS Physical Link Signaling, part of Layer 1 that encodes and decodes transmissions, e.g. Manchester coding (IEEE 802).
PM Performance Monitoring, function in ATM.
PMA Physical Medium Attachment, electrical driver for specific LAN cable in MAU, separated from PLS by AUI (802.3).
PMA Primary Market Area, metro area as served by MAN.
PMD Packet Mode Data, ISDN call type.
PMD Physical layer, Medium Dependent; a sublayer in layer 1 (below PLS) of LAN protocols; also PMA (802).
PMP Point to MultiPoint, broadcast connection (ATM UNI).
PNNI Private Network-Network Interface, between ATM switches in public and private networks.
PO Primary Out, FO port that sends light into the main fiber ring (FDDI).
POF Plastic Optical Fiber, for short distances rather than glass for long haul.
POH Path OverHead, bytes in SDH for channels carried between switches over multiple lines and through DCCs .
POP Point Of Presence; end of IXC portion of long-distance line at central office (Tel).
POS Point of Sale.
POTS Plain Old Telephone Service, residential type analog service.
PPDU Presentation (layer) PDU (OSI).
ppm Parts Per Million, 1 ppm = 0.0001%.
PPP Point to Point Protocol, non-proprietary multi-protocol serial interface for WAN links.
pps Packets Per Second, switch capacity.
pps Pulses Per Second, speed of rotary dialing dial pulses.
PR Page Reservation, SMDS header field.
PRC Primary Reference Clock, GPS-controlled rubidium oscillator used as stratum 1 source.

PRA Primary Rate Access, via PRI for ISDN.
PRBS Pseudo-Random Bit Sequence, fixed bit pattern, for testing, that looks random but repeats.
PRI Primary Rate Interface; 23B+D (T-1) or 30B+D (CEPT).
PRM Performance Report Message, error history from CPE over FDL.
PRM Protocol Reference Model.
PROM Programmable Read Only Memory; non-volatile type chip.
PRS Primary Rate Source, stratum 1 clock.
PS Power Status, 2-bit control field at BRI.
PS Presentation Services, level 6 of SNA.
PSC Public Service Commission, telecom regulator in many states, also PUC.
PSDN Public Switched Data Network, national collection of interconnected PSDSs.
PSDS Public Switched Digital Service, generic switched 56K intra-LATA service.
PSI Primary Subnet Identifier, part of address in network level header (MAP).
PSK Phase Shift Keying, modem modulation method.
PSN Packet Switched Network.
PSN Public Switched Network.
PSPDN Packet Switched Public Data Network.
PSTN Public Switched Telephone Network, the telco-owned dial-up network.
PT Payload Type, field in frame or cell header.
PTAT Private Trans-Atlantic Telephone, cable from US to U.K., Ireland, and Bermuda.
PTE Path Terminating Equipment, SONET nodes on ends of logical connections.
PTI Payload Type Identifier, control field in ATM header.
PTT Postal, Telephone, and Telegraph authority; a monopoly in most countries.
PU Physical Unit, SNA protocol stack that provides services to a node and to less intelligent devices attached to it.
PU2 Cluster controller or end system.
PU4 Front End Processor.
PUB AT&T technical PUBlication, Bell System de facto standard, most from before divestiture.
PUC Public Utilities Commission, state body that regulates telephones, also PSC.
PVC Permanent Virtual Circuit (Connection), assigned connection over a packet, frame, or cell network, not switchable by user.
PVN Private Virtual Network, VPN.
PWB Printed Wiring Board, PCB.

Q

Q Quiet, line state symbol (FDDI).
QA Queued Arbitrated, portion of packet traffic that contends for bandwidth (DQDB).
QAM Quadrature Amplitude Modulation, high speed modem, also used in CAP.
QFC Quantum Flow Contrl, way to manage ABR service (ATM) based on buffer usage in switches.
QLLC Qualified Logical Link Control, a frame format (SNA).
QoS Quality of Service, performance specification for network, like delay, jitter, or error rate.
QPSX Queued Packet Synchronous eXchange, old name for DQDB; QPSX Systems Inc. originated it in Australia.
Q.921 CCITT recommendation for level 2 protocol in signaling system 7.
Q.931 CCITT recommendation for level 3 protocol in signaling system 7.

R

R Interface reference point in the ISDN model to pre-ISDN phone or terminal.
R Red alarm bit in synch byte (TS 24) of T1DM (Tel).
R Reserved, bit or field in frame not yet standardized, not to be used.
R Ring, one of the conductors in a standard twisted pair, 2-wire local loop (the one connected to the 'ring,' the second part of a phone plug) or the DTE-to-DCE side of a 4-wire interface.
R1 Ring, or R lead of the DCE-to-DTE pair in a 4-wire interface.
RACE Research for Advanced Communications in Europe, program to develop broadband.

RACF Remote Access Control Facility, security program (SNA).

RAI Remote Alarm Indication, (yellow alarm) repeating pattern of 8 ones and 8 zeros in EOC of ESF T-1 line (also in ATM).

RAID Redundant Array of Inexpensive Disks.

RAM Random Access Memory; volatile chip.

RARE Reseaux Associes pour la Recherche Europeene, European Organization of Research Networks.

RARP Reverse ARP, Internet protocol to let diskless workstation learn its IP address from a server (see BOOTP).

RBHC Regional Bell Holding Company, one of the seven "baby Bells."

RBOC Regional Bell Operating Company, one of about 22 local telephone companies formerly part of Bell System.

RBS Robbed Bit Signaling, in PCM.

RD Receive Data, lead on electrical interface.

RD Request Disconnect, secondary station unnumbered frame asking primary station for DISC (Layer 2).

RDA Remote Database Access, service element (OSI).

RDT Remote Digital Terminal, advanced channel bank functionality on fiber or copper loop.

REJ Reject, S-format LLC frame acknowledges received data units while requesting retransmission from specific errored frame (Layer 2).

REL RELease, signaling packet on disconnect (SS7).

RELC Release Complete, packet to acknowledge disconnect (DSS1 and SS7).

RELP Residually-Excited Linear Predictive Coding, voice encoding scheme (8-16 kbit/s).

REN Ringer Equivalent Number, the load presented to CO line during ringing, compared to one analog phone.

REQ Request (OSI).

RF Radio Frequency.

RFC Request For Comment, documents that are modified then adopted by IETF as Internet standards.

RFH Remote Frame Handler, FR switch or network accessed over CS links.

RFI Radio Frequency Interference.

RFP Request For Proposal.

RFT Remote Fiber Terminal, equivalent to SLC96.

RH Request/response Header, 3 bytes added to user data in format for first upper layer frame (SNA).

RHC Regional Holding Company, one of the 7 telco groups split from AT&T in 1984, see RBOC.

RI Ring Indicator, digital lead on modem tells DTE when call comes in (phone rings).

RI Routing Indicator, bit in LAN packet header to distinguish transparent- from source-routed packets.

RIM Request Initialization Mode, layer 2 supervisory frame.

RIP Routing Information Protocol, method for routers to learn LAN topology (TCP/IP).

RISC Reduced Instruction Set Computer.

RJ Registered Jack, connector for UNI; RJ11 is standard phone, RJ45 for DDS and terminal, RJ48 for T-1.

RJE Remote Job Entry, one form of BSC.

RL Ring Latency, time for empty token to traverse full ring with no load (FDDI).

RLL Radio Local Loop.

RM Rate Management, flow control (cell in ATM connection).

RM Reference Model.

RMN Remote Multiplexing Node.

rms Root Mean Square, form of V average related to power in a.c. circuits.

RN Redirecting Number, DN of party that forwarded a call via the network (ISDN).

RNR Receiver Not Ready, S-format LLC frame acknowledges received data units but stops sender temporarily (Layer 2 HDLC).

RO Receive Only.
ROH Receiver Off Hook, signal from CO switch that finds line off-hook but not in use; "howler."
ROLC Routing Over Large Clouds, IETF study.
ROM Read Only Memory; non-volatile chip; applied to CD holding data.
ROSE Remote Operation Service Element (OSI).
RPOA Recognized Private Operating Agency, X.25 interexchange carrier (ISDN).
RR Receive Ready, S-format LLC frame acknowledges received data units and shows ability to receive more (Layer 2).
RS Radiocommunications Sector, part of ITU, 1993 successor to CCIR.
RSET Reset, layer 2 supervisory frame to zero counters.
RSL Request and Status Link, same as PHI or SCAI.
RSP Response (OSI).
RSU Remote Switch Unit, multiplexing equipment outside CO that serves a CSA.
RT Remote Terminal, CP end of multiplexed access loop, a mux.
RTS Request To Send; lead on terminal interface.
RTS Residual Time Stamp, control information in ATM to support CBR service.
RTT Round Trip Time, twice total transmission latency.
RTU Remote Terminal (Test) Unit.
RU Request/response Unit, unframed block of up to 256 bytes of user data (SNA).
RVI Reverse Interrupt, positive ACK that lets station take control of a BSC line.
RZ Return to Zero; signal pauses at zero voltage between each pulse, when making zero crossings.

S

S Status, signaling bit in CMI.
S ISDN interface point between TA and NT-2.
S Supervisory frame, commands at LLC level: RR, RNR, REJ, SREJ (Layer 2).
s Second (unit of time).
S0 European notation for BRI.
S2 European notation for PRI (30B+D).
SA Source Address, field in frame header (802).
SA Synchronous Allocation, time allocated to FDDI station for sending sync frames (802.6).
SAA Systems Application Architecture, compatibility scheme for communications among IBM computers.
SAAL Signaling ATM Adaption Layer, for Q.2931 messages.
SABM Set Asynchronous Balanced Mode, connection request between HDLC controllers or LLC entities (Layer 2).
SABME SABM Extended, uses optional 16-bit control fields.
SAFER Split Access Flexible Egress Routing, service at one site from two toll offices over separate T-1 loops (AT&T).
SAI S/T Activity Indicator, BRI control bit.
SAP Service Access Point, logical address of a session within a physical station, part of a header address at an interface between sublayers (802).
SAP Service Advertising Protocol, periodic broadcast by LAN device (Netware).
SAPI Service Access Point Identifier, part of address between layers in protocol stack; e.g., subfield in first octet of LAP-D address.
SAR Segmentation And Reassembly, protocol layer that divides packets into cells.
SAR-PDU SAR Protocol Data Unit, segment of CS-PDU with additional header and possibly a trailer (e.g., a cell in ATM).
SARM Set Asynchronous Response Mode, unnumbered frame connection request (layer 2 HDLC).
SARME SARM Extended, uses optional 16-bit control field.
SARTS Special Access Remote Test System, the way telcos test leased lines.
SAS Single-Attached Station, FDDI node linked to network by 2 optical fibers (vs. DAS).
sB Signal Battery, second lead to balance M lead in E&M circuit.

SBC Sub-Band Coding, compressing voice into multiple bit streams.
SCADA Supervisory Control and Data Acquisition, in nets for oil and gas producers, factory automation, etc.
SCAI Switch-to-Computer Applications Interface, link between host CPU and voice switch to integrate applications; also PHI and RSL.
SCAMP Single-Channel Anti-jam Man-Transportable, DoD project for LBRV terminal.
SCCP Signaling Connection Control Part, upper layer 3 protocol (SS7).
SCIL Switch Computer Interface Link, PHI by Aristacom.
SCP Service Control Point, CPU and database linked to SS7 that supports carrier services (800, LIDB, CLASS).
SCPC Single Channel Per Carrier, analog satellite technology (telephony).
SCR Sustainable Cell Rate, traffic parameter (ATM).
SD Starting Delimiter, unique symbol to mark start of LAN frame (JK in FDDI, HDLC flag, etc.).
S/D Signal to Distortion ratio.
SDDN Software Defined Data Network, virtual private network built on public data net.
SDH Synchronous Digital Hierarchy, digital multiplexing plan where all levels are synched to same master clock,
SDLC Synchronous Data Link Control; a half-duplex IBM protocol based on HDLC.
SDM Subrate Digital (Data) Multiplexing, a DDS service to put multiple low-speed channels in a DS-0; also Multiplexer.
SDN Software Defined Network.
SDS Switched Digital Service, generic term for carrier function.
SDSL Symmetric DSL, DSL with same bit rate in both send and receive directions.
SDU Service Data Unit, information packet or segment passed down to become the payload of the adjacent lower layer in a protocol stack.
SEND clear to Send, signaling bit in CMI.
SEP Signaling End Point.
SES Severely Errored Second, interval when BER exceeds 10-3, >319 CRC errors in ESF, frame slip, or alarm is present (see index).
SEV Self-Excited Vocoder, form of CELP voice compression.
SF Single Frequency; form of on/off-hook analog signaling within telcos.
SF Subfield (SNA).
SF Super Frame, 12 T-1 frames.
SFET Synchronous Frequency Encoding Technique, a way to send precise isoc clocking rate as a delta from system clock.
sG Signal Ground, second lead to balance E lead in E&M signal circuit.
SHR Self-Healing Ring, topology can survive one failure in line or node (802.6, etc.).
SI Sequenced Information, LAP-D frame type.
SI Secondary In, FO port that receives light from secondary fiber ring (FDDI).
SIF Signaling Information Field, payload of a signaling packet or MSU (SS7).
SIM Set Initialization Mode, layer 2 supervisory frame.
SIO Service Information Octet, field in MSU used to identify individual users (SS7).
SIP SMDS Interface Protocol.
SIPO Signaling Indication Processor Outage, alarm on failure of processor that receives signaling packets ("indications") (SS7).
SIR Sustained Information Rate, average throughput; basis for SMDS access class.
SIT Special Information Tone, audible signal (often three rising notes) preceding an announcement by the network to a caller.
SITA Societe Internationale de Telecommunications Aeronautiques, operator of worldwide airline network.
SIVR Speaker Independent Voice Recognition.

SLC Subscriber Loop Carrier, usually digital loop system.
SLIC Subscriber Line Interface Card (Circuit), on a switch.
SLIP Serial Line Internet Protocol, older PPP for IP only.
SLS Signaling Link Selection, field in routing label of SU that keeps related packets on same path to preserve delivery order (SS7).
SMAP Systems Management Application Process, all the functions at layer 7 and above to monitor and control the network (SS7).
SMB Server Message Block, a LAN client-server protocol.
SMDR Station Message Detail Recording, keeping list of all calls from each phone, usually by PBX or computer.
SMDS Switched Multi-megabit Data Service, offered on a MAN by a carrier; service mark of Bellcore.
SME Subject Matter Expert.
SME System Management Entity, process in ATM hardware that supports remote NMS.
SMF Single Mode Fiber, thin strand that supports only one transmission mode for low dispersion of optical waves.
SMP Simple Management Protocol, newer and more robust than SNMP.
SMR Specialized Mobile Radio, for fleet management and dispatching.
SMT Station ManagemenT, NMS for FDDI.
SMTP Simple Mail Transfer Protocol (TCP/IP).
S/MUX Workstation software to allow UNIX daemons to talk to SNMP manager station.
SN Sequence Number, transmission order of frames or cells within channel or logical connection.
SNA SDH Network Aspects, evolving standards for VC payloads and network management (SDH).
SNA Systems Network Architecture, IBM's data communication scheme.
SNADS SNA Distribution Services, communication architecture for electronic mail and other applications.
SNAP Sub-Network Access Protocol, identifies encapsulated protocol and user (802.1, ATM).
SNI Subscriber-Network Interface, the demark point.
SNMP Simple Network Management Protocol, started in TCP/IP, but extending to many LAN devices (Layer 4-5).
SNP Sequence Number Protection, CRC & parity calculated over SN field in header (AAL-1).
SNR Signal to Noise Ratio, in dB.
SNRM Set Normal Response Mode, unnumbered command frame (layer 2).
SNRME SNRM Extended, uses optional 16-bit control field.
SO Secondary Out, FO port that sends light into the secondary fiber ring (FDDI).
SO Serving Office, central office where IXC has POP.
SOH Section OverHead, bytes in SDH for channels carried through repeaters between line terminations like DCC or switch.
SOH Start of Header, control byte in BSC.
SOHO Small Office Home Office.
SONET Synchronous Optical Network.
SP Structure Pointer, field in AAL-1 cell (ATM).
SPAG Standards Promotion and Applications Group, has same function as COS.
SPCS Stored Program Controlled Switch, CO switch (analog or digital) controlled by a computer.
SPDU Session (layer) PDU (OSI).
SPE Synchronous Payload Envelope, data area in SONET/STS/SDH format, with POH.
SPF Shortest Path First, LAN router protocol that minimizes some measure (delay) and not just "hops" between nodes.
SPID Service Profile IDentifier, DN or DN plus unique extension (ISDN in N.A.).
SQPA Software Quality Program Analysis, Bellcore process to evaluate vendors.
SREJ Selective REJ, layer 2 frame that requests retransmission of one specific I frame.
SRL Singing Return Loss.

SRT Source Routing Transparent, variation of source routing combined with spanning tree algorithm for bridging (802).
SS7 Signaling System 7, CCS within PSTN; replaced CCIS or SS6.
SSA Systems Applications Architecture, SNA plan to allow programs on different computers to communicate.
SSAP Source Service Access Point, field in LLC frame header to identify the sending session within a physical station (802).
SSCF Service Specific Coordination Function, maps SSCOP functions to lower layer (Q.2130, ATM).
SSCOP Service Specific Connection Oriented Protocol, provides assured transport for Q.2931 PDUs (ATM).
SSCP System Services Control Point, host software that controls SNA network.
SSCS Service Specific Convergence Sublayer (ATM).
SSM Single Segment Message, frame short enough to be carried in one cell.
SSN SubSystem Number, local address of SCCP user (SS7).
SSP Service Switching Point (ISDN).
ST Stream, network layer protocol for very high speed connections.
STDM Statistical Time Division Multiplexer.
STE Secure Terminal Equipment.
STE Section Terminating Equipment, SONET repeater.
STEP Speech and Telephony Environment for Programmers, Wang's PHI.
STM Synchronous Transfer Mode, one of several possible formats for SONET and BISDN.
STM-1 Synchronous Transport Module-1, smallest SDH bandwidth; = 155.52 Mbit/s, STM-n = n x 155.52 Mbit/s.
STP Shielded Twisted Pair, telephone cable with additional shielding for high speed data and LANs.
STP Signal Transfer Point, packet switch for SS7.
STS-1 Synchronous Transport Signal, level 1; electrical equivalent of OC-1, 51.84 Mbit/s.
STS-N Signal in STS format at N x 51.84 Mbit/s.
STSX-n Interface for cross-connect of STS-n signal that defines STS-n.
STX Start of Text, control byte in BSC.
SU Segmentation Unit, info field of L2-PDU, <= 44 octets of a L3-PDU (SMDS, 802.6).
SU Signaling Unit, layer 3 packet (SS7).
SV Subvector, part of NMVT (SNA).
SVC Switched Virtual Circuit (Connection), temporary logical connection in a packet/frame network.
SVD Simultaneous Voice and Data.
SWG SubWorking Group, part of a technical committee or forum.
SWIFT SWItched Fractional T-1, telco service defined by Bellcore, includes full T-1.
SWIFT Society for Worldwide Interbank Financial Telecommunications, global funds transfer network of 2000 banks.
SW56 Switched 56 kilobit/s, digital dial up service.
SX.25 Simplified X.25, layer 2 (LABB) without source address, etc.; in FDL.
SYN Synchronization character, 16h ASCII.
sync Synchronous.
SYNTRAN Synchronous Transmission, byte aligned format for an electrical DS-3 interface.

T

T Interface between NT-1 and NT-2 (ISDN).
T Non-data character in 4B/5B coding, ending delimiter (802.6).
T Measurement interval, seconds, = Bc/CIR.
T Tip, one of the conductors in a standard twisted pair, 2-wire local loop (the wire connected to the 'tip' of a phone plug) or one of the DTE-to-DCE pair of a 4-wire interface.
T Transparent, no robbed bit signaling in D4/ESF format.
T-1 Transmission at DS-1, 1.544 Mbit/s.

T1 The standards committee responsible for transmission issues in US, corresponds to ETSI (Europe) and the Telecommunications Technology Committee (Japan).

T1 Tip or T lead of the DCE-to-DTE pair in a 4-wire interface.

T1DM T-1 Data Multiplexer, brings DS-0Bs together on a DS-1 (Tel).

T1D1 TSC of T1 for BRI U interface.

T1E1 TSC of T1 for SNI.

T1M1 TSC of T1 for NMS and OSS.

T1Q1 TSC of T1 for ADPCM, voice compression, etc.

T1S1 TSC of T1 for ISDN bearer services.

T1X1 TSC of T1 for SONET and SS7.

TA Technical Advisory, a Bellcore standard in draft form, before becoming a TR.

TA Terminal Adapter, matches ISDN formats (S/T) to existing interfaces (R) like V.35, RS-232.

TABS Telemetry Asynchronous Block Serial, M/S packet protocol used to control network elements and get ESF stats.

TAC Technical Assistance Center, network help desk.

TAPI Telephony Applications Programming Interface.

TAPS Test and Acceptance Procedures, telco document for equipment installation and set up.

TASI Time Assigned Speech Interpolation; analog voice compression comparable to DSI and statistical multiplexing of data.

TAT Trans-Atlantic Telephone, applied to cables, as TAT-8.

TAXI 100 Mbit/s interface to ATM switch.

TBD To Be Determined, appears often in unfinished technical standards.

TC Terminating Channel; local loop.

TC Transport Connection.

TC Transmission Control, level 4 in SNA.

TC Trunk Conditioning, insertion of various signaling bits in A and B positions of DS-0 during carrier failure alarm condition.

TCA TeleCommunications Association.

TCA Threshold Crossing Alert, alarm that a monitored statistic has exceeded preset value.

TCAP Transaction Capabilities Application Part, lower layer 7 of SS7.

TCC Telephone Country Code, part of dialing plan.

TCP/IP Transmission Control Protocol (connection oriented with error correction) often runs on Internet Protocol (a connectionless datagram service).

TD Transmit Data.

TDD Telecom Device for the Deaf, Teletype machine or terminal with modem for dial-up access.

TDM Time Division Multiplexing (or Multiplexer).

TDMA Time Division Multiple Access, stations take turns sending in bursts, via satellite or LAN.

TDS Terrestrial Digital Service, MCI's T-1 and DS-3 service.

TDSAI Transit Delay Selection And Indication, way to negotiate delay across X.25 bearer service (ISDN).

TE Terminal Equipment, any user device (phone, fax, computer) on ISDN service; TE1 supports native ISDN or B-ISDN formats (S/T interface); TE2 needs a TA.

TEI Terminal Endpoint Identifier, subfield in second octet of LAP-D address field (ISDN).

TEST Test command, LLC UI frame to create loopback (Layer 2).

TFT Thin-Film Transistor, pixel in display panel.

TFTP Trivial File Transfer Protocol, simpler than FTP.

TG Transmission Group, one or more links between adjacent nodes (SNA).

TH Transmission Header, 2 bytes in framing format for layer 4 protocol (SNA).

TIA Telecommunications Industry Association, successor to EIA, sets some comms standards.

TIFF Tagged Image File Format, for graphics files.

TIRKS Trunk Inventory Record Keeping System, telco computer to track lines.

TIU Terminal Interface Unit, CSU/DSU or NT1 for Switched 56K service that handles dialing. 61330
TLA Three Letter Acronym.
TLI Transport Level Interface, for UNIX.
TL1 Transaction Language 1, to control network elements (TR482); CCITT's form of MML.
TLP Transmission Level Point, related to gain (or loss) in voice channel; measured power - TLP at that point = power at 0 TLP site.
TM Traffic Management (ATM).
TMN Telecommunications Management Network, a support network to run a SONET network.
TMS Timing Monitoring System.
TN TelNet, remote ASCII terminal emulation (TCP/IP); also TN-3270 for SNA over Ethernet.
TN Transit Network, IEC (ISDN).
TN3270 Remote emulation of IBM 3270 terminal.
TO Transmit Only; audio plug for a channel bank without signaling.
TOA Type of Address, 1-bit field to indicate X.121 or not (X.25).
TON Type Of Number, part of ISDN address indicating national, international, etc.
TOP Technical and Office Protocol; for LAN's.
TOPS Task Oriented Procedures, telco document for equipment operation and maintenance.
TOS Type Of Service, connection attribute used to select route in LAN (SPF).
TP Transaction Processing, work of a terminal on-line with a host computer.
TPEX Twisted Pair Ethernet Transceiver.
TP-N Transport Protocol of Class N (N=0 to 4), OSI layer 4.
TP-0 Connectionless TP (ISO 8602).
TP-4 Connection oriented TP (ISO 8073).
TPDU Transport Protocol Data Unit (OSI).
TPF Transaction Processing Facility, IBM host software for OLTP.
TPSE Transport Processing Service Element (OSI).
TR Technical Reference (Requirement), a final Bellcore standard.
TR Token Ring, a form of LAN.
TS Terminal Server, allows async terminal to talk over a LAN (Telnet/IP, e.g.)
TS Time Slot, DS-0 channel in T-1, PRI, etc.
TS Transaction Services, top level (7) of SNA protocol stack, on top of LU 6.2.
TS Transport Service (OSI).
TSAPI Telelphony Server API.
TSB Telecommunications Standardization Bureau, formed by ITU in 1993 from merger of CCITT and CCIR.
TSC Technical Subcommittee, for standards setting.
TSDU Transport Service Data Unit (OSI).
TSI Time Slot Interchange(r); method (device) for temporarily storing data bytes so they can be sent in a different order than received; a way to switch voice or data among DS-0s (DACS).
TSK Time Shif Keying, modulation method applied to xDSL.
TSS Telecommunications Standardization Sector, a variant on TSB.
TSY Technology Systems, Bellcore group renamed Network Technology (NWT).
TTC Telecommunications Technology Committee, Japanese standards body.
TTL Transistor-Transistor Logic; signals between chips.
TTR Timed Token Rotation, type of token passing protocol (FDDI).
TTRT Target Token-Rotation Time, expected or allowed period for token to circulate once around ring (802.4, 802.6).
TTY Teletypewriter.
TU Tranceiver Unit, active device at end of DSL.
TU Tributary Unit, virtual container plus path overhead (SDH).
TUC Total User Cells, count kept per VC while monitoring, field in OAM cell.

TUG TU Group, one or more TUs multiplexed into a larger VC (SDH).
TUP Telephone Users Part, ISDN signaling based on MTP without SCCP, used outside N.A. only.
TWX TeletypeWriter Exchange, switched service (originally Western Union) separate from Telex.

U

u English transliteration of Greek mu (μ), for micro or millionth; prefix in abbreviation of units like us, um.
U Interface between CO and CP for ISDN.
U Unnumbered format, command frames, same as UI (Layer 2).
U Rack Unit, vertical space of 1.75 inch.
UA Unnumbered Acknowledgement, LLC frame to accept connection request (Layer 2).
UART Universal Async Receiver Transmitter, interface chip for serial async port.
UAS Unavailable Second, when BER of line has exceeded 10^{-3} for 10 consecutive seconds until next AVS start.
UBR Unspecified Bit Rate, service with no bandwidth reservation (ATM).
UDLC Universal Data Link Control, Sperry Univac's HDLC
UDP/IP Universal Data Protocol or User Datagram Protocol over Internet Protocol; UDP is a transport layer, like TCP.
uF Microfarad, one millionth of the unit of capacitance.
UI Unnumbered Information, frame at LLC level whose control field begins with 11: XID, TEST, SABME, UA, DM, DISC, FRMR (802).
UID User-Interactive Data, circuit mode digital transport (ISDN).
ULP Upper Layer Protocol.
UNI User-Network Interface, demark point of ATM, SDH, FR, and B-ISDN at customer premises.
UNMA Unified Network Management Architecture; AT&T's umbrella software system.
UNR Uncontrolled Not Ready, signaling bit in CMI.
UOA U-interface Only Activation, BRI control bit.
UP Unnumbered Poll, command frame (Layer 2).
U-Plane User Plane, bearer circuit for customer information, controlled by C-Plane.
UPC Usage Parameter Control, flow control of ATM cells into network (I.555).
UPS Uninterruptable Power Supply.
us Microsecond; 10^{-6} second.
USART Universal Sync/Async Receiver Transmitter, interface chip for sync and async data I/O.
USAT Ultra-Small Aperture SATellite; uses ground station antenna less than 1 m diameter.
USB Universal Serial Bus.
USOC Universal Service Order Code.
UTC Universal Coordinated Time, the ultimate global time reference.
UTP Unshielded Twisted Pair, copper wire used for LANs and local loops.
UUSCC User to User Signaling with Call Control, ISDN feature that passes user data with some signaling messages.

V

V Volt, unit of electrical potential.
V5 ETS for interface between AN and PSTN.
V.25bis Dialing command protocol for modems, CSUs, etc.
V.35 CCITT recommendation for 48 kbit/s modem that defined a data interface; replaced by V.11 (electrical) and EIA-530 (mechanical and pinout on DB-25).
VAD Voice Activity Detection, silence suppression, VOX.
VAN Value Added Network; generally a packet switched network with access to data bases, protocol conversion, etc.
VBD Voice Band Data, ISDN terminal mode that may include a modem or fax.
VBR Variable Bit Rate, packetized bandwidth on demand, not dedicated (ATM).
VC Virtual Container, a cell of bytes carrying a slower channel to define a path in SDH; VC-n corresponds to DS-n, n = 1 to 4.

VC Virtual Circuit (Channel), logical connection in packet network so net can transfer data between two ports.
VCC Virtual Circuit (Channel) Connection; between terminals (SMDS, SONET, ATM).
VCI Virtual Circuit (Channel) Identifier; part of a packet, frame, or cell address in header (802.6, ATM).
VCL Virtual Channel Link (ATM).
VCX Virtual Channel Cross-connect, device to switch ATM cells on logical connections.
VDSL Very-high-speed DSL, up to 52 Mbit/s over short loops (1000 ft).
VDT Video Display Terminal, often applied to any type of "tube" or PC.
VESA Video Electronics Standards Assoc., defined the VESA-bus for personal computers.
VF Voice Frequency, 300-3300 Hz or up to 4000 Hz.
VFRAD Voice FRAD, one with voice port(s) for VoFR.
VG Voice Grade; related to the common analog phone line; 300-3000 Hz.
VGA Video Graphics Array, 640x480 display.
VGPL Voice Grade Private Line, an analog line.
VHF Very High Frequency, radio band from 30 to 300 MHz.
VLAN Virtual LAN, term for logical LAN connectivity based on need rather than physical connection.
VLSI Very Large Scale Integration, putting thousands of transistors on a single chip.
VMTP Versatile Message Transport Protocol, designed at Stanford to replace TCP and TP4 in high-speed networks. Commweek 12Nov90
VNL Via Net Loss, related to TLP.
VOD Video on Demand.
VoFR Voice Over Frame Relay.
VOX Voice Activation, in voice over FR, silence suppression by not sending frames when audio level is below threshold.
VP Virtual Path, for many VCCs between concentrators (ATM).
VPC VP Connection.
VPI VP Identifier, VCI in ETSI version of ATM; applies to bundle of VCCs between same end points (ATM).
VPT Virtual Path Termination (ATM).
vPOTS Very Plain Old Telephone Service; no switching; ARD etc.
VPC Virtual Path Connection; between switches (SONET).
VPL Virtual Path Link; between switches, may carry many connections (ATM).
VPN Virtual Private Network, logical association of many user sites into CUG on PSTN.
VPX Virtual Path Crossconnect; SONET device like a DACS.
VQC Vector Quantizing Code; a voice compression technique that runs at 32 and 16 kbit/s.
VQL Variable Quantizing Level; voice encoding method.
VR Receive state Variable, value in register at receiver indicating next NS expected (Layer 2).
VR Virtual Route (SNA).
VRU Voice Response Unit, automated way to deliver information and accept DTMF inputs.
VS Send state Variable, value in register of sender of NS in last frame sent (layer 2).
VSAT Very Small Aperture Terminal, satellite dish under 1 m.
VSELP Vector-Sum Excited Linear Prediction, compression algorithm used in some digital cellular systems.
VT Virtual Tributary, logical channel made up of a sequence of cells within SONET or similar facility.
VTAM Virtual Telecommunications Access Method, SNA protocol and host communications program.
VTE Virtual Tributary Envelope, the real payload plus path overhead within a VT (SONET).
VTG Virtual Trunk Group, pseudo TDM channels over ATM.
VTNS Virtual Telecommunications Network Service.

VTOA Voice and Telephoney Over ATM, a working group.
V.35 Former CCITT recommendation for a modem with a 48 kbit/s interface on a large 44-pin connector, being replaced by EIA-530 pinout on DB-25.

W

WACK Wait before transmit positive Acknowledgment, control sequence of DLE plus second character (30 ASCII, 6B EBCDIC).
WAN Wide Area Network, the T-1, T-3, or broadband backbone that covers a large geographical area.
WATS Wide Area Telephone Service, large-user long distance, includes 800..
WBC WideBand Channel, one of 16 FDDI subframes of 6.144 Mbit/s assignable to packet or circuit connections.
W-DCS Wideband Digital Cross-connect System, 3/1 DACS for OC-1, STS-1, DS-3, and below, including T-1 (see B-DCS).
WDM Wavelength Division Multiplexing, 2 or more colors of light on 1 fiber.
WIRE Workable Interface Requirements Example, definition of interface between protocol layers (ATM).
WWW World Wide Web, information service nodes linked over the Internet.

X

X X class central office switch is in HPR net but not linked to NP.
X Any digit, 0-9.
X.25 CCITT recommendation defining Level 3 protocol to access a packet switched network.
xDSL Unspecified Digital Subscriber Line method, any of various ways to send fast data on copper loops (ISDN, HDSL, xDSL, ADSL, etc.).
XGMON X-Windows (based) Graphics Monitor, IBM net management software for SNMP.
XID Exchange Identification, type of UI command to exchange parameters between LLC entities (layer 2).
XNS Xerox Network Services, a LAN protocol stack.
X-off Transmit Off, ASCII character from receiver to stop sender.
X-on Transmit On, ok to resume sending.
XTP eXpress Transfer Protocol, a simplified low-processing protocol proposed for broadband networks.

YZ

Y Yellow alarm control bit in sync byte (TS 24) of T1DM, Y=0 indicates alarm.
Z Impedance, nominal 600 ohm analog interface may be closer to complex value of 900 R + 2 uF C.
Z2 LOH byte used in ATM to return FEBE value.
ZBTSI Zero Byte Time Slot Interchange, process to maintain 1's density.

Numeric

1Base5 1 Mbit/s BASEband signaling good for 500 m, STARLAN standard (802.3).
27** 2 raised to the 7th power; exponential notation.
2B+D Two Bearer plus a Data channel, format for ISDN basic rate access.
2B1Q 2 Binary 1 Quaternary, line code for BRI at U reference point.
2W 2-Wire, analog interface with send and receive on same pair of wires.
4B/5B Coding that substitutes 5 bits for each 4 bits of data, leaving extra codes for commands (802.6 & FDDI). See also DMC, NRZ, AMI.
4W 4-Wire, analog or digital interface with receive and send on separate wire pairs.
5E8 Software release for 5ESS that supports National ISDN-1.
5ESS Trademark for class 5 (end office) ESS made by AT&T.
10Base5 10 Mbit/s BASEband signaling good for 500 m, LAN definition of Ethernet (802.3).
10BaseT 10 Mbit/s BASEband signaling over twisted pair, Ethernet (802.3).
23B+D 23 Bearer Plus a Data channel, ISDN primary rate T-1 format.
30B+D 30 Bearer channels plus a Data channel, ISDN primary rate E-1 format.

800 Area code for phone service where called party pays the carrier for the call (Free Phone); also 888.
802.x IEEE standards for LAN protocols.
802.1 Spanning tree algorithm implemented in bridges.
802.3 Ethernet.
802.4 Token Bus architecture for MAP LAN.
802.5 Token Ring, with source routing.
802.6 Distributed Queue, Dual Bus MAN.
802.10 LAN security.
802.11 Spread spectrum local radio for LANs.
900 Area code for phone service where calling party is charged for the call plus a fee that the carrier pays to called party.
54016 Specification for ESF; AT&T Pub.
62411 Basic description of T-1 service and interface; old AT&T Pub.

New Additions

You are invited to send your own list to the author for inclusion in a future edition.

Index

1's density 87
1000 Hz tone 73
15 zeros 82
2B+D 132, 226
2B1Q DSL 111
4ESS 91
56 kbit/s digital service 213
56/64K multiplexers 215
5ESS 91, 129, 242
62411 85, 87
64 kbit/s channel 287

A

A-law 38
Access 77
- test 119, 255
- to the DACS computer 125
- via a DACS 121

ACCUNET
- RESERVE 124, 272
- T1.5 with CCR 92

Adaptive DPCM 45
Add and drop 152
Administering a network 241
ADPCM 45
- M44 197
- signaling 197
- switching 199
- T-1 circuit 126
- transcoder 197

Aggregate input 279
Aggregate throughput 169
Alarm Indication Signal 85
Alarms 244, 253
- CO 82
- reporting 236

Algorithm routing 262
Aliasing 36
All 1's 85, 244
Alternate configurations 252
Alternate mark inversion 79, 288
Alternate routing 267
Amplification 25, 281
Analog
- input 281
- line 182
  - from RT 203
- modem technology 223
- signal 26, 34
- tail circuits 211
- voice 118, 142
- voice grade facilities 221

Architecture
- bus 169
- center-weighted 243
- nodal 168

ARPANET 237
Asymmetrical DSL 115
Async data circuits 230
Asynchronous data 211
Audible frequencies 24
Automatic
- alternate routing 6, 151, 157, 167, 169, 188, 236, 249
  - ways to reconnect 250
- backup protection 192
- fallback 193
- protection 163
- ringdown 143, 156

Availability 71, 190
Average ones density 82

B

B channel 131, 225
B8ZS 96
Backbone network 214
Backplane 169
- throughput 155

Bandwidth
- allocation 5
- assignment 157
- cable 107
- of the BRI 111
- spent on signaling 66
- voice channels 24, 36
- wide 8

Basic rate
access 225
interface 111, 132, 226
Basic system reference frequency 91
Battery 61
BCM 46, 126, 197
Bearer channel 131, 225
Bell
Alexander Graham 22
operating companies 103
System 103
Bell Communication Research Corp. 218
Bellcore 218
Binary 8-zeros suppression 96
Bipolar
format 92
signal 79
violation 72, 93, 95, 289
Bit
integrity 141
interleaved 64, 281
interleaved multiplexers 127
rate 140
robbing 66
stuffing 216
time 63
Bit compression multiplexer 46, 57, 126, 146, 197
Bit error rate 71
Blocking 155
Blue alarm 85
Break even 3, 175
Bridged taps 227
Bridging
analog 182
digital 184
Digital T-1 193
multiple LANs 228
Broadcasting 100
BSRF 91
Budget 16
Buffer 102, 167
Buffer memory 91
Buffering 280
Bulk bandwidth 30
Bumping 6, 251
Bundles, M44 58
Bursty seconds 70
Bus Architecture 169
Busy indication 55
Busy out 147
Bypass 31
technologies 105
Byte interleaved 64, 283
Byte interleaving 282

C

C bit Parity 218
Cable
international 128
telephone 29
Cable connections 205
Cable systems 227
Carrier Clocking 91
Carrierless Amplitude/Phase DSL 112
Carriers, leased T-1 102
Category 5 UTP 109
CCIS 130
CCITT 131
CCR 123
Center-Weighted Control 169, 243, 249
Central
control station 253
controller 258
network control 271
processor 243
Central office 1, 21, 119
connections 104
equipment 131
Functions 195
layout 118
Multiplexing 195
Centralized
architecture 241
control 242
processor 250
Centrex 61, 130
CEPT
Format 287
frame 288
Cesium atomic clock 91
Channel 24
Clear 56, 95, 101, 289
Channel associated signaling 291
Channel bank 27, 29, 30, 39, 56, 60, 63, 134
drop and insert 152
format 57
intelligent 135
plugs 40
Channel Routing 148
Channel Service Unit 83
Charges, usage 130
Circuit
priority 249
switching 159
Class of service 248
Clear channel 56, 95, 101, 289
Clipping
DSI effect 53
Clock
master 90
one for all switches 91
precision 166
recovery 81
signal 79
source 164, 167
station 167
stratum 167
Clocking 140, 163
DS-2 217
fallback 164
isochronously 167
Strata 167
Cluster controller 182
CMI Line Coding 294
CMOS chips 41
Coax 107
Code Excited Linear Prediction 50
Codec 28, 36, 39
Coding
AMI 96
B8ZS 96
line, Japanese 293
ZBTSI 96
Coding rule violation 293
Common channel interoffice signaling 130, 241
Common channel signaling 60
Companding 38

Compatibility
- ADPCM 57
- D4 138

Compatibility Bulletin 119 82
Computer Inquiries 2 and 3 94
Computer-PBX Interface 144
Conditioning VG lines 7
Configuration control 4
Connection
- management 229
- priorities 251
- request 248
- set up 232, 236

Connectorized wiring 119
Constant current source 77
Consultants 17
Contention for resources 232
Continuously Variable Slope Delta Modulation 47
Contracts 17
Control
- card 170
- Center-weighted 243
- Centralized 241
- Distributed 238
- Functions 246
- leads 231
- signals 142

Control point
- transferable 253

Controlled slip 72
Conversion
- between analog and digital 36
- of E-1 to T-1 287

Coordination of installation 182
Copper pairs 109
Cost
- network 177
- of T-1 nodes 274
- Reduction 9
- savings 265, 270

Cost-based pricing 31
CPE 94, 133
CPI 144
CRC errors 256
Cross connect 119
- clocking needs 217
- device 120
- frames, DSX-0 120
- panel 121

Cross connections 41
Cross-connects 119
CRV 293
CSU 70, 83, 109, 255
- clear channel 96
- integral 41
- power source 85
- redundancy 86

CSUs 202
Custom chips 41
Customer Controlled Reconfiguration 123
Customer Premise Equipment 70, 94, 133, 211
Customer premises equipment 75, 133
CVSD example 267
Cyclic redundancy check 69, 102

D

D channel 131, 225
D4 144
- format and high efficiency 136
- frame 282
- Framing 63

DACS 91, 121
- centralized control 241
- functionality 121
- limitations 125
- regenerates the ESF 70
- transparent connection 127
- with CCR 199

Daisy Chain 186
Dark fiber 77, 108
Data
- bypass 152, 187, 249, 274
- channel 126
- channel, D 131
- communications 238
- compression 47
- distribution 211
- Interfaces 139
- multiplexer 153
- PBX 127, 159, 232
- switching 156

Data oriented viewpoint 153
Data Over Voice 222
Data switch 233
Data termination unit 226
Datascope 257
DB-15 79
DDS
- channels 120
- subrates 40

Dedicated T-1 line 79
Delay 278, 282
- DACS 121
- existing cables 78
- introduced by a TDM 278
- propagation 282
- statistical multiplexing 280
- time slot interchangers 278
- ZBTSI 97

Delta channel 58, 197
Demarcation point 88
Demark
- Japanese T-1 292
- T-1 CPE 80

Design philosophy 169
Design the Network 17
Design Tools 236
Desktop connectivity 221
Determinant routing 247
Diagnostic Control 4
Diagnostics 4, 94, 150, 236, 253
- remote 41

Dial pulses 67
Dial Tone 172, 202
Differential Pulse Code Modulation 43
Digital
- backbone 5
- bridges 184
- CO switches 217
- facilities 24
- Hierarchy 28
- milliwatt 73
- Multiplex Interface (DMI) 144
- nodes in a CO 91
- PBX 144
- regeneration 28

Ruler 38
service unit 92
signal 26
signal level zero 28
signal processor (DSP) 51
signals 34
Speech Interpolation 52
switched services 118
switches 120
transmission 24
Trunk Interface (DTI) 145
values 25
Voice 33
Digital Access and Cross-connect System 60, 121, 184, 199
Digital signal processor 114
Digital Subscriber Line 24, 109
Digital subscriber loop 109
Digital To Analog Conversion 39
Digitizing distortion 37
Digroup 64
Discrete MultiTone DSL 113
Distortion 37, 77
Distributed
Control 238
AAR 249
intelligence 239
network 169
processing 168
Diverse routing 78
Divestiture 4, 103
CPE 94
DMI 144
Documentation 18, 206
Dorgan's dilemma 90
DoV 223
DPNSS 145, 291
Drop and insert 149, 152, 187
Drop circuit 182
Dropped call 251
DS-0 28, 287
connectivity in a DACS 121
Loop Back 84
DS-1 2, 136
circuits 77
DS-1C 29
DS-3 216
connections 155
formats 216
lines 219
DSI 52, 53
DSL-based services 110
DSP chips 109
DSU 92
DSX-1 88, 100
port 172
specification 88
DTI 145
Dynamic bandwidth allocation 148
Dynamic clock selection 167

E

E lead 60
E&M signaling 142, 194
Earth ground 206
Earth station 100, 106
Economics 178
Economies of scale 271
Efficiency 135
EIA signal leads 142, 257
Elastic store 91, 102
Electrical noise 25
Electronic mail 170
Embedded operations channel 72
Encoding techniques, voice 42
Encryption 141, 271
T-1 79
End office 130
Error
bursts 7, 250
checks 280
correction 278
history 260
level 256
rate 78
via satellite 102
Errored seconds 70
ESF 67, 70
ANSI standard 71
Examples
CAD/CAM group 8
Dallas and Denver 9
price analysis 11
Response times 8
teller productivity 5
Exchange Carriers Standards Association 218
Extended framing 68
Extended Superframe 67, 94

F

F bit 65
comparison table 69
ESF 69
Japanese 294
Facilities 77
coax 107
Infrared 107
microwave 105
Optical fiber 107
Private media 105
satellite 106
twisted pair 108
Facilities data link 69, 70, 83, 98, 289
Facsimile 9
machine 159
Fade 25, 102
Failed seconds 71
False Framing 73
Fast packet switches 281
Fault isolation 254
Faults 4
FCC 31
licensing of microwave 106
FDL 70
FDM 24, 281
FEP 182
Fiber optic networks 31
Filtering bridges 228
Finger pointing 4, 94, 182, 214
Fixed frame 284
Flexibility 1
Flexible
frame in TDM's 284
framing 137
subframes 137
Flow control 142, 231, 280
Foreign exchange, office 61
Foreign exchange, subscriber 62

Forward error correction 102
Fourier 113
  transform 114
FR access devices 202
Fractional DS-3 129
Fractional E-1 122
Fractional T-1 122, 175, 196
  Access 171
Fractional T-1/T-3 121
FRADs 202
Frame
  in a channel bank 63
  slip 91, 167
  supplies organization 63
  synchronization (CEPT) 289
Frame relay
  Access 201
  false framing 73
Framing 57, 277, 283
  Bit Patterns 69
  bits 29, 64
  CEPT format 288
  D4 format 63
  error 72
  migration from D4 to ESF 94
  overhead 137
  pattern 219
  pattern sequence 69
  requirement 73
  SUMMARY 284
Frequency division multiplexing 23, 281
Frequency domain 114
Frequency standards 217
Front end processor 182
FT-1 CSUs 172
FXO 61, 143
FXS 62, 143

G

G.703 interface 288
Give clock 163
Global Positioning System 89, 165
GPS 89
Grooming 59
Ground station 102
Group, channel 64
Group bypass 151
Growth planning 235

H

Handoffs between countries 128
Hard error 101
HCV 51
HDB-3 Coding 288
Hierarchical philosophy 247
Hierarchy, digital transmission 28
High Capacity Voice 51
High speed side 170
High-speed DSL 117
Hybrid network 174, 178

I

I/O port card 134
ILC 103
In-band characters 280
Incremental cost 145
Infra-red transceivers 107
Inputs
  Analog Voice 142
  Asynchronous 139, 194
  Digital Voice 144
  Isochronous 141
  many kinds 138
  Synchronous 140, 194
  voice 194
Inside wiring 84
Installation 102, 123
Integral CSUs 86
Integrated Access 195, 196, 203
Inter-LATA T-1 265
Interexchange carrier 103
Interface
  analog 142
  async 140
  bipolar 92
  digital voice 145
  DS-3 215
  DSX-1 88
  E&M 59, 142
  How Many 134
  Isochronous 141
  leads 142
  PBX 175
  port variety 138
  satellite 100
  synchronous 140
  to analog switch 61
  to phone 62
Interleaving 277, 281
Internal expertise 178
Internal oscillator 89, 166, 167
International traffic 128
Internet Access 13, 170, 201
Internet Service Provider 13, 170, 220
Interoffice trunks 22
Intra-LATA channel 103
Inventory programs and features 259
Inverse Fourier transform 114
Inverse Multiplexing 129
ISDN 130
  basic rate interface 146
  central office 225
  chips 226
  demonstrations 131
  DSL 117
  Is Inevitable 132
  outlook 131
  primary rate 145
  signaling 292
  technology 41
  Terminal Adapter 225
Isochronous Data 141
IXC 103

J-K

Jitter 77, 81, 90, 102, 217
Justification 216
Keep alive 84

L

LAN 227
LAN traffic 171
Land line 77
LAP-D 72
Large networks 213
LATA 2, 102
  distribution channel 103
LDC 103
LDM 222, 226
Leased backup circuits 79
Leased line 77, 119
Lightning protection 227

Limited Distance Modem 222
Line
- coding 288, 292
- driver 222
- installation, 56K 158
- interface 172
- Loop Back (LLB) 83
- Monitor Unit (LMU) 86
- Noise 25
- repeater 82

Linear Predictive Coding 55
Link efficiency 135
Load sharing power supplies 162
Loading coils 108
Local
- access coordination 104
- access lines 265
- area networks 227
- bridge, LAN 228
- connectivity 232
- distribution of data 215

Local Access and Transport Area 2, 102
Local Area Network 13
Local loop 2, 21, 29, 62, 79, 103, 109, 131, 132, 191, 227
- for T-1 11

Logical circuit 244
Long distance carrier 103
Lookup table 50
Loop
- current 61
- start 143, 194
  - phones 62
- timed 163
- timing 90, 92, 172

Loopback 83, 93, 255
- control at T-1 interface 79

Loran 165
Low speed side 170
LPC 55

M

M lead 60, 159
M12 multiplexer 216
M13 89, 216
M13 multiplexers 219
M23 217
M24 195
M24 multiplexer 64
M28 218
M44 46, 58, 126, 146, 197
- format 57

M55 197
Mainframe computer networks 202
Maintenance 4, 235
Management
- network 235
- objectives 190

Manholes 28, 77
Manual ringdown 62, 143
Manual Routing 246
Manufacturers guarantee 17
Map
- DACS 123
- network 123

Master timing source 163
Matrix switch 194
Medium-speed DSL 117
Megacom 67, 92, 130, 146
Mesh 189
- network 245

Metallic path 24
Microwave
- relay 78
- systems 105
- transmission 24

Migration path 154
Modem 182, 221
- management system 221
- signal 43, 281
- traffic 50
- transmission requirements 52

Modem signal DSL 113
Modem Signals 52
- on compressed voice channels 51

Modulation, DSL 110
Monitor mode 257
Moves and Changes 260
Mu-law 38
Multidrop line 182, 202
Multiframe 290
Multiframe Alignment Signal 291
Multiplexer 129
- exemption 94
- Frequency Division 24
- interfaces 133
- ports 134
- techniques 277

Multirate Calling 128, 171

N

N bit 289
National use bits 289
NCTE 94, 132
Network
- access
  - SONET 219
- all-digital 212
- changes configuration 90
- control 157
- design 173, 234
  - responsibility 178
- Design Tools 261
- Example 272
- expansion 169
- hybrid 178
- installations 258
- integration 266
- large node count 237
- management 86, 235, 258
  - voice interfaces 143
- map 123
- Number of nodes 211
- optimization 137
- partition 243
- planning 211
- private 177
- public 177
- size 154
- Termination 225
- topology 6, 236, 249
- voice and data 56

Network channel terminating equipment 70, 83, 94
Network Control 3, 168, 235, 237, 258
Network management system 6, 157, 211, 214, 215, 226, 258

Networking
  At DS-3 219
  functions 169
New Networking 215
New services 8
Nodal Architecture 168
Node
  configuraters 236
  configurations 262
  count in a network 169, 211
Noise 24
  electrical 25
Nyquist theorem 36

O

Octet 99
OCU 88, 89, 109, 132
Off hook 46, 55, 67
Off premise extension 61, 194
Office Channel Unit 88, 132
On hook 67
One for N sparing 163
One's density 82
Ones density 56, 82, 92, 290
Optical fiber 85, 106, 107
Optimization
  network 137
OPX 61
Oscillator 89
Out of frame 70
Outage time 71
Outlook 131
Outside plant 22, 109
Outside wiring 84
Overhead, framing 137
Owning versus renting 177

P

Packet
  data 132
  networks 237
  nodes 282
  Switching 279
Packetized voice 53
Parameter settings 232
Parity 232
  C bit 218
  conversions 231
Part 68 87
Partitioning 179
Patch panel 119
Payback calculation 137
Payload Loop Back (PLB) 83
PBX
  DS-1 interface 145
  fall back to PSTN 267
  management 214
  networking 146
  upgrade to T-1 175
PCM voice
  encoding 135
  samples per second 37
  with frame slips 92
Perceived voice quality 7
Performance report message 72
Phase locked loop 81, 163
Phase shift 113
Phase shifts 81
Plain old telephone service 118
Planning 19
Pleisiochronous 163
Point of presence 103
Point of service 103
Point to Point Protocol 171
Point-to-point topology 184
Pointer 99
Polled protocols 202
Port expansion 231
Postal, Telephone, and Telegraph authority 2
Postalization of rates 202
POTS 118
Power
  consumption 41
  failure 85
  supplies 162
Practical answers 265
PRI 145
Primary Rate 287
  Interface 132, 145
  service at 1.92 Mbit/s 291
Private
  communications network 133
  data networks 238
  network 5, 174, 274
  T-1 network 61, 211
Processing delay 282
Program audio channel 46
Programmable switches 247
Project Management 205
Propagation delay 278, 282, 283
Proprietary signaling 60
Protection switching 41, 244, 250
Protocol analyzer 257
Provisioning 92, 123, 260
PSTN 33, 129
PTT 2
Public Network Option 174
Public Switched Telephone Network 33
Pulse amplitude modulation 36
Pulse code 36
Pulse Code Modulation 35, 42
Punch block 119

Q

Quality 7
Quantizing noise 37
Quaternary 111

R

Rate-Adaptive DSL 117
Reconfiguration, automatic 251
Recovery after power loss 253
Redundancy 161, 192, 239, 271
  IXC 78
  on the long-haul 6
  on the low speed port 161
  Power supply 162
  within the multiplexer 6
Redundant
  module 161
  T-1's 78
Regenerator 26, 77, 83
Reliability 1, 5, 160, 190, 235, 271
Remote bridges, LAN 228
Remote terminal 30, 203

Repeaters 77, 89, 109
Rerouting of connections 3
Retransmission 102, 280
Return to Zero 79
Ring topology 187
Ringing 61
voltage 62
RJ-48 connector 59
RJ-48C 79
Robbed bit signaling 56, 65, 92, 127
Route diversity 190
Router 228
for Internet access 201
Routing 123, 246, 248
of connections 137
functions 201
options 180
table 251, 258
in every node 240
Routing Information Protocol 201
RS-232 cables 231

S

Sa4 289
Sample 99
Sampling rate 36, 63
Satellite
facilities 100
movement 102
transit time 100
Savings
money 1
on line charges 267
SDN 129
Seize the line 55
Self checks 254
Serial bus 169
Serial Line Internet Protocol 171
Service contracts 19
Service on demand 159
Severely errored seconds 70
Signal processing 109
Signal to noise ratio 25, 281
Signaling
bit-oriented 291
bits, C and D 68
CEPT 290
D3/D4 57
data 142
in-band 142
information in every sixth frame 66
ISDN 131
M44 126
message-oriented 291
out of band 142
Signaling bits 291
Signaling System #7 60, 131, 241, 291
Silence suppression 52
Simple Network Manatgement Protocol 235
Simplification 3, 5
Simplified X.25 70
Single point of control 3, 9, 253
Single-Ended T-1 Access 200
Site Preparation 205
Slip, frame 72, 167
SLIP accounts 171
Slot limitations 155
Slow switch 121
Smart Jack 84
SNA circuit 101
Soft-configured device 157
Software Defined Data Network 130
Software Defined Network 129
SONET 216, 218
Spare capacity 191
Spectrum analysis 114
SRDM 222
Staff 16
Staggered pins 162
Staging 206
Star
configurations 184
topology 179
Start delay 283
Start-stop protocol 139
Station clock 91, 165, 167, 217
Statistical multiplexer 215, 230, 279
Statistics 256
Gathering 245
historical 260
of normal activities 257
Stratum 1 165
Stuff bits 217
Sub-band Coding 54
Sub-rate
data multiplexing 222
multiplexing 126
switching 222
Subnets, intra-LATA 265
Super-Rate 122
Super-rate channels 123, 137
Superframe 65
E-1 multiframe 290
ESF 57
Japanese 294
Supervisory channel 58, 222, 226, 289
Supervisory messages 254
Supervisory Port 254
Surface mount technology 41
Switch
ADPCM voice channels 126
central office 91
subrate channels 126
Switchboard 21
Switched digital services 118, 199
Switched T-1 128
Switches
in the public network 81
single clock 91
Switching 155, 194
architecture 160
as a value added feature 118
at tandem nodes 281
distributed vs. centralized 159
flexibility 157
function of T-1 Backbone 234
statistical multiplexers 127
voice and data 179
Symmetrical DSL 118
Synchronization 65

Synchronous
- Data 140
- format 218
- Optical Network 218

Syntran 216, 218

T

T carrier 109
T-1
- access circuits 105
- bit rate 88, 166
- circuits 1, 21, 77
- connector 79
- cross-connect specification 88
- Data capacity 12
- Defined Historically 21
- DS-1 rate 29
- economic incentive to use T-1 3
- economics 33
- facilities everywhere 29
- facility 77
- for Me? 14
- in Japan 292
- Installations 205
- interface 59
- local loops
  - exceeded 1 million 200
- multiplexer 41, 133, 277
- network justification 176
- nodes 211
- outlook 131
- Pricing 265
- reasons to install 265
- regenerators 96
- repeaters 29
- service 1
- Switched 128
- transmission 33
- unframed 79
- What it is 1

T-1C 216
T-2 216
T-carrier 109
T-span 77, 96
T1 committee 218
T1DM 65
Tail circuits 221
Take clock 163
Tandem connections 158
- data PBX 234

Tandem nodes 250
Tariff 11, 103
- data base 137
- files 236, 261
- services 261
- trends 262

Tariff #10 104
Tariffed costs 213
Tariffed rates 175
TDM 25, 278
Tear down 236
Tech Pub 62411 88
Telcos 29
Telephone
- service 21
- Signaling 55
- signals 34

Telephony 238
Terminal adapter 225
Terminal server 229
Terminating channel 103
Terrestrial
- delay 78
- microwave 102
- routing 78
- service 77
- T-1 line 79

Test 254
- access 255
- instruments 257
- pattern generator 256
- signals 255
- tone 73

Thick film hybrids 41
Tie
- line 7, 59, 148
- trunks 142

Time
- delay, land lines 78
- delay, satellite 100
- Division
  - Multiplexer 27, 135
  - multiplexing 278
- domain 114
- of day 8
- out 101
- primary reference 91

Time Shift Keying DSL 114
Time slot 28, 63, 290
- Sixteen 291

Timing 79, 217
- source 91

Toll
- office 130
- Quality 35

Topology 179, 236, 258
- change 240
- design 262
  - tool 262
- for DS-3 and faster backbones 220
- linear 186
- mesh 189
- network 6
- point-to-point 180
- ring 187
- star 180

Total costs 177
Trade-off 34, 173
- between bit rate 36

Trader turret 62
Traffic
- Analysis 261
- analyzers 236
- volume 266

Training 18, 206
Transcoder 46, 57
Transfer switch 192
Transit delay 91, 278
Transit time 282
- for a T-1 circuit 78

Transmission, voice and data 179
Transparent
- DACS connection 127
- multiplexer 278

Tri-state devices 162
Trouble ticketing 259
Troubleshooting 4
Trunk Conditioning 244
Trunks, wanted 202
Turning Up a Node 207
Twisted pair 2, 29, 108, 226
- for T-1 108

U

U interface 225
Unclassified National Security Related (UNSR) information 271

Unframed 79
Uniform numbering plan 130
Unifying control 214
Uninterruptable power
 for the CSU 85
Unshielded twisted pairs 109
USAT 106
User
 Clocking 89
 Ports 134
 records 260

V

Vacuum tube 24
Variable Quantizing Level 54
Vector Quantizing Code 50
Vendor files 260
Very Small Aperture Satellite 106
Very-high-speed DSL 118
Video
 encoders 141
 teleconference 5, 8, 157, 253
Virtual circuits 281
Voice
 Applications 59
 bandwidth 24, 36, 281
 Channel 24, 52
 Channel speed 56
 circuit 248
 compression 11, 52, 135, 199, 266
 conversations 52
 cross-connects 120
 digital interface 144
 digital T-1 interface 145
 Digitization Methods 34, 47
 dominates communications 266
 encoder 38
 encoding techniques 42, 144
 frequencies 42
 inputs 194
 networking 146
 packet type 280
 packetized 53
 ports 134
 quality 1, 7, 34, 52
 sample rate 36
 signal 25, 42
 switched by PBX 211
 switching 156
 tie lines 9, 266
 transmission 214
 trunks on a T-1 203
 volume range 37
Voice and data 33, 179
Voice oriented 152
Voice-grade lines 7
Voice/Data Switch 157
Volume range 43
VQC 50
VQL 54

W-Z

Wander 217
War room 4
Wave form coding 51
Wide area network 228
Wink 67
Wire
 center 119
 wrap 119
Wire plant 109
Wiring 108
 bays 89
 frame 119
Wobble 102
World Wide Web 13
Yearly saving 11
Z bits 98
ZBTSI 96
Zero-byte time slot interchange 70, 96